THE COMPLETE BOOK OF HOUSEPLANTS

ROB HERWIG
MARGOT SCHUBERT

THE COMPLETE BOOK OF

HOUSE PLANTS

Translated by Marian Powell

LUTTERWORTH PRESS
GUILDFORD AND LONDON

ISBN 0 7188 2550 0
First published in Great Britain 1975
Second impression 1980
New revised edition 1982
English translation by Marian Powell © 1975, 1982 Lutterworth Press
© 1974, 1982 Zomer & Keuning Boeken B.V., Wageningen and
BLV Verlagsgesellschaft mbH, Munich
No part of this book may be reproduced in any form, by
print, photoprint, microfilm or any other means without
written permission from the publisher.

Printed in Holland

CONTENTS

1. Examples and illustrations

Ideal and reality 9
Interaction – Vigour – A question of attitude – Fixed rules – A plant's individuality – Age records – The problem of space – The golden mean

The plant as a starting point 15
Environment – Forms of growth – Variegated foliage – Overall formation and type – Origin – The pleasure of collecting – Know your limitations

Plants away from the window-sill 24
Dangers for hanging plants – Essential precautions – Ugly and attractive plant tables – Indoor propagators – A plant case is practically indispensible – Where to find good examples

Views on the plant window 30
Essential preparations – Plants feel at home in the plant window – Take advice – The plant window: a bridge to the open air – Instructions on design – The indoor propagator in addition to the plant window

Plants in offices 36
Deadly decor – The use of lasting plants – What is an 'office'? – What everyone should know about the care of plants – Factors which must be taken into consideration – Changes in architecture – Colour provided by flowering plants – Preference for undemanding plants – Plants as presents – Problems in winter – The art of restraint – 'Office landscape' and working climate

Soilless cultivation and artificial light 43
From Aristotle to von Liebig – Plants do not eat soil – Fresh vegetables on Ascension Island – Initial problems and how to overcome them – The drawback of pumice – Possibilities practically unlimited – For domestic use: what plants require and what they tolerate – The best possible living conditions – Correct doses of artificial light

2. The practical care of houseplants

Light, air, situation 49
Sunshine and shade – Assimilation or photosynthesis – Indoor plants supply oxygen – A plant's right to live – Lack of light is obvious – To move or not to move – Sidelight or toplight – Growth in artificial light – Little light for the bud formation of shortday plants – Artificial light or daylight? – Out of doors in summer – Typical hothouse plants remain indoors – Borderline cases: exceptions, etc.

Water and heat 56
Hard water is harmful – The system of pH values – The best pH value for your plant – Degrees of hardness – How to soften water – Rain water is no longer always suitable – Water must not be too cold – Dry air is harmful – Plants dislike having cold feet

Potting composts and feeding 60
Quality of compost is important – Tips for mixing – Attention for the roots

The correct pot for your plant 62
Choice of pots – Drainage hole

When and how to repot 65
A notebook with repotting record – Crocks for good drainage – Roots growing through the hole – Plastic pots require a different method – Preparation of old rootball – From repotting to rerooting

The pleasures of propagation 70
Sexual and vegetative propagation – Cuttings – Quick method of shortening an overgrown rubber plant

Twelve months of plant care 76
Care of the compost – Dust – Washing the windows – Dried-up compost – Hints on pruning

Bottle gardens 81
Relation between available space and choice of plants – In the long run foliage plants are more suitable

What exactly is a plant window? 85
Height of parapet with width of container – Flower window built into the room or as a bay? – Improving design with less branches – Forest moss rots too quickly – Do not water individually – Choosing plants – It need not be a whole tree – A permanent plant community

Growing plants in hydroculture 91
Filling – Containers – Nutrients – Changing over to hydropot cultivation – Cultural troubles and their remedies – Pest attack and cure – Fungal diseases

3. Survey of plant families discussed in this book

Classification of plant families and their relationships   100

4. Houseplants in colour

A huge, 240 page illustrated ABC of houseplants   113

5. For quick reference

Technical terms – Light requirements and winter temperature – Plants for the plant window   353

6. Index

358

INTRODUCTION

If statistics are to be believed, no living thing is as popular as the houseplant. Neither dogs nor cats, tropical fish nor guinea pigs approach the astronomical numbers of indoor plants decorating millions of living rooms. It is the azalea, the rubber plant, the African violet that are man's greatest friends in his desire to bring nature into his environment. Clearly people feel a tremendous need to compensate for the vastly increased tensions of modern living and the frequently sterile and unattractive surroundings in which they work. While some seek relaxation in nature outdoors, others prefer to bring nature into their homes.

There are many books on the subject of houseplants. However, we intend to strike out in new directions. To begin with, this book exceeds any previously published on the subject of indoor plants, both in scope and, as a result of international co-production, in size of edition. Consequently it offers a great deal of information and numerous photographs at a relatively low price. Second, we have made the book so comprehensive that practically all houseplants are included. And third, we wish to show you what a plant can do for your environment. For this reason we have not merely photographed individual plants, but have gone a step further in showing a number of fine, full-grown specimens correctly placed. We hope that these illustrations will give you ideas for your own surroundings. The text will provide the necessary information for growing the plants successfully.

It has given the two authors—Margot Schubert wrote the text and Rob Herwig took the photographs—great satisfaction to compile this book.

Margot Schubert and Rob Herwig

1. EXAMPLES AND ILLUSTRATIONS

Ideal and reality

Plants fulfil an entirely different function today than twenty or thirty years ago; our whole style of living has changed and with it the interiors of our homes and our attitudes to plants. Everyone, at one time or another, has seen photographs of old-fashioned rooms in which a clivia stands, between trim curtains, as an integral part of the decor. In addition there is usually a small table with a trailing plant, but that is all. Modern houses present greater opportunities for houseplants. Our much enlarged windows let through a great deal of light, and as a result a far wider range of plants can be grown. Their situation is also no longer a matter of tradition; today, plants can be used in a much more light-hearted fashion. Particularly the larger species have become an important element in interior decoration and are often used simply as decorative aids or to form an airy partition. Climbers have also found new freedom and are allowed to cover entire walls.

All kinds of accessories for growing houseplants have recently appeared: varied types of hanging pot, tables with sunken plant-troughs, tall cylinders, magnificent tubs, etc. In fact it seems as if nearly all branches of modern industry are intent on helping our plants adapt successfully to their new domestic environment. So we now have a wide range of unusual possibilities for hanging or placing our plants, there are even fully automatic glass cases which will provide us with an enclosed area of tropical jungle. The once dull office or bank building is increasingly becoming the home of tastefully assembled groups of plants, and this has obviously greatly improved the atmosphere for the people who work there.

This baroque jardinière is placed on the moderately lit landing of a large house. The plants must be replaced from time to time.

This development has one drawback, however, for it is possible to have too much of a good thing. In face of so many plants and so many ways of arranging them it is becoming ever more difficult to know when one is making a right decision. One needs not only a good deal of knowledge but also a good deal of aesthetic sense. This book is designed to help on both these counts.

It is not always recognised that correct positioning is an important point. Not all plants fit every interior; one has to choose with care and in accordance with one's personal taste. Only then will the plant truly suit the atmosphere of the interior, be it home or public building, for as a rule the occupant's personality is reflected in his or her choice of furniture and colours.

Interaction

Since dracaena quickly loses its lower leaves, a virtue has here been made of necessity, by growing a separate standard plant. This *Dracaena marginata* has been planted in an ordinary clay pot, encased in marble chippings.

So why should this personal taste not also be expressed in the choice of plants?

It is for this reason that a book on houseplants which claims to be complete must pay attention to the subject of the positioning of plants in the home. Nevertheless, however attractive a plant looks in a certain spot in your living room, this will not necessarily mean that it is happy there.

To combine these two factors and achieve a good result is surely an art worth cultivating.

Vigour — A characteristic of any happy, well cared for plant is that it will quickly produce a rich and vigorous growth. This, of course, is something from which we can derive not only a great amount of aesthetic pleasure but also considerable benefit. We can, in fact, use the beauty of our plants to hide ugly corners for, thanks to radio, television and similar modern appliances, most houses today have more than their fair share of unsightly installations. Here then is one practical way in which our plants can repay us for our care and devotion to them.

A question of attitude — It is, naturally, commonplace that plant lovers are of various kinds and fall into quite distinct categories. Many people regard the 'greenery on the window-sill' merely as a modern form of bric-a-brac. Such people buy a flowering plant and if, after a week or a couple of months, it has lost its beauty it is carelessly thrown into the dustbin. They apparently regard their 'purchase' as just another disposable object, which is surely an abominable attitude to plants.

Fortunately there are also other people with a different attitude and, thank goodness, they are in the majority. These people truly love their plants. They worry if a plant is ailing or if it does not thrive; and are delighted by any signs that it is vigorous and healthy. Sometimes such people are surprised and disappointed to discover that a plant is also bound by immutable laws of nature and after a certain period of time comes to a natural end, depending on the life-span of its genus. In the plant world, also, the rules of nature are inescapable.

Fixed rules — We must accept the fact that on the window-sill, as in the open air, there are annual, biennial and perennial plants, of which the last (large) group can be sub-divided into herbaceous and woody plants, bulbous and tuberous plants, shrubs and trees, evergreen and deciduous trees, needle-leaved trees and grasses. All these can be further divided into numerous other groups. But when we know that an annual plant *cannot* exceed its allocated life-span, or that a biennial plant will not flower until the second year, after which it will inevitably die, we need not anxiously search for possible errors in our care of these plants. It is a fact of life that a slipper plant, a begonia, or a cineraria can hardly be expected to survive after its first flowering, which unfortunately, is also its last. And we have to accept that such plants are marketed as a more or less factory-grown mass product and, through no fault of their own, fulfil the detestable role of 'throwaway-décor'.

A plant's individuality

In addition to its life-span, which is determined by its species — a subject of which amateurs frequently know far too little, especially where houseplants are concerned — each plant has its individual capacities: the vigour or potential energy given to the plant in your living-room as a result of greater or lesser care in its cultivation and of correct treatment. The more our houseplants become the victims of supermarkets and exaggeratedly rational methods of cultivation, the less favourable are the prospects for their further development and health in the living-room. The amount of adaptation required is just too much.

Age records

It is becoming rare that a plant cultivated by these methods is able to fulfil the ideal of every true plant lover, namely become a kind of family heirloom which can be passed on from one generation to another. We have done our best to capture some of these veterans in our photographs, since most people only get a chance to see youthful specimens on sale in their local florists or garden centres.

The large monstera illustrated on the facing page, may be counted among these, even though it is less than twelve years old. To get plants to grow to a ripe old age it is not sufficient to look after them reasonably well; they have also to be loved. A particularly fine specimen of a veteran is the passion-flower illustrated on p. 285: because of its exceptionally favourable situation, a sheltered, south-facing wall in a Dutch garden, it has managed, with one root, to cover six metres of wall. This must be regarded as an exception, and in any case such an enormous plant can hardly be termed a houseplant.

I, myself, have in the course of time seen many old, indeed very old, plants. So far the record is held by a hoya of more than seventy years, which has now been passed on by a Swiss family to a fourth generation. In the house of Danish friends I was privileged to admire a forty-two year old 'queen of the night' *(Selenicereus grandiflorus)* in full flower

This huge *Monstera deliciosa*, shows how an older plant can be used to enhance a sparsely furnished hallway.

In the author's study this fairly recently bred streptocarpus hybrid 'Constant Nymph' can be seen growing in an antique stand from eastern Asia.

and some time ago I was invited to the thirty-fifth birthday of a *Crassula arborescens*. Although not flowering, this specimen could hardly be surpassed for gnarled irregularity. And when a *Cyclamen* 'Thusnelda', grown from seed by its owner, flowered for the thirtieth time with a wealth of soft pink blooms, I made a special journey to see it. All this beauty came from a tremendous, shapeless corm, almost bursting out of its pot.

But make no mistake: these honourable age records have their drawbacks, for in the course of time these old plants will create space problems. This applies not only to the tall, upright species such as ficus or sparmannia, but also to the various kinds of hanging plant. It is a law of nature that houseplants, even when they grow slowly, and are regularly pruned and divided, eventually become too large

The problem of space

for ordinary indoor use. Everyone must decide for himself how to solve this problem in his own special circumstances. In some cases the solution will inevitably be the cause of deep regret.

The golden mean The examples cited over the past two or three pages indicate two totally different attitudes to houseplants. On the one hand we find the people who throw their plants into the dustbin when they have finished flowering and, on the other, the idealists, who pamper their plants into perfection. But surely there is a via-media? Indeed there is. It is that narrow path trodden by truly successful plant lovers: those happy people who are blessed with 'green fingers' or who have learnt a great deal from books; people who give their houseplants exactly what they need, without demanding the impossible or the *un*natural. In other words, the way of sensible people who know how to curb their desires and only purchase those plants which will feel at home in the environment they can offer. In my opinion this self-imposed restriction, which does, however, require a certain knowledge, gives rise to the greatest satisfaction, and is the only way to achieve that happy balance between the ideal and the possible.

The plant as a starting point

The art of successfully growing houseplants begins with the realisation that pot plants are confronted with many difficulties in an average living room. They have arrived here from distant places as beautiful strangers, and instead of the blue skies and warm air of their native countries all we offer them is a life in a small flower-pot on the window-sill. There can be no doubt that they have a much harder life than plants out of doors which, in addition to enjoying a more suitable climate, have practically unlimited space for their roots. Nobody can pretend that houseplants do not also require plenty of space for their roots, so why is one always warned not to plant them out in the garden during the summer months? The answer is that, even if our houseplants were suited to such an operation, they would eventually suffer. You see, once their urge to grow is unrestricted, they will abandon themselves to their new found liberty. The problems arise in the autumn, when you have to try to get their greatly enlarged root system back into a pot. Inevitably, this will cause damage.

Environment

In actual fact a houseplant requires exactly the same living conditions as its more fortunate relations out of doors: plenty of light and air, a suitable position, and alternating sun and shade. It, also, makes similar demands with regard to soil structure, nourishment, and correct temperature during its periods of growth and dormancy, all of which are modified by its native habitat and country of origin. Man, on the other hand, surrounds himself with plants of varying character and, as a rule, thoughtlessly takes it for granted that they will adapt to his surroundings and share his life like willing slaves. In reality our plants will not tolerate such treatment.

This is why we should always ask ourselves: what goes where? And if, being sensible plant lovers, we are not satisfied with a few basic facts and refuse to act merely on fragmentary information and skimpy instructions, the question — what goes where? — is immediately

As many plants like to take deep root, plastic cylinders such as the one illustrated have definite advantages. The attractive standard tree is a *Dracaena deremensis* 'Warneckii', one of the most vigorous kinds.

followed by another question: on what does good positioning depend? In many cases the answer is supplied by the appearance of the plant itself.

Forms of growth The various characteristics of growth pattern, such as branch structure, the position of the foliage, flowering or non-flowering, all provide indications for the classification of plants and help to prevent many errors in cultivation. For example, smooth leaves and soft, succulent stems indicate great evaporation, which means that soil humidity and a shady situation are desirable. The fact that both leaves and flowers of some typically shade-loving plant (such as gloxinia and saintpaulia) are more or less velvety, does not invalidate this classification, for their velvety surface is of an entirely different kind to that, for instance, of those felty white plants which grow quite happily in dry areas with full sunlight throughout the day. In the latter case the hairy surface clearly restricts evaporation, and does so extremely well because the epidermis below, which is often very thick and hard and has a store of moisture, also helps to counteract evaporation.

But nature knows many more ways of helping its plants to maintain themselves in their natural environment, and is able, therefore, to assist us in cultivating a very wide assortment of houseplants. For example, all succulents, whether globe- or column-shaped cacti from the Mexican desert, or the honest houseleek growing on a thatched roof, are miniature reservoirs; they store water caught during rainy periods so as to be able to survive those months of summer drought when evaporation is at its height. Thus, we have to attempt to imitate these conditions on our windowsills as best we can.

However, plants with firm, leathery, evergreen leaves and of moderate vigour, such as dracaena, palms, philodendron and various forms of ficus, indicate a need for little water. On the other hand, plants which rapidly develop new shoots and a great deal of foliage, as do the hoya and the passion flower during their periods of growth, basically require a great deal of light.

You will find that the photographs both immediately preceding and following this section have been specially chosen not only to give you some useful ideas on positioning but also to illustrate plants which have proved most successful indoors. The best indoor growers are: dracaena, palms, philodendron, ficus, aspidistra, chlorophytum, pandanus, and sansevieria. From this list you will be able to build up an assortment of plants which will never disappoint you.

Variegated foliage Plants with white or yellow variegated leaves are an entirely different matter. Their 'shading' is the result of either a partial or complete absence of chlorophyll. This gives rise to a number of fascinating and quite distinct markings, the leaves being striped, streaked, marbled, or even stippled. However, an absence of chlorophyll as a result of cultivation is certainly a fault: plants with variegated leaves are, in general, more sensitive, rarely or never flower, and grow slowly (the latter, of course, is not always a fault).

Overall formation and type

Finally, a plant's general formation provides us with some useful clues on its care. Ground-covering plants want to grow in a different manner from hanging plants while shrubs and trees from the onset demand the roomiest possible situation, where they are not hampered by other plants. Climbing and trailing plants, which wind themselves around a post or are trained against a wall and whose leaves grow naturally only on one side of the stem, have yet different requirements. The language of plants cannot be learned from books alone — it also requires a keen eye.
Another important point to be considered, in our research into the question 'what goes where?', is the plant's origin, in other words, we must consider the plant's mother country with its particular climate, soil and landscape.
From the knowledge thus gained we can deduce what a plant's natural requirements are and, regardless of the methods used in its cultivation, these still remain vitally important. They represent the plant's natural limitations and to ignore them is to cultivate disaster.

Origin

If, therefore, one possesses a bromeliad, a stag's-horn, an orchid or a Christmas cactus, one should know that these are epiphytes from the tropical rain forests. This origin imposes certain requirements as regards their situation on window-sill or in flower window and means that we must ensure the correct soil mixture and humidity as well as seeing that there is no direct sunlight. It is this kind of practical advice that you will find in the descriptions of individual plants beginning on p. 113.
It must be remembered that in all enclosed spaces warmth and atmospheric humidity are concentrated at the top. This must certainly be taken into consideration in the case of highly placed hanging plants or of large epiphyte branches in the flower window. The least sensitive plants, requiring the least water, should always be placed in the highest positions. *Oplismenus hirtellus*, described on p. 276, is a textbook example for such positioning.
Even if you use a special watering can for your hanging plants, watering in these upper regions is no easy matter. On page twenty-four,

Whether you like the rather stiff foliage of *Chamaerops humilis* or not is purely a question of personal taste. In any case its dark green colour certainly suits this cool and rather sombre environment.

which deals with plants that do not live on the window-sill, we shall consider this subject in greater depth.

Here, however, are some further points which may be considered. Reflection on the origin and native habitat of plants frequently leads us to try and combine plants which match one another in every detail. In this way a plant lover, without any particular preferences, may grow into a genuinely enthusiastic specialist collector. For instance, he may turn his attention to cacti or orchids, but there are numerous other fields to which he can turn should he so wish. Think, for example, of the many other succulents which, botanically, form part of the huge cactus family.

The pleasure of collecting

Medinilla magnifica can only remain ▷ in the living-room for a certain period, and it is therefore necessary to have a hot house where it can live for the rest of the year. Judging from the photograph, the owner of this proud specimen has had no problems in this respect.

Since plastic plant cylinders are available in a variety of shades, one can usually find one to suit the plant. One such possibility is shown in this photograph of a *Ficus elastica* 'Schrijvereana'.

In a sunny situation wonderful results may be obtained with a well-grouped bowl of cacti. It is advisable to keep them cool and dry in winter.

A fine small collection of *Aizoaceae* — living stones, fenestraria, argyroderma, *Faucaria tigrina* and glottiphyllum — can effectively be combined in a tray or plant-bed on a widened window-sill. We should also mention the bryophyllum (in all its forms) which has recently been classified among the kalanchoë genus. The many species of peperomia, not forgetting the various tradescantia species, can also be recommended as subjects for collecting. In larger rooms, where plants can be cared for all through the year, philodendrons might be considered. An essential requirement in this case is a large window, and if we start by growing relatively easy flowering species in favourable conditions, the result will be a valuable group of plants of great beauty. In addition there are the ficus species, the indoor ferns, foliage and shrubby begonias, bromeliads and indoor ivy in its many forms.

In my opinion, the pleasure of living with flowers becomes all the greater if, instead of being satisfied with the chance success of some plant chosen at random, we introduce an element of purpose into our collection and try to be rather more systematic when combining the various plant 'personalities' at our disposal. Remember, you will still be able to cheer up your living-room with seasonal flowering plants. It suddenly occurs to me that I have so far forgotten to mention one of the most delightful subjects for collecting, namely the African violet, nowadays available in this country in many varieties. In the United States this has been top favourite for many years: there are specialist societies and new cultivars are constantly being developed — to date, more than a hundred — as well as new shades, varying between red and blue, and new shapes of flower and foliage all of which enable one to assemble a truly splendid collection.

Know your limitations

One thing must be stressed — you should never attempt the impossible! Someone living in a constantly heated bed-sitting room cum kitchen must choose different plants from someone who owns a villa with rooms facing in all directions and in which the heating can be constantly controlled.

When planning our collection we can take it as a general rule that plants which the grower has cultivated in an unheated greenhouse will only thrive in cool rooms, barely heated in winter. Plants grown in a moderately heated greenhouse are, with some restrictions, those that will succeed in a living-room. Specimens cultivated in hothouses are less suitable for simple indoor use and thrive best in a plant window or in a tropical section of the greenhouse. Details about plant windows can be found on p. 30.

There is nothing against temporarily placing a flowering *Clivia miniata* in a central position away from the light. Afterwards however, it must be returned to its usual position near the window.

Plants away from the window-sill

It is not unusual today to find people knocking down walls, and making three rooms into one, or converting their lofts into spare bedrooms. Indeed, the fact of the matter is that nowadays most of us live in extremely cramped conditions. It is hardly surprising, therefore, if there is now little space left for plants on our window-sills, let alone for a proper plant window. Yet, there are still thousands of people who care about their houseplants and insist on finding a suitable solution to this problem. Suitable, in this context, means finding positions in which our plants can find the correct amounts of light and air. In the centre section of this book (the green pages) this

In modern entrance halls, with an open staircase, there is often sufficient space and light for one to suspend a hanging plant. The plants illustrated here are *Iresine herbstii, Peperomia glabella* and *Alloplectus vitattus.*

point is covered in more detail, namely in the paragraph on light, air and situation (p. 49).

Dangers for hanging plants In this connection the pots used for hanging and climbing plants play an important part.

As has already been said on p. 18, these plants should be watered freely, because the temperature is higher in the upper part of the room. It must also be said that hanging pots and wall containers are usually too small for their purpose, which is particularly harmful where the plant is put straight into the container. As there is a danger of the soil-ball rapidly drying out the plant is watered frequently and because there are usually no drainage holes the moisture on the bottom cannot be properly checked. Inevitably, too, water will be

Dr Ward's nineteenth century Wardian case was the forerunner of our bottle-garden.

spilled, causing ugly stains. Even soilless cultivation does not solve all the problems of hanging plants, since it is often not a simple matter to replace the nutrient solution.

From this point of view, standing containers have the advantage. In the first place we should mention the bottle-garden, as this form of cultivation — provided it is done correctly — is independent of conditions in the room and, in any season, demands a minimum of care — it will even survive long periods of absence by its owner. How to achieve the correct balance in the cycle of vegetative functions within a sealed bottle is fully described on p. 81. Admittedly a bottle-garden will always remain a kind of curiosity, almost an abstract form of 'living with flowers'. Not everybody, therefore, will care for it.

The filigree-like, tall and dense growing fronds of *Cyperus haspan*, a tropical plant with a wide area of distribution, here look almost like a bouquet of dried flowers. The plant resembles a small-scale *Papyrus* and its close relationship to the *Cyperus* grasses shown on p. 185 is obvious.

If you do not have a greenhouse, this fairly inexpensive indoor propagator with built-in heating is ideal for growing indoor plants from seed or from cuttings.

Essential precautions

People who value personal contact with their houseplants and care for them lovingly, will clearly have to find other means of compensating for the lack of space on the window-sill. Provided the light coming in from outside is sufficient, or can be supplemented by artificial light, there are many places in which to grow pot-plants other than the traditional window-sill.

Ugly and attractive plant tables

Using the circular plant table with upper bracket, plus the 'what-not' from grandmother's day, inventive manufacturers have developed numerous small pieces of furniture. However, not all of them are quite as useful as their makers would have us believe and opinions on the designer's taste may also vary considerably. But if we accept the principle that there is no accounting for taste, a standard lamp with flower basket attached has a right to exist as have those small indoor fountains in the middle of a plant trough, illuminated by coloured lights. Apart from these inevitable examples of 'bad taste' there are, of course, many excellent designs, all the more useful for being inconspicuous. That earlier generations also realised that simplicity makes for good taste, is confirmed by the illustration on p. 323 of an original plant table with copper tray dating from 1780. As we see it now forms as charming a setting for African violets as it did, two hundred years ago, for the forget-me-nots or resedas so beloved by Frenchmen of the time.

Indoor propagators

The heated propagator illustrated on p. 27 provides one very simple solution to the problem of growing plants away from the window — it is a kind of indoor greenhouse or plant case. On this page we see a considerably more elaborate version — a version that is modern, tasteful and technically well-nigh perfect — and its ample size allows us plenty of freedom to select the plants we want. The simpler, plastic models on sale at the florist's will require careful handling in order to give lasting pleasure. Usually they do not admit air and are too small to allow plants to thrive in the correct climatic conditions but they are eminently suitable for growing plants from seed or from cuttings. Such a small 'mobile garden' will also, temporarily, form a kind of improved and colourful plant container. However, one should be aware of its limitations and not expect the impossible.

A plant case is practically indispensable

Needless to say a large plant case, with its various automatic installations for looking after the plants, offers much greater opportunities.

If you wish to be entirely independent of light and temperature, this enclosed plant case is ideal for most plants. Light, temperature and even atmospheric humidity can be controlled at the turn of a knob.

At the top of the case are the thermostatically controlled heating system, the watering and spraying vents complete with hydrostat, the carefully adjusted lighting system and, finally, good ventilation. Other electrically operated apparatus, connected to the mains, is placed at the bottom of such a case — in the elegant light metal structure illustrated these are hidden by the wide frame. Nothing could be more convenient.

For plant lovers living in rented houses or flats (and not only for them!) this is a truly ideal solution. In a rented house one is often not allowed to make any structural alterations, either because the landlord refuses to give permission, or because it is not architecturally justifiable to construct a permanent plant window. It should also be borne in mind that the latter would become an inherent part of the house and would be very expensive, if not impossible, to move.

The indoor greenhouse on castors is, therefore, a much more convenient proposition. When you move house there can be no problem concerning the right of ownership: it is your property wherever you go. Nor will the housewife have to be careful when drawing the curtains and she will be able to open the windows wide when washing them, for the small glass house filled with flowering orchids can easily be pushed aside.

Where to find good examples There are then many ways of keeping plants other than on the windowsill. In the section on plants you will see a photograph of an old plant table, the left front leg of which has been somewhat bent as a result of its many years of burden but which, nevertheless, provides an excellent method of living with flowers. It is for this reason that such photographs of indoor settings have been included in this book. The first example will be found on p. 140, where the tropical fern *Blechnum gibbum* stands proudly on an antique piece of furniture. An entirely different atmosphere is illustrated on p. 142, showing the unusual bulbous plant *Bowiea volubilis* placed on a modern telephone table made of white plastic. Another modern environment is that of the magnificent dieffenbachia on p. 189 — this is a truly perfect specimen.

This book also contains two examples of palms in a contemporary setting: the first shows a large howeia or kentia palm (p. 235) which flourishes in lighting from above and may be regarded as a textbook example of a well cared for office plant (see section on 'Plants in Offices' p. 36). The other example is a coco palm promoted to desk plant — its situation may well horrify some solicitous plant lovers (p. 261). In addition there are the firmly grounded pandanus on p. 281 and the schefflera on p. 327. The latter, which is certainly thriving, is being grown by the soilless method.

Views on the plant window

It's a long road from the window-sill with a row of separate pot plants to a fully developed plant window. Unless everything has been properly considered and correctly built, and unless the installation and the choice of plants have been preceded by careful preparation, a flower window will inevitably lead to disappointment. And this is doubly unpleasant since all the work you have put into it, manually as well as mentally, is aimed at achieving a permanent garden in your living-room.

Why do we build our flower windows in this way? Why do we attempt to create such a structure, either with or without a partition

Essential preparations

There are plenty of opportunities for adapting plant containers to the style of a modern interior — witness this window-sill filled with succulents.

on the inside, to house a community of carefully attuned houseplants? Are we chiefly concerned with its decorative function or do we regard it as a kind of status symbol, to be cared for by a florist under contract? Genuine plant lovers will indignantly reject all such suppositions.

They are, however, most enthusiastic about the invention of the genuine plant window with its many technical gadgets because, quite apart from its attractive shape and decorative function, it will provide better conditions for their plants. They know, often as a result of years of experience, that their plants frequently regard it as a doubtful pleasure to be offered the traditional situation in our living-rooms. A properly constructed and correctly planted plant window can largely fulfil the plant's essential requirements, whereas the

Plants feel at home in the plant window

The so-called flower window is becoming more and more popular. In this case it is a kind of niche, with a water-tight tray at the bottom, the whole surrounded by bookshelves.

limited opportunities for care provided by the normal living-room environment only too frequently lead to the plants' downfall.

This leads us to a number of familiar complaints: of over-dry air and a winter temperature indoors which is unbearable for many plants; of too little light during the dark months; of chill behind drawn curtains during periods of strong night frost; of the struggle to achieve the correct climatological balance in summer, to say nothing of the owner's absence on holiday and the problems this involves. There are surely quite enough problems in the care of plants without all these? So do remember, with a plant window most, if not all, of these problems can be solved without difficulty.

Take advice

Space is lacking to discuss all the necessary preparations. Such details are properly the work of the expert who builds the plant window, the interior decorator, and the specialist florist if you yourself lack sufficient knowledge of plants. There are not many professionals in this field who can help you to make the correct choice of plants, but with a little effort, or perhaps with a florist's cooperation, you will doubtless be able to find one. Naturally you will have to pay for such advice but, if your own choice of plants should prove wrong, your large investment will be a financial loss. A balanced choice of plants will be to the plants' benefit, since they will feel most at home in a plant window if they are allowed to form a family.

The plant window – a bridge to the open air

In whatever form the plant window is built, our first consideration must be the well-being of the plants themselves, houseplants already have a considerably more difficult life than their relations out of doors. Nor should we forget that a properly designed and correctly cared-for plant window introduces nature into our homes much more effectively than do the separate pot plants on our window-sill – a point of particular significance to flat dwellers in large cities. Seen in this context, as a bridge to nature, a plant window is much more than merely a hobby on the part of its owner. If we approach the matter from this ethical point of view we will certainly conclude that only the best will do for our houseplants. However modest the scale of our plant window, each new day will bring with it the satisfaction of a job well done and the pleasure derived from thriving plants.

Further advice will be found from p. 85 onwards, under 'Plant windows in practice', where the chief requirements are all discussed. If in addition you read the instructions concerning light, air and situation beginning on p. 49, you will have gathered a fair amount of information for planning such a window. I should now, however, like to add a few directives resulting from my own experience (my most

A floor-level plant window is most suitable for larger plants. This solution is recommended for entrance halls in offices, public buildings, etc.

recent plant window was constructed fifteen years ago when my house was built):

Instructions on design

1. If you have little time available to look after your plants, simply widen a north-facing window-sill and do not glaze the inside. Buy only strong plants requiring a shady position, which will be able to tolerate the atmosphere and temperature of winter. You will be able to introduce some colour in any season by occasionally adding a flowering plant.

2. Your starting point should always be the knowledge that a plant window, however simple, must offer the plants better living conditions than a window-sill with a few pot plants and a radiator below.

3. You should realise that a plant window without a plant trough big enough to hold the larger pots is not really a plant window. But at the same time remember that taking the plants from their pots and planting them straight into the trough is the greatest mistake a beginner can make. The result will be unchecked root growth; the only plants that can be treated in this way are cacti and other succulents.

4. Always take into account that each individual plant has its growth- and rest-period and requires a certain rhythm in the provision of light, heat and fresh air. Design your plant window with the idea of filling it with a previously determined plant community which can be of a permanent nature.

5. Avoid having a radiator, or other form of heating, below the plant window. Even the mildest heat may be too much for your plants to bear.

6. Be careful with the hardness of your water: check the local degree of hardness and if necessary use a water softener. Use soft water for the electric humidifier or spray.

7. Better to spend a little more money at the outset — it will save added expense later on.

It seems to me that after this enumeration of problems and possibilities, in which I have tried not to make matters appear more favourable than they really are, you will be able to enjoy the examples selected all the more. Any plant lover with a modicum of knowledge of the subject

An indoor propagator in addition to a plant window

will realise that our illustrations are not carefully arranged studio photographs, but well-functioning windows in private ownership. In the case of the magnificent tropical plant case on p. 35 it should be added that the owner also possesses a tropical hothouse in which he brings his orchids to flower before putting them into the plant case for a few weeks. Outside their flowering season such plants are not so attractive.

Hot house conditions can be created by separating the flower window from the room by means of a sliding window. The atmosphere can be maintained entirely automatically. Even orchids are relatively easy to grow in this way.

Plants in offices

This decorative *Ficus lyrata* in a polyester tub has been placed on the mezzanine floor of a bank building. It receives an extra dose of artificial light from the ceiling.

Deadly decor

In our grandfathers' day the most that one could expect to find in an office by way of floral decoration was a prickly cactus or perhaps one of those laurel wreaths gracing the bust of some local dignitary. The overall impression was, if not of death then, certainly of sober propriety.

The use of lasting plants

It is really quite remarkable that, notwithstanding his general lack of time, modern man has exchanged these easily cared for objects and taken on 'living' plants which are much more difficult and time consuming to look after. Following a period, during which offices and hotels were decorated with short-lived, cut flowers, the demand arose for more lasting plants to provide a living decor. The fact that today this need is recognised by society in general, even by boards of directors and government authorities, shows what a vast subject the care of houseplants has become.

What is an 'office'?

Naturally we are not only talking about plants in offices but also in workshops, schools, post offices, petrol stations, chemists and waiting rooms.
Today many magnificent plants are to be found in all these places and this inevitably makes for a number of problems. The plants in a boardroom, for instance, lovingly tended by the manager's secretary enjoy a happy fate, as do 'contract plants' looked after by professionals. Plants on a living-room window-sill, on the other hand, whose owner, though often full of goodwill, frequently lacks sufficient expertise, are not so fortunate. This section is addressed, therefore, not merely to people in offices, shops and hospitals, but plant lovers in general.

What everyone should know about the care of plants

As we have already said, it is essential that all growers of indoor plants possess a sound knowledge of the basic principles of care and cultivation. This being so, let me say at once that there are five essential factors for providing reasonable living conditions for all plants rooted in a narrow pot. These pre-requisites are the same whether plants are in the home or in a working environment.

Factors which must be taken into consideration

The five factors are: light, humidity, temperature, water and nourishment. Naturally, these also embrace the wider issues of situation, comparative room temperature, periods of rest and growth and other details with which one must be familiar to enable the plants to thrive.

If one of these factors is lacking the plant may suffer; for instance, where we force a plant used to shade to live in a sunny situation or another, whose origin was in a soft-water area, to endure hard and often excessively cold water. Another instance of cruel treatment is where a plant which requires cool surroundings during its dormant period, is forced always to remain in a normal room temperature with its dry atmosphere.

To this must be added the fact that working areas have changed even more radically than other features of our modern homes. Modern architecture with its functional character, sober shapes, large windows and centrally heated interiors, provides greater opportunity for plant growth than was the case in earlier decades. Up to a few years ago long weekends with too much sun and too little heat created a problem for office plants, but this is no longer the case, today's problem is that of a constantly maintained temperature. On the whole it is better to turn the thermostat down a few degrees at night.

Changes in architecture

In a work corner plants soften the ▷ business-like effect. A strong plant with a relatively good life-span will give you more pleasure than a bunch of flowers fading after a few days.

The right choice is important

It is therefore more important than ever to choose the right plants for the surroundings in which we work. The appearance of the plant should not be the first criterion. It is much more important to ask oneself which plants will thrive best in given circumstances. Everyone knows that foliage plants are attractive all the year round and also that they are easier to look after than flowering plants with their dependence on a certain growth rhythm. The photograph on the facing page gives a good example of the friendly atmosphere which can be created by colourful plants in the work corner of a living-room. It is possible, moreover, to introduce colour into the greenery, by placing flowering plants in between, thus providing flowers all the year round. Any reasonably priced pot plant is suitable, usually an annual bought only for its flowering period and then, since it is difficult to keep this kind of plant in an office, thrown away.

Colour provided by flowering plants

My advice after many years of experience is, therefore, to select only lasting, undemanding plants, those, for instance, which do not require too strict a rest period or any other form of treatment which is difficult to provide in an office. In this connection one should also bear in mind possible absence during weekends and holidays.

Preference for undemanding plants

In many firms it has become the custom to club together to give colleagues a pot plant for their birthday. To avoid mistakes in such cases, it is useful to consider first of all which plant would be best

Plants as presents

suited to the spot for which it is intended. Since there is no lack of choice in the relatively large number of suitably undemanding plants, the recipient might be asked to state his preference in advance.

A difficult question: which plants will go where?

Certain plants are unsuited for use in offices, reception rooms and other places where it is difficult to look after them. These are primarily species which will flourish only in a thermostatically controlled tropical environment, demand a high degree of atmospheric humidity as well as special water and soil, and which, furthermore, require regular spraying. I also regard those plants as unsuitable which require a rest period in a moderate or cool temperature; among these are the myrtle, the passion flower, the hoya, the bougainvillea and clivia. In a normally heated room these plants develop long and weak roots and are soon attacked by greenfly with the result that they will not flower a second time. For this reason a number of cacti are also unsuitable for leaving all the year round in a functional environment; they definitely demand dry and cool conditions in winter which, of course, runs contrary to the human desire for greater warmth. In addition it should be remembered that few pot plants are sun-lovers. Plants subjected to the heat of the sun through an unscreened window will perish not only in summer, but even in the bright weather of winter and spring.

Problems in winter

In rooms where people work it is also sensible to limit one's 'horticultural' efforts. The best thing to do is to create simple groups of plants which will eventually develop into harmonious communities. For a sensible selection of plants we clearly cannot do better, initially, than turn to those veterans of the 'green front' which, over the years, have more than proved their powers of endurance. For instance, a bowl of sansevieria in a south-facing window will not only provide a decorative effect with its yellow-edged leaves but will also produce fragrant flowers every spring. Or we could take a *Collinia elegans* (syn. *Chamaedorea elegans*), that elegant mountain palm from Mexico. This is an undemanding, slow-growing plant which often produces mimosa-like flowers in its second season and, although preferring light, will tolerate a great deal of shade.

The art of restraint

Office plants must be vigorous and should never be placed in small pots, since this would result in lack of moisture during weekends or holiday periods. In the photograph *Monstera deliciosa* has been combined with dieffenbachia and other plants.

'Office landscape' and working climate

Now that architects and designers have become interested in what they call 'office landscape', plants have gained even more in spatial and decorative importance. Examples of such landscaping may be seen in nearly all large buildings, and now we have come to realise that plants make a working environment more agreeable, we can hardly imagine being without them. An interesting development, however, is the search for practical solutions to the problems created by an absence of natural light. Technical advances in the field of 'botanical light-physiology' are almost limitless. Naturally plants growing in artificial light must be provided with perfect conditions. This will be discussed in the following section on hydroculture.

Soilless cultivation and artificial light

From Aristotle to von Liebig

Justus von Liebig, who was born in 1803 in Darmstadt and died in Munich in 1873, is still regarded as one of the most important chemists and researchers in history. He is sometimes called 'the father of modern agricultural science', because he discovered several processes in organic chemistry. He also invented a method of silvering glass to make mirrors, made an important contribution by inventing and introducing a method to make meat extract and, last but not least, he once and for all put an end to the dispute about the way in which plant-roots take nourishment from the soil. Up to that time not a single scientist had been able to answer this question, which had been raised as early as 284 B.C. by the Greek philosopher Aristotle. Justus von Liebig, sometimes scornfully referred to as the 'father of artificial fertilisers' in his *Die Chemie in ihrer Anwendung auf Agrikultur und Physiologie* published in 1840, established irrefutably that 'inorganic nature is the sole source of plant nutrition'.

Plants do not eat soil

It has therefore been known only for the last hundred years or so that a plant does not 'eat' soil, but draws its nourishment, such as nitrogen, potassium and phosphorus from the soil in a watery solution. This quickly led scientists to the logical conclusion that plants would be able to live without soil provided one could supply the roots with the necessary elements in a dissolved form. However, proving this theory was not such a simple matter. Although a Swiss Botanical Garden had apparently grown a tropical tree in a nutrient solution, around the beginning of the century, this was merely regarded as a curiosity.

The actual principle appears most convincing. Food tablets or concentrated liquids to be diluted with water are made in accordance with exact calculations. When plant roots are immersed in such a solution, they have the same elements at their disposal as in good potting soil. The degree of concentration, moreover, is constant. Obviously such a pill for soilless cultivation or concentrated nutrient solution, cannot be compared with ordinary artificial fertilisers in solid or liquid form. The latter usually lack trace elements, though this does not matter as they are generally present in sufficient quantities in the soil.

◁ Provided there is sufficient light, an intimate corner can be created in an industrial canteen by means of a screen of plants. The photograph shows *Ficus benjamina, Syngonium auritum* and *Schefflera actinophylla*.

You will be able to read more on this subject from p. 91 onwards. It may, however, be stated that the various products necessary for soilless cultivation have nowadays reached an unprecedented degree of perfection; witness the fact that nearly all plants flourish when grown by this system.

But let us continue with the history of hydroponics: which is every bit as gripping as a novel. An interesting development took place during the last war. In the middle of the Atlantic, 1500 kilometres (937 miles) from the continent of Africa and 1300 kilometres (812 miles) from St. Helena, lies Ascension Island where volcanic ash, lava, sand and rats are found, but no vegetation to speak of. In 1942 this dreary place had become an important military base with an exacting American garrison. Fresh vegetables were important for maintaining the GI's morale and since they were difficult to obtain, even by air transport, some clever engineer had the idea of trying hydroponics. Enormous flat tanks were filled with a nutrient solution and covered with a metal grating through which the vegetables could grow. It had, after all, long been proved that this form of cultivation was practicable. The transport problem was thus rapidly solved, for instead of frequent transports of large, heavy crates of vegetables, it was now only necessary to fly in the installations and an occasional quantity of germinating plants. Later the same method was introduced into Hawaii and the Philippines, where certain bacteria in the soil made locally grown fruit indigestible for white people. These bacteria

Fresh vegetables on Ascension Island

Where a partition is required at some distance from the window, artificial light can come to the rescue. In this photograph you can see a special plant table with fluorescent lighting.

naturally did not occur in the nutrient solution and the taste of the fruit was improved. In colder climates the same system is used under glass.

From commercial growing to the amateur plant window

Initially, growing vegetables under glass by the soilless method appeared to have commercial possibilities even in Europe. It was hoped that with this new system the growing of lettuces, cucumbers and tomatoes could be completely rationalised. Growers also experimented in their greenhouses with indoor plants, arranged separately in small glass jars containing nutrient solution. However, after a few years of experimenting it was found that, because of all the technical installations required, the vegetables became far too expensive. Attention was now centred on the houseplants which originally were only of secondary importance. In the United States, Latin America and Switzerland the cultivation of houseplants by the soilless method has already made triumphant progress and looks like continuing to do so.

Initial problems and how to overcome them

In the beginning problems arose, chiefly in connection with the supply of oxygen to the under-water roots. I clearly remember huge pumps being installed into commercial greenhouses, while in domestic surroundings a bicycle pump was used to 'aerate' the solution. This sort of thing did not exactly stimulate the development of the hobby, especially as the 'time saving' which was claimed to be one of the chief advantages of the system was thus completely nullified. And since larger plants, such as the rubber plant or philodendron, could hardly be grown like lettuce in a small metal frame or in sphagnum moss, or wood shaving soaked in nutrient solution, and since, also, window-sill containers cannot, like an aquarium, be equipped with a complete oxygen supply, people began to experiment with materials which would be able to give the plants support as well as air. The first requirement for such materials was that they should on no account be affected by the nutrient solution.

The drawback of pumice

Conversely the material should not cause chemical changes in the solution or stimulate root-rot. From the first, various grades of basalt splinters were used. People who feared that the fine roots would be damaged by the sharp edges of the stone were, like so many critics of later famous inventions, proved wrong. But since there was at first obviously little appreciation of the connection between growth and the pH of the nutrient solution, and especially of the supporting material, it happened that, apart from basalt splinters, ground pumice was often used — to such an extent in fact that the system was sometimes called the 'pumice method'. True, this ground pumice had the excellent properties of being sterile and much lighter in weight than the heavy basalt splinters, and inspired confidence because of its soft and porous nature, but nevertheless it was like bringing in the Trojan horse, for untreated (as it was mostly used), pumice contains a great deal of lime — generally recognised to be extremely harmful to some plant roots. Yet, when confirmed lime-haters did not thrive when grown by this 'pumice method', nobody at first traced this fact to the obvious cause. In such cases it was said that 'these plants are unsuitable for soilless cultivation . . .'

Practically all houseplants can be ▷ grown in a nutrient solution in these special watertight containers with a filling of compacted clay granules. This new method of soilless cultivation was developed in West Germany.

Most houseplants are suitable

It is now realised that in principle all houseplants can grow without soil. However, there are, of course, certain plants which, because of their powers of adaptation, do better than others. To begin with, one should select plants which can remain in the living room all the year round, without losing their beauty. The photograph on the facing page gives a good example. In addition excellent examples of single or grouped plants grown by this method are to be found on p. 125 (anthurium), p. 233 (hippeastrum or amaryllis) and p. 327 (schefflera).

It is hardly within the scope of this book to describe how much the soilless system has meanwhile spread. In many countries special nurseries have been established for growing young plants in a nutrient solution. Plant lovers are thus relieved of the laborious change-over from soil cultivation to water cultivation. In addition, the plants enjoy the advantage of more rapid and healthier growth. Good results were also obtained by Luwasa, originally a Swiss firm, who invented a system in which the plants are placed in a kind of expanded clay medium. The nutrient solution fills the bottom half of the containers.

Possibilities practically unlimited

A new form of artificial lighting for plants where daylight is lacking

For domestic use: what plants require and what they tolerate

In addition it was thought that plants which lacked enough daylight could be encouraged if they were given artificial light. In this respect many plants both in living rooms and in offices have a difficult time, particularly in winter. There are even places where one would like to grow plants but in which there is no daylight at all.

Thanks to the latest systems, mistakes in illumination can be avoided and no undesirable side effects occur, such as lank shoots and incomplete development of flower buds.

As you may know, electric light can be divided into various wavelengths. To begin with eight groups were identified; later numerous experiments were undertaken to find the 'kindest' radiation. The results of these experiments led to the introduction of fluorescent tubes and high pressure mercury lamps which have properties favourable to plant growth. It would be perfectly feasible to adjust artificial light to the individual needs of shade-loving plants from the tropical rain forest or of natives of sunny deserts and steppes. However, for a window-sill with a mixture of houseplants, or for an office interior with an equally varied selection such individual adjustment would be quite impractical. Attempts were therefore mainly aimed at obtaining a lamp which averaged good results for a large number of indoor plants.

This modern style of decorative plant illumination has great attractions for many people. The photograph on p. 44 shows how it can look in practice. However, success is assured only where plants grown in artificial light enjoy maximum conditions in other respects as well. This perfect combination of growing conditions is rarely enjoyed by soil-grown plants; these can hardly do without the helping hand of daylight. Plants grown by the soilless method, on the other hand, tolerate a lack of daylight much better. Both at home and abroad it has long since been proved that soilless cultivation is essential to the health and development of plants grown in artificial light.

The best possible living conditions

Finally a few practical hints for plant illumination at home and in the office. If fluorescent tubes are used – these have the advantage of being inexpensive and giving off little heat – the best spectral distribution is given by a 'daylight tube'. The distance between a light source consisting of two 40 watt tubes and the top of the plants should be 40–80 cm (16–32 in). A ceiling fitting is therefore not adequate. The strength of light can easily be established with a light meter. From 1000 lux upwards is sufficient for undemanding plants; flowering plants such as begonias and African violets need about 5000 lux.
A high pressure mercury lamp gives off more light, but also more heat and is considerably more expensive, particularly because of the transformer required. Distance from the plants should be 1–2 m (3–6 ft). The period of illumination (never more than 16 hours) can be controlled with the aid of a time switch.

Correct doses of artificial light

2. THE PRACTICAL CARE OF PLANTS

Light, air, situation

A plant, like man, is an extremely complicated creature. Its various aspects of growth are exactly attuned one to the other and are greatly influenced by outside factors.
While it is beyond the scope of this book to enter into great technicalities, we shall nevertheless have to explain a number of points in order to achieve a better understanding of the correct care of our house plants.

There is risk in sunshine and in shade

'Light is the soul of a plant's life' said an early Victorian writer on the art of cultivating indoor plants. In prosaic language this means that no green plant can live without light. The amount of light needed by different plants, however, varies considerably. Some require a great deal; others are happy in moderate shade. In any case light is not the same thing as full sunlight, which can make a window very hot and scorch the plants. On the other hand, lack of light has a very detrimental effect on all plants — and especially on pot plants, which frequently suffer from too little living space; in fact, it may in some case be fatal, since the main source of energy is absent.

Assimilation or photosynthesis

It is well known that without light a plant cannot start its growing process. This process is controlled by light and by chlorophyll and is called assimilation or photosynthesis. The leaves absorb carbon dioxide from the air through microscopically small pores. Under the influence of light, this carbon dioxide (CO_2), together with water supplied by the roots and the mineral salts it contains is converted into organic matter (carbohydrate), releasing oxygen in the process.

The fact that plants release oxygen means that they play an enormous part in creating a healthy atmosphere in the living room. Cut flowers do not produce oxygen.
So why it is that, knowing that light is essential for the plant's very existence they are so often robbed of it? We should realise that the human eye is a bad light-meter. We often consider there is sufficient light at some distance from the window, whereas the hoya, the clivia and the sparmannia would find the darkness at such a spot debilitating.
Just a remark in passing concerning the word photosynthesis. Any amateur photographer who knows how to use a light-meter can measure the light objectively. He will find that, with every step away from the window, the marking pointer will move rapidly backwards. Tables and graphs are available which are more detailed than the diagrams given below, which have purposely been kept simple.

Indoor plants supply oxygen

All houseplants can be divided into three groups. The first, and oddly enough the smallest group, consists of plants growing in deserts, on steppes, on high mountains, moors and other barren areas entirely devoid of shade. They tolerate this natural situation, for there is a real

An objective measurement of degrees of light and shade in a room will give results very different from those estimated by the human eye.

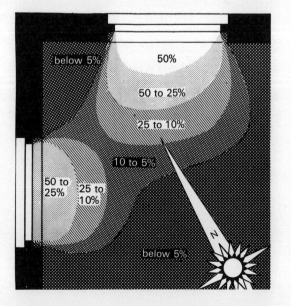
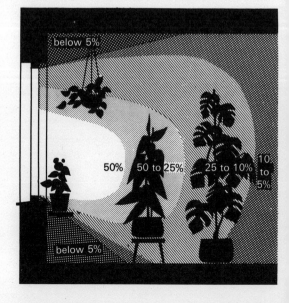

difference between outdoor positions in full sunlight and those behind heated glass, as is the case with windows facing south.

The second, and by far the largest group is formed by plants tolerating unshaded daylight, but requiring protection against fierce sunlight around noon, or against direct radiation from the afternoon sun.

The third group consists of so-called shade plants. In addition there are numerous groups in between. Among the third group, some do not object to some morning sun; others are endangered by direct sunlight. To this latter group belong, among others, ground covering plants originating in tropical rain forests, such as foliage begonias, ferns and mosses. Cyclamen and azalea do not like sunlight, but are nevertheless good examples of plants which require more light when in flower (and more water as well!)

A plant's right to live

In practice rules, alas, again and again prove to be merely theories. Quite often we see our modern interior decorators using flowers and plants as if they were pieces of furniture or pictures. For instance: an empty spot beside the bookcase in the corner? It might not be a bad idea to put a little table there with a rubber plant ... and at the same time we could move that large fern forwards a little on the window-sill, more in the light so that we can see it better ... But no, I am afraid that is not the way to go about it. We are responsible for our house-plants — this can never be sufficiently stressed. It is not necessary to become enslaved, and to impose restrictions on oneself for the sake of one's flowers and plants, making the living room less comfortable in the process; but one should try to take a plant's needs into account, for instance by giving it the correct position.

Lack of light is obvious

It is easy to see from a plant's appearance whether it gets sufficient light or not. An only too obvious indication of lack of light is provided by the rank, and almost white, growth of a pelargonium kept in a cellar in winter. Long sections of bare stem and pale green leaves show that a philodendron, which actually does not require all that much light, has, nevertheless, too little. Moving plants from a dark to a light situation, or from one room to another, can cause damage by excess of light. We know, for instance, of someone who wondered why her camellia had lost nearly all its flower-buds and even some leaves, although she had placed it in the sun every morning to get it to flower, moving it from its permanent position facing north to a window-sill facing east to catch the morning sun.

To move or not to move?

The answer is simple and connected with phototropism, in other words, the fact that plants naturally turn towards the light. This frequently distorts a plant somewhat, so that it is often the less beautiful side, with fewer flowers, which faces the room. As a rule houseplants do not like being moved or turned in order to make them grow in a different direction. In the case of vigorous green plants it does little harm to turn them carefully to encourage them to grow regularly. On occasion it is essential to move a plant: for instance, if we want it to spend the summer out of doors and bring it back inside in autumn, some deterioration is inevitable. A dormant period, moreover, should never begin in a fully heated room.

Among our favourite indoor plants a number are particularly sensitive to moving and turning — especially the azalea, the gardenia, many cacti (Christmas cactus, Easter cactus!), the camellia and the hoya during the period when the buds develop and open. Indoor rose plants, too, when almost completely in flower, may lose petals and even leaves if they are moved or turned at an early stage of their development. The clivia, in fact, will not tolerate movement at all. This problem is discussed in greater detail in the individual plant descriptions.

Marking pot and position.

Marking the pot and its situation in red or blue will prevent a change of the angle of light. Alternatively, tiny coloured labels can be used, or perhaps you can come up with a good idea of your own. These marks will ensure that the plants are not placed in a different position when they have to be moved for a short time, for instance while airing the room or washing the windows. Clever people will meanwhile have noted that all these separate pots are rather inconvenient. It is much better to place the pots together in peat-filled troughs or trays. If for some reason this is impractical, the pots could at least be combined in shallow bowls (see also p. 27).

If one trough should be too heavy, since it may contain a layer of coarse pebbles or gravel for drainage, you could have two or three separate ones.

Sidelight and toplight.

Of course the problems caused by the direction of the light occur only when the light comes from one side, as is the case of the window-sill. Plants surrounded on all sides by daylight — as they are in their natural state — need not twist and curve in all directions in order to obtain some extra light and will consequently not grow ridiculously long and weak shoots. Greenhouses provide the maximum amount of light available under artificial growing conditions and the fact that the light comes from above will ensure that the plants are not harmed as a result of phototropism.

Situation in relation to the sun

North

(north-north-west to north-east)

Advantages. Always 'slight shade' without having to screen, and constant temperature. The best situation for all plants naturally growing in shade and for plants from tropical rain forests. Little maintenance: no damage from scorching or drying if one forgets to water or to screen. Screening in fact not necessary at all.

Disadvantages. Too much shade from neighbouring tall trees or houses can be harmful to the plants by robbing them of too much light.

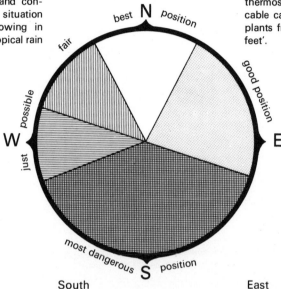

Artificial light can overcome this problem. Danger of light frost is greater in this position than anywhere else, but can be prevented by good insulation and double glazing; rolling shutters can be recommended as protection against wind and radiated cold. To prevent too severe cooling of the filling material in plant trays a thermostatically controlled heating cable can be introduced. Protect pot plants from damage caused by 'cold feet'.

West

(west-south-west to north-west)

Advantages. Positions towards the north can be very suitable for plants requiring a great deal of light and heat, as the rays of the late afternoon and evening sun enter here.

Disadvantages. In periods of strong sunlight a position facing directly west or slightly south-west may create similar problems through overheating of the window and cooling at night as does a situation facing south. Wind and rain often come from the north-west as well as from the north.

South

(east-south-east to west-south-west)

Advantages. Prolonged and abundant light.

Disadvantages. Without effective screening this side is dangerous to plants except when the sky is cloudy. Even cacti or other succulents originating in a desert climate cannot tolerate strong sunlight. An insufficiently damp atmosphere and incorrect watering (drying soil, scorching foliage) are naturally detrimental to the plants. In winter the contrast between day and night temperatures is considerable.

Important: screening should be outside the house. Remember to regulate the amount of light let through. Prolonged and intensive screening is more harmful than a position in the north, with deep shade.

East

(north-north-east to east-south-east)

Advantages. Friendly morning sun comes in by the window and encourages the growing process. Positions slightly towards the north or slightly towards the south are very suitable for arranging large, lasting groups of plants. Also good for plants enjoying slight shade. Little danger of scorching or drying out.

Disadvantages. The north-east side may suffer from rain, wind and cold. In strong sunlight screening is essential. Screening against the winter sun is necessary if deciduous vegetation outside, which gives shade in summer only, is unable to provide this.

Growth in artificial light

By now every plant lover knows that artificial light can come to the rescue where there is lack of natural light. Once one has some experience one can even influence bud-formation in certain plants with the aid of extra light. With regard to light we distinguish between neutral, short-day and long-day plants.

Little light for the bud formation of short-day plants

Short-day plants will produce flower buds only in the dark season, when there are less than 12 hours of daylight. A convincing example is *Euphorbia pulcherrima*, the poinsettia. Although it may appear that this plant continues to flourish, it does not flower for the second time, or rather, does not produce flower-like red bracts with very insignificant true flowers, to the great disappointment of its owner. The reason is that the plant gets too much light; even the electric light burning in the room at night is sufficient! This encourages vegetative growth, causing excessive development of foliage, while generative growth (i.e. the development of flower buds) is absent. On the other hand a correctly cultivated poinsettia can be made to flower at any time of the year by ensuring strict alternation of 10 hours of light and 14 hours of darkness, while otherwise caring for it normally. But do not move the plant backwards and forwards for this purpose: leave it in its customary place! Long-day plants need the opposite, that is, at least 12 hours of daylight. If we connect this with the course of the seasons, this naturally applies to summer-flowering and early autumn-flowering plants. We need therefore not be surprised when plants flower at unusual times; it isn't magic, merely the results of expertise and technical perfection in the establishments of various growers. The scientific word for plant's dependence on the rhythm of light/dark is photoperiodism.

Artificial light or daylight?

The fact that in greenhouse cultivation the growth of indoor plants can be influenced with the aid of artificial light has long been known, even among amateurs. It used to be said that artificial light alone did not give good results, but once more it has been proved that nothing is further from the truth. Incredible technical advances are continually being made in the development of better and better lamps for extra illumination, cyclical illumination, etc. It is useless to mention names and types: they will have been superseded tomorrow. Rules for the care of plants in artificial light only will be found in the chapters on soilless cultivation on pp. 46 and 91.

Out of doors in summer

These few notes on plant positions would be incomplete if we did not mention plants which can be placed out of doors in summer. Which

Surrounding the plunged pot when taken outdoors with a layer of gravel will prevent harmful insects or an excess of moisture from entering.

plants can go outside, and when, depends on the climate. Once the 'Ice Saints' have departed (mid May), you might begin to think about it; possibly the plants cannot go out of doors until early June — that depends also on what the spring has been like. Before giving the plants a definite position in the garden it is desirable — and this applies in particular to plants requiring a great deal of light — to institute a certain hardening-off period, for the 'climate' outside is definitely different from the 'climate' in the living-room. A transitional period in mild and rainy weather is very suitable — to begin with possibly in a shady and sheltered position.

Under no circumstances should plants be taken from their pots when placed outside. We have said this before in a different context. This restriction of too great a liberty for the root system is a kind of long-term precautionary measure, for in autumn you would be faced with the consequences: the plant, with its large and unruly cluster of roots, would no longer fit into any pot. It would be badly damaged by being confined to a too small pot and soon nothing would be left of its healthy condition. All this is much worse than the reverses occurring when the first cold nights, with temperatures of 8–10°C (46–50°F) force you to take up the plants.

The foregoing paragraphs, of course, give general rules only, which must be adapted to each plant. The point of time at which a plant should be put outside in summer and brought back inside in autumn depends on various factors, temperature being the most important one. It is also important to prevent damage to a plant by rapid cooling; this should be taken into account in bringing a plant through the winter.

Typical hothouse plants remain indoors

To begin with we should consider whether a houseplant is suitable for passing the summer out of doors at all. Naturally enough, hothouse plants, originating in tropical rain forests, are unable to face the short western European summer anywhere but indoors. Nevertheless, among plants which normally spend the entire year in the living-room, there are some which feel perfectly at ease on a balcony, a terrace, or in the garden.

This applies to a fairly large number of cacti, such as aporocactus (rat's-tail cactus), large globe and column-cacti, and the large epiphyllum cactus, formerly called phyllocactus (do not confuse them with the Christmas cactus or the Easter cactus).

Borderline cases: exceptions, etc.

Not all houseplants require to live permanently indoors. Plants such as the araucaria, the fatsia, ivy, billbergia, azalea, abutilon and others enjoy being placed out of doors. Others, such as the sparmannia, feel differently and would already abhor a short excursion to the balcony. Opinions differ as to the rubber plant. According to some plant lovers, this is a border-line case, like many other stalwart houseplants, with which one may have happy results or not. This also applies to chlorophytum and tradescantia species (except, of course, species which grow wild); at most these may be given a very sheltered position on the balcony.

The sturdy *Passiflora caerulea* (only this particular species!) and *Saxifraga stolonifera* (mother-of-thousands) adjust very well to outside conditions and, in a mild climate, may even pass the winter out of doors if given some protection.

Among houseplants which definitely ask to spend the summer outside we must certainly count the camellia, in addition to such balcony and tub-plants as aucuba, oleander, fuchsia, pelargonium, etc. Other plants which enjoy being out of doors have been mentioned above; they can be found in the garden in company with the camellia. It has only recently been found that the cyclamen, too, likes to join the party.

A subject such as this cannot be treated in depth here, since all sorts of variable climatic factors and characteristics of the individual plants have to be taken into account. It is therefore hardly feasible to give exact instructions for achieving the best results. You would probably do best to experiment a little.

A holder like this prevents large plants overturning; the legs can be shortened as required.

Water and heat

One of the major problems in the care of houseplants is the present condition of the tap-water in many areas. From more than half the taps at our latitude flows a liquid which, if not yet harmful to man and beast, is nevertheless fatal for many plants. The high content of calcium and magnesium cause important physical changes in the compost, which becomes calcareous, and this is poison to many plants. The less chalky a plant's original habitat, the more rapidly it will decay if cultivated in such conditions.

Symptoms of this poisoning by calcium are: white blotches on the outside of clay pots (see drawing on p. 60), white chalk spots and blotches on the foliage, caused by spraying with hard water, and quite often on the pot and pot-soil a mass of acid-smelling blue-green algae. To avoid these troubles it is necessary to know the degree of acidity of our drinking water.

Hard water is harmful

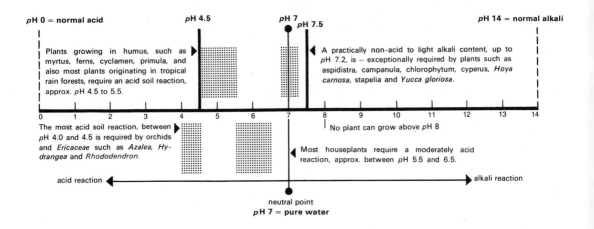

| | The system of pH values | Officially this value is determined by the chemical measurement of the pH, a rather complicated method for laymen. As you will see from the diagram, the neutral point is pH 7; to the right, up to pH 14, are the areas which are harmful to plants and within which no plants can grow; to the left of the neutral point you will find the lower pH values, representing a more or less acid soil reaction, suitable for plants. |

It is the best pH value for your plants   The majority of our houseplants — including most cacti and other succulents — are satisfied with pH values between 5.5 and 6.5. Only avowed lime-haters such as azalea, camellia, orchids and other plants from the tropical rain forests demand pH values to a minimum of pH 4.

Degrees of hardness   The lime and magnesium content of water, but not its chemical reaction, may be more simply determined by telephoning the laboratory at your local water department to enquire about the degree of hardness of your water supply. This will normally be expressed in parts per million (ppm) (i.e. parts of lime and magnesium in one million parts water) and varies throughout the country. In the course of the years this figure has increased enormously and it is quite likely that, if you had enquired a few years ago as well, you would find that, in the meantime, it has risen by several points. The water department has maps available, showing the degree of hardness in different parts of the country. There are few places where the water is still nice and soft. In general it may be said that degrees of hardness above 150 ppm endanger the growth of acid-loving plants.

A pratical pH indicator with colour scale on the lid.

If the water in your area is between 100 and 120 ppm, it is advisable to leave the water for your plants standing overnight. The simplest way is to fill the watering can immediately after watering. Above 120 ppm the water may be boiled; part of the hardness will be deposited in the form of scale. After cooling the water can be used for watering and spraying. A degree of hardness of 120–150 ppm can also be treated with the aid of peat fibre. Put a few pounds in a linen (or better still muslin) bag, hang this in a small bucket and pour the hard water over the peat. Leave the bag in the water overnight; the softened water can then be used. The peat can be used several times.

If the tap-water is harder than 150 ppm, more drastic measures will have to be taken. Some people think that a water-softener based on ion exchange, such as is often used domestically, will do the trick. You know them – those appliances which have to be rinsed with salt. The water thus produced is 'soft' water, these people argue, so why should it not be suitable for plants? This is easy to explain: the principle of ion exchange is based on exchanging the 'hard' calcium and magnesium ions for sodium ions derived from the kitchen salt added. The water is now called soft, because no chalk is formed in the washing machine. But the hardness which is harmful for plants is now merely a different form of hardness, based on the bicarbonate-ion, and this so-called carbonate hardness, or temporary hardness, of the water can only increase as a result of the exchange of ions. This method of water softening is therefore useless.

Bicarbonate can be removed from water by the addition of sulphur, but the quantities must be measured so exactly that this method is unsuitable for domestic use. A fraction too much sulphur and your plants will die.

To solve all your problems you will need a demineralisation apparatus. As the name indicates, it removes all minerals, leaving distilled water. Nearly every garage has such an apparatus for making distilled water for filling batteries. For houseplants it is possible to buy small filters, but these are expensive in use. Personally I have for years had good results with an inexpensive demineralisator consisting of a perspex tube filled with special, minute grains of a synthetic material. The water runs in at the top through a rubber tube and runs out at the bottom completely distilled. When the filling is used up it discolours and can easily be replaced. To make such an apparatus easy to use, I would advise you to have a special tap installed in the kitchen or in the utility kitchen, to which it can be permanently attached by means of a rubber tube. Every day you quickly run off the required quantity of water. If you do not want to demineralise the water completely (this is in any case not necessary), you can mix the distilled water with

How to soften water

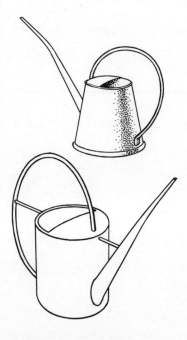

Watering cans should be comfortable to handle; the spout should be wider at the bottom.

Plant sprays moisten the atmosphere for a short time, but create water drops.

ordinary water. Such an apparatus is also very useful for making coffee and tea; it removes any taste of chlorine and even radio-activity.

Rainwater is no longer always suitable

Why take all this trouble you may ask, when plenty of soft rainwater falls from the sky. By catching it in a barrel, you solve the problem. Yes, indeed, it used to be so, before we had air pollution. If you look in the rain-water butts in urban areas nowadays you will find a thick oily layer floating on top. Don't give such water to your plants unless you want to kill them off. On the other hand, if you live in an area where the rain is still pure, you really have no problem and can do without the water-softener.

Water must not be too cold

When you read the plant descriptions in this book you will note that under the caption 'watering' you are often warned against using the water too cold. It is a fact that all houseplants detest icy cold water, even while resting in a cool place. Get into the habit of always filling your not too small watering can with tap- or rain-water immediately after watering and standing it above, or at least near, a radiator; in this way you can be sure that the water will always be at the right temperature.

Dry air is harmful for your plants

The automatic humidifier saturates the atmosphere without drops being formed. It should ideally be filled with distilled or demineralised water.

You will be aware of the fact that the air in centrally heated rooms can be extremely dry, especially in winter. It is, of course, very useful to spray the plants frequently, and this does some good; the question is whether you will always remember to do this in time. For this reason the purchase of an electric humidifier or spray (see illustration) can be recommended. Moister air is more pleasant for people as well. But remember that such an apparatus should ideally not be filled with tapwater, especially if it is hard. Here, again, the demineralisator comes to your aid: the pure water it produces is excellent for all types of humidifiers.

Plants dislike having cold feet

A sufficiently high soil temperature is one of the most important conditions for bringing a plant successfully through the winter. Unfortunately many plant lovers err in this respect: they may check the temperature of the environment, but disregard the soil temperature. It has, however, repeatedly been proved in practice that plants will tolerate a low air temperature better than a lack of soil heat. This is understandable when you think of the situation in nature. The problem is made even worse if, in addition, the plant is watered too liberally and with water that is too cold.

It is instructive to check the differences in air and soil temperature with the aid of a simple thermometer. Behind a badly insulated window the difference may be as much as 8–10°C (16–20°F). Result: severe damage to the roots, often followed by the entire plant dying. Plants originating in the tropics, in particular, easily suffer from cold feet. Even in the so-called dormant period, the soil temperature should really not drop below 18°C (65°F). All these problems may easily be solved with the aid of modern heating cable, which can be used for many other purposes as well (see p. 70). Naturally you should make sure that the soil does not dry out as a result of heating, and that the roots do not become *too* warm. This can be avoided by regularly checking (push your finger in the soil) or by the use of a simple thermostat.

White crust on the flower pot indicates an agglomeration of mineral salts in the potting compost.

Potting composts and feeding

Even in the days of our grandfathers it was considered 'a law of the Medes and Persians' that garden soil should never be used for potting houseplants. There were, in fact, special potting compost mixtures; what is more, old text-books contain pages and pages describing dozens of special mixes, produced to comply with various points of view. And wasn't it fun searching for a nice soft mole-hill to be used for one special plant, or, after years of work, managing to produce a really good compost. And what about those determined plant lovers, who resolutely tracked down old pollard willows in a search for the small amount of dark brown humus occasionally hidden in the interior?

Meanwhile a great deal has changed in this field. Instead of the almost mystical emotions connected with the subject of potting compost, we now see the housewife in the supermarket carelessly dropping a bag of pre-packed compost into her trolley. Needless to say many manufacturers at first took advantage of this uncritical method of buying. Nowadays, fortunately, standards have been laid down to which good potting compost for general use must adhere. Most manufacturers voluntarily subject their own brands for testing; however, there are

Quality of compost is important

still some large firms who ignore the rules, and if you suspect that a compost is not up to the required standard, help can be obtained under the Trade Descriptions Act.

A commonly available compost is the John Innes potting compost, which was originally worked out at the John Innes Horticultural Research Institute in the 1930s. It has a standard formula, consisting of seven parts sieved loam, three parts peat and two parts coarse sand (parts by bulk), together with 21 g ($\frac{3}{4}$ oz) chalk and 112 g (4 oz) base fertiliser to every bushel of the mixture. This is J.I. potting compost No 1; No 2 has twice as much chalk and fertiliser, and No 3 three times as much, per bushel, and each is used according to the size of plant. For acid-loving plants there is the J.I. potting compost (acid), in which the chalk is substituted by 21 g ($\frac{3}{4}$ oz) flowers of sulphur.

Other potting compost is based on a mixture produced from domestic refuse. This soil is naturally rich in lime and contains a wide range of nutritious elements. It can be particularly recommended for plants which like a little more lime, for example members of the *Vitaceae* family.

Tips for mixing If you prefer to make your own mixture — something to be encouraged, since it is such a pleasant job — you can make a choice among the following more or less easily available components.

Leafmould is the decayed foliage of beeches, limes, alders, etc. You can collect these leaves yourself. Leafmould contains many nutritious elements.

Compost from your own domestic or garden refuse, properly made, can form an excellent basis for all kinds of potting soil.

Rotted cow manure is stable manure left on a heap for a number of years and, if well rotted, smells nice; only then can it be used as an addition to potting soil mixtures.

Rotted turves consist of soil taken from the upper layer of meadowland based on clay or loam. If possible it should be composted for a year. Contains easily absorbed minerals.

Sharp sand refers to lime-free river sand, roughly sieved. Mountain sand and very fine sand are not lime-free nor suitably coarse grained, and are therefore better avoided.

Peat fibre is very acid, contains practically no nutritious matter and is used as a water-retaining additive.

Sphagnum moss and *fern roots* are used particularly in mixtures for orchids and other epiphytes.

Dried blood, hoof and horn, and *bonemeal* are introduced in potting soil mixtures in spoonfuls only, to provide a lasting organic fertiliser.

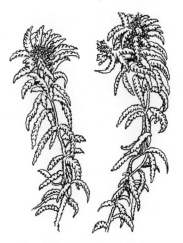

Sphagnum moss, still a widespread plant, which thousands of years ago turned into peat.

Crocks (pieces of clay pots) are indispensable at the bottom of clay pots. They ensure good drainage.
Perlite is a kind of expanded volcanic rock which improves the drainage of soil mixtures. Used especially in cactus growing.
Artificial substances such as polystyrene granules serve the same purpose: they lighten the soil.

In the preceding pages we have already stressed the importance of feeding houseplants. It might be advisable to repeat that, with few exceptions (for example indoor primulas) houseplants should be fed only during their growing season. There are many proprietary compound fertilisers available. Good brands contain nitrogen, phosphorus and potassium in the correct proportions, often with the addition of other elements which encourage the growth of your plants. A good fertiliser should not contain too much nitrogen; this leads to foliage development, but does not strengthen the plant, let alone help it to flower well. A correct ratio of the quantities of nitrogen, phosphorus and potassium might be 6:4:6. In the case of cactus fertilisers, however, the ratio should be 4:7:9, for an excess of nitrogen is exceptionally dangerous for succulent plants.

Fertilisers may be bought in liquid or in powder form. Both are suitable, provided you do not exceed the indicated doses. Proprietary organic fertiliser is also available, usually in the form of concentrated dried cowdung or chicken manure. Strangely enough the excellent fertilisers — dried blood, hoof and horn and bonemeal — are not very often marketed in small quantities. Another special fertiliser is foliage feed, which is dissolved in water and sprayed on to the leaves. Its special composition ensures that the nutritious elements can be absorbed directly through the foliage.

The correct pot for your plant

Unlike other living creatures, plants have to spend their entire lives in one and the same spot. This applies particularly to houseplants. It follows that the pots in which they are placed plays an extremely important role in the lives of our indoor plants. Visible growth, in fact, is far less important than what is happening in the soil: the root development is certainly no less multiform than the life of the plant

The reward for good care: after having been cherished for twenty years, this dracaena has at last flowered.

above ground.

We learnt at school that there are two main groups of roots: in the first place the tap-roots, which penetrate deep into the soil in search of the necessary water, and secondly the much divided fibrous roots just below the surface. In the case of houseplants — and especially of cacti — there is a third group: the tuberous roots. It would be easy to make a more detailed classification, for there are numerous transitional forms, but this subject is outside the scope of this book.

Attention for the roots

It can be easily understood that all roots assume entirely different forms in a flower pot to those of their natural state. If you have ever repotted a plant you will know that the fine fibrous roots, in particular, tend to combine in a dense, felt-like mass which nestles against the wall of the pot. It is here that most of the moisture and nutritious matter is concentrated — at least in clay pots — and this is the reason for such unnatural growth.

In some cases the growth underground does not much resemble roots in the ordinary sense of the word: we are referring here to the rootstock or rhizomes. These are not so much concerned with absorbing food, but rather serve to retain it. Rhizomes can also grow into bulbs or tubers. In that case the closely joined fleshy scales are actually a metamorphosis of leaves, while the basal plate underneath this remarkable phenomenon can form its own roots.

After this small excursion into the field of root systems, you might do well to remember that the kind of pot in which a plant has to grow is by no means unimportant. I mention this because in general the individual needs of roots are rarely taken into account.

Main shapes of plant pots: left normal pot, centre shallow azalea pot and right deep palm pot

Choice of pots

Although in theory there is a choice of different types of terra cotta clay pots, such as the tall, narrow palm pots and the wide, shallow pots for azaleas, in practice only the ordinary pots, neither specially wide nor deep, are available in the sizes 5 to 23 (the figures refer to the inside measurements expressed in centimetres; the equivalent figures in inches are 2 to 9). Larger sizes, up to 35 cm (14 in), are available from well-stocked garden centres. There are also plant bowls in various shapes, such as round, oval and rectangular. In addition there is a tremendous choice in plant troughs, with or without legs, in tubs in various materials, pendant pots, ornamental pots, etc. Some have a form of drainage system built in, so that the plants can be placed directly into the container. Many ornamental pots are intended as a cover, the plant remaining in its original clay or plastic pot.

When the first plastic pots appeared, everyone thought that the old, heavier and more expensive clay pots would soon disappear from the scene. But before long the plastic pots proved to be murderers, the main reason being overwatering. In the terra cotta clay pots a fair quantity of water is evaporated; the rest of the moisture is exuded by the foliage. In plastic pots evaporation naturally takes place only through the leaves, and where the plant is rather small in proportion to the pot, or where we are concerned with succulents which do not exude much moisture, a daily dose of water soon creates a mudbath. In general we may take it that a plant in a plastic pot requires about half the amount of water given to one in a clay pot. Of course it remains advisable to check the dampness of the compost constantly with your finger. The idea that a porous wall, as provided by clay pots, is essential to the roots of a plant has proved to be a myth. The soil undoubtedly receives the necessary air from above. The fact that the roots like to move towards the wall is due merely to the movement of the soil moisture through the wall and the resulting direction of the nutrients. The hungry roots automatically try to follow.

Almost cube-shaped containers provide the maximum capacity. They can be combined in rows, possibly placed on castors.

A sensible plant lover will attempt to determine which plants will grow best in plastic, and which in clay pots. The choice will depend, among other things, on how heavy-handed the watering is, so it is difficult to lay down definite rules. If using clay pots, it is important to soak them in water, preferably rain-water, for twenty-four hours before potting. Not only may the clay contain elements dangerous to the plants, but in addition an unsoaked, porous pot-wall will take so much moisture from the compost, that the plant itself may get into difficulties. There is nothing against using old clay pots, provided they are not covered in thick white crusts or green slime. In any case, these pots should be thoroughly washed, perhaps with hot water containing soda, in order to destroy possible pathogens.

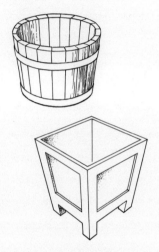

Two large wooden plant containers. It is advisable to apply a wood preserver before planting.

Regardless of whether the pots are of plastic or of clay, good drainage is of enormous importance. For that reason there should always be a drainage hole; more than one in the case of larger pots. Drainage remains the best precaution against excess watering, one of the greatest sins committed in the care of houseplants. Excess water can thus find a natural way out. If the pot stands in a saucer, the layer of crocks at the bottom of the pot should be slightly higher than the edge of the saucer. In this way the plant cannot get wet feet. If a pot is placed in a tall ornamental cache-pot, special attention must be paid, since the water level may imperceptibly rise. Some pot-covers fortunately have drainage holes themselves; some of them appear in the photographs. Other ornamental pots, tubs and containers for

Drainage hole

Plant trough on legs, square tub and circular tub, suitable for larger groups of plants. Clay pots can be placed in peat fibre.

large indoor plants illustrated in this book have no drainage holes at all. Provided great care is taken in watering, plants will nevertheless thrive in these.

If you want to make absolutely certain that a plant placed in such a watertight ornamental container does not get a footbath, you can place a plastic tube of 5–8 cm (2–3½ in) diameter vertically in the container before putting in the plant. By looking in the tube you can see whether there is water in the bottom of the pot.

When and how to repot

As a rule houseplants are kept purely for pleasure and looked after any old how, without much regard to their origin. Regrettably this applies not only to watering and other aspects of daily care, but also to the important matter of repotting. The individual plant descriptions in this book contain separate paragraphs with more or less detailed instructions on repotting; it would therefore be a relatively simple matter for you to maintain a kind of repotting record for all your plants.

A notebook with repotting record

It can be extremely useful to keep a notebook recording the necessary details about each of your plants. Under each name you might, for instance, begin by stating how much light the plant needs, which window it prefers, which are the growing and resting seasons, what temperature is required during those periods, and so on. You can then write down your experiences: whether they agree with what you have read about the particular plant, how many flowers it produced in any one year, in short, all those details which seem of value in the care of the plant. Naturally these notes should also contain an indication concerning the correct date of repotting, when this was actually done, in what compost, etc. You might also add something about the size required of the new pot, whether it should be a clay one or a plastic one, and so on. It would also be fun to make an annual chart so that you can see the repotting date of each plant at a glance.

If you followed such a system it would be a simple matter to plan the entire time-consuming repotting operation in advance. Everything necessary is assembled and even if, on closer examination of all your plants, it should appear that a couple more require repotting, this will involve little more effort. In practice it will certainly be sufficient if you organise such a repotting session two or three times a year.

When should one repot? *Left:* too soon; the insufficiently rooted soil-ball will easily break up. *Centre:* This is how the soil-ball should look — it will not break up when being repotted. *Right:* Repotted too late; too much root formation; hardly any soil left.

In the preceding chapter I mentioned the necessity of soaking new clay pots for twenty-four hours before using them. The same applies to the pots to be repotted: it is advisable to get them thoroughly wet a few hours in advance by watering them. One might also place the plants in a large bowl of lukewarm water the evening before. The next morning the soil-ball will have softened sufficiently to be easily detached from the pot. Occasionally the fine fibrous roots will have attached themselves so firmly to the wall of the pot that it will be advisable to run a sharp knife between soil and pot, as shown in the drawing.

New clay pots should be soaked for twenty-four hours; used pots must be scrubbed clean inside and out.

Crocks for good drainage

Before we can proceed to the actual repotting, the pots have to be prepared for their task. As soon as they are taken from their bath, they are provided with the necessary crocks. These crocks, obtained by breaking up old clay pots, are placed in layers over the drainage hole, in such a manner that the water can run away unhindered, but not too rapidly, through the hole. The crock immediately over the drainage hole is usually placed with the rounded side up. If no crocks are

This is how the crocks should cover the drainage hole. A 4-cm (1¾in) layer of pebbles may be used instead.

available, fairly large pebbles may be used instead, but these are clearly less suitable.

If the old pot already contains crocks (newly bought plants never do) a recalcitrant soil-ball may be moved by pushing the handle of a wooden cooking spoon, or a blunt-ended stick, through the drainage hole. Roots which may have grown through the hole will be discussed in a later paragraph.

The current method for taking a soil-ball from a pot is usually described as follows: 'hold the plant upside down and firmly bang the rim of the pot on the edge of the table'. Where the plant is not too large this may be an effective method, but in the case of large, mature houseplants, or of thorny cacti, I would not agree. You could, of course, wear heavy working gloves. But what if you find a bundle of roots growing underneath the pot? Plenty of people, both professional and amateurs, are sufficiently barbaric to cut off such healthy, strong roots, in order to remove the plant from the pot more easily. This causes, entirely unnecessary, large wounds, which are slow to heal and make it even harder for the plant to grow in its new pot. This kind of amputation is particularly irresponsible in those cases where an attempt is being made to transfer the plant from soil to soilless cultivation.

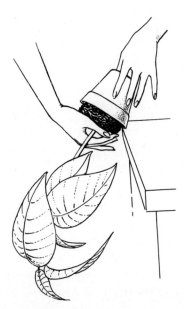

Left: loosening the soil-ball with a long, sharp knife.
Centre: a knock on the table is not always the best method.
Right: a well-aimed blow with a hammer at any rate saves the roots.

In cases such as described above it is therefore much better to forget about the table and instead to use a light hammer. With a couple of firm blows the pot can easily be broken in such a way that the roots hanging from the drainage hole are spared.

Roots growing through hole

Of course plastic pots cannot be broken with a hammer. In this case a pair of pliers or heavy scissors may come to the rescue. Naturally there will be exceptional cases, for instance when an ailing plant, or a dropped one, must be dealt with immediately, without regard to the repotting schedule. Apart from this no plant should be repotted after late July–early August, since shortly before — or during their dormant period — plants develop little or no root. Damage to the root system might therefore be so great that they will not be able to manage the transitional period to the next growing period.

Plastic pots require a different method

The state of health of the root system can be determined by close examination. If the roots on the whole are brownish in colour, with white tips, everything is in order. If you come upon black-brown, rotting patches, these must be cut back to healthy tissue. 'Healthy' in this case means that the root section must be free of dark lines or other discoloration.

Preparation of old root-ball

In general a dissected root is creamy white in colour; other possible colours will be indicated in the plant descriptions. It is, moreover, advisable to remove not only diseased, but also damaged parts of roots. This should be done by shaping the healthy, undamaged part of the root to a slight point with a sharp knife. Do this only in the case of the thicker roots.

Whether or not it is necessary to remove the old soil from the roots should be determined for each individual plant. You might carefully investigate the interior of the soil-ball with the aid of a pointed stick. Old soil that has turned sour may be got rid of by 'kneading', but if a plant is thoroughly rooted there will be little soil left inside. As much of the upper part of the soil as possible is removed, since this contains most of the waste matter. At the bottom of the ball you will often find some crocks which have become encased by the roots; try to remove these as well. If in the case of older plants it becomes essential to remove part of the roots, this should always be accompanied by pruning a proportional amount of growth above ground. Disproportionate growth in relation to the root system always leads to difficulties.

Sometimes plants are divided when being repotted, in order to produce new specimens. We only mention this in passing, since it is a question of propagation, rather than of repotting.

Loosening and roughening compacted compost.

From repotting to rerooting

Cutting down the soil-ball with a knife.

The correct proportion between the existing soil-ball and the new pot.

In principle new pots should be 2 cm ($\frac{3}{4}$ in) larger in diameter than the old ones. Of course there are exceptions, details of which will be given in the individual plant descriptions. Some plants have such a genuine dislike of over-large pots that they refuse to make new roots. Others, on the contrary, are only too happy with a little extra compost. It goes without saying that the vigour of the plant concerned plays an important part in this. This vigour will have been partly demonstrated by the old root system (did the contents of the old pot consist largely of white roots, or was there a lot of compost left?); apart from this, experience is the best teacher.

Among all the other measures to be taken when repotting a plant one is particularly important: the plant must be placed at exactly the same level in the new pot as in the old one. One of the greatest mistakes made in repotting is to plant too deep; for many plants this will create grave difficulties. We should also mention that firmness of planting varies from case to case. An oleander in heavy compost demands different treatment from a plant with surface roots planted in light humus. As a rule the new compost, placed carefully around the old root-ball, should be pressed down moderately hard with the fingers or with a blunt stick. The plant should in any case stand firm. It should be placed 2–3 cm ($\frac{3}{4}$–$1\frac{1}{2}$ in) below the edge of the pot to facilitate watering.

After repotting a little luke-warm water should be given; at the same time the leaves can be rinsed with the same water. Keep the plants out of the sun for the time being and ensure maximum atmospheric humidity. For this purpose the entire plant is sometimes placed in a half open plastic bag.

Planting in the new pot.

Press down well.

Check if it is centred.

The pleasure of propagation

'Living with plants' is not complete until one begins to enjoy the many opportunities provided by plant propagation. Practical considerations alone are sufficient argument for home propagation; there are, for instance, many well-known and well loved plants which last for only a very short time. For example, tradescantia as well as coleus should be increased at least every second year, since otherwise they outgrow their strength and become unsightly. You can increase them quite easily yourself and save the expense of buying new specimens. It also happens that plants grow too large and are given away. Take a cutting for your own window-sill first! At other times a plant may have been pruned and no true plant-lover could bear to throw away the good shoot tips; thus a number of cuttings become available. When you see small bulbs or plants appearing from the soil or on the leaves, it is difficult to resist the desire to propagate.

There are, of course, many good arguments for attempting propagation, but initially it may be pure curiosity which sets us off. However, in some instances, propagation is the only way to obtain a plant. Those in the know are aware of the fact that the most attractive houseplants are never for sale, but can only be obtained by other means. Can we blame the plant-lover, if his passion for collecting plants sometimes leads him to go too far? "I was in the Botanical Gardens at Easter and quite by chance a tiny shoot of *Sedum morganianum* stuck to my finger. You have no idea what a magnificent plant that produced last summer. . . .' This song is heard among collectors in every possible key.

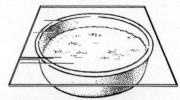

Simple clay seed pan with glass cover. Holes covered with crocks; a layer of polystyrene granules provides drainage. Gently firm the soil. Sow coarse seed singly; fine seed in rows. Cover thinly with sifted sand and moisten

Heated seed pan. The heating cable, which can be thermostatically controlled, is attached with sellotape. Drainage holes must be made in plastic pan. Sow as described above.

Sexual and vegetative propagation

As you may know, there is a distinction between generative propagation, that is, increase by a sexual process from seed (among which we count the propagation of ferns from spores) and vegetative propagation, which includes all such methods as cuttings, division, air-layering, etc. In general one might say that garden plants are usually grown from seed and houseplants on the whole from cuttings, a practice

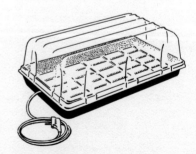

Indoor propagator with or without thermostatic control. Also available without heating. Ensures humid atmosphere and warm soil. Sow in pots or directly into the tray.

△
Tip cutting: two well formed leaves suffice (hoya).

Two leaf-cuttings: left African Violet, ▷ right cissus.

Left: a leaf-cutting of the variegated sansevieria invariably produces plain green plants.
Right: leaf-cutting from a ficus must have an eye in order to develop.

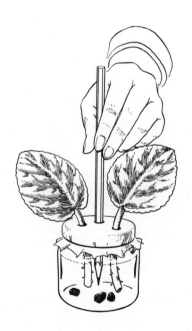

Rooting African violet leaves: stem twice the length of the leaf. Charcoal to purify the water.

followed by the average plant-lover. This is not, however, always the best method, for there are many attractive plants easily grown from seed, a pleasure which you will not readily give up once you have experienced it.

From a technical point of view the process of germinating various exotic seeds — fairly easily obtainable from seed merchants at home or overseas — presents a number of problems, since it is necessary to provide a constant degree of humidity and a fairly high temperature, both difficult to realise without special aids.

In any case we usually face problems when growing plants from cuttings, for without a sufficiently high temperature and a high degree of humidity, no 'callus' will form, and it is in this scar tissue that the roots are produced.

To solve all these problems at once, we strongly advise you to buy a seed-propagator for indoor use, together with an electric heating cable. Unfortunately not all seed merchants or garden centres stock them, but they are nevertheless obtainable. The heating flex, which can be plugged into the lighting circuit, is adapted to the correct temperature increase of 10–15 °C (20–30 °F).

Another disadvantage of propagation from seed which should not be overlooked is the fact that plants obtained by this method often flower less profusely than plants grown from cuttings, especially if the latter have been taken from a freely-flowering mother-plant. The opposite often applies to cacti, which, in some cases, flower magnificently and after only two or three years. The method of propagation to be preferred for each plant is mentioned in the plant descriptions under the heading 'propagation'.

Rooting cuttings and sowing seed should preferably not be done in ordinary potting compost, which contains too many nourishing elements and will therefore burn the young roots. Special composts for sowing seeds and for cuttings are available; these are based on either the John Innes type, or the soilless kind. If you want to mix your own compost for growing cuttings, the soilless kind will contain equal quantities of peat fibre and sharp sand, thoroughly mixed. This mixture contains no nourishing elements, but these are unnecessary for root formation. When the cuttings have rooted they are transplanted into ordinary potting compost.

Compost for growing seeds, on the other hand, should contain some nourishment. Ordinary potting compost may be 'thinned' with at

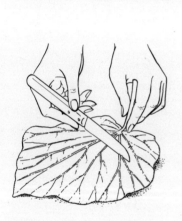

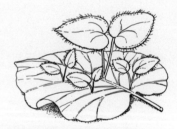

Begonia rex — leaf cuttings.

least half the quantity of sharp sand and possibly some peat fibre, to produce the correct mixture. There is also the J.I. seed compost, which contains two parts loam, one part peat and one part sand (parts by bulk), together with 21 g (¾ oz) chalk and 42 g (1½ oz) superphosphate per bushel; this is available from chain stores and garden shops.

Cover the seeds with a very thin layer of pure sharp sand. This is best done with the aid of a strainer (for instance a tea-strainer), so that coarse grains of sand and grit are kept back.

A seed pan should always be well drained. To dampen the compost the entire pan is placed in water until a change in the colour of the compost indicates that the water has reached the top. The pan should then be allowed to drain thoroughly. It is unlikely that you will need to water again until germination takes place, but should this become

Left: the thickest veins are cut below a junction.
Centre: place leaf face upwards on damp sand and peg down. Cover with glass and keep warm and damp.

Right: where you have cut, roots and plants will develop. Harden carefully, removing glass. Transfer to small pots containing good soil when they are 5 cm (2 in) high and keep in moist atmosphere.

Date palm growing from stone.

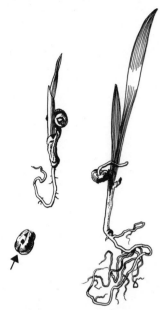

Sandpaper or file hard stone lengthwise and place 2 cm (¾ in) below surface in fine soil in small palm pot. Temperature 25°C (78°F). Will soon germinate and form roots.

A wide pot with roots growing through the hole is fatal, since repotting is practically impossible.

All will go well in a tall palm pot. The leaves are smooth-edged. Only after two years does the first incised leaf appear.

Pineapple plant grown from crown.

Excise the crown to a depth of 2–3 cm (¾–1¼ in); remove flesh. Dry for two to three days and place in sandy soil under glass at soil temperature of 26–28°C (79–83°F). If leaves develop it proves that roots are growing.

necessary, you should increase the moisture slightly by means of a plant spray.

As most seeds will germinate in the dark, the seed-pans or indoor propagators may be screened in many cases with a newspaper or greaseproof paper. This will prevent damage caused by the sun (too high a temperature!). As soon as there is any sign of plant life the paper should be removed, although even the pans must be screened from too strong sunlight. It is best to place them behind net curtains, then nothing can go wrong even in your absence. In a covered seed-pan the temperature will be maintained.

Hardening off is a different matter, however. As a rule this starts when the first two true leaves become visible — after the seed leaves. The glass cover should be removed during the daytime, or the lid of the propagator propped up. Now you should beware of the compost drying out. After a few days, up to a week, you can begin to prick out the seedlings into small pots. Use only the strong plants.

The illustrations show the major operations for taking cuttings. As a general rule, tip cuttings should have three to four leaves or pairs of leaves and should be cut obliquely just below a bud. The two lower leaves are then removed, together with their stem if present. Exceptions are: cissus, hoya (see sketch p. 71), philodendron, scindapsus and stephanotis which must always be cut 2–3 cm ($\frac{3}{4}$–$1\frac{1}{4}$ in) below a leaf-bud. The shorter the cutting the sooner it will strike. Until then it must be protected against evaporation. This is best done in a propagator, under glass, or if necessary in a plastic bag, but be careful that the cutting does not rot. It should be placed in a well-lit situation,

Grafting columnar cactus. Straight cut grafting: cut off top with sharp knife; trim lower part obliquely. Blot escaping liquid. Pressing lightly, place graft on top with a turning movement and hold in place with rubber bands.

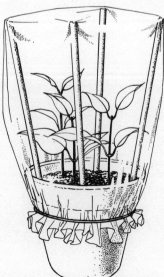

◁ Indispensable: polythene cover.

Left: Polythene bag over bent wire; it should not touch cutting.

Right: Four canes keep polythene away from cuttings.

but out of the sun. When shoots appear we know that roots have developed. Now the little plant can be hardened off by gradually removing the glass or the plastic. Further hints will be found under the heading 'propagation' in the plant descriptions. The best time for taking cuttings is usually in spring, but often also after the plants have been pruned. Geranium *(Pelargonium)* cuttings are taken in autumn. Cuttings from moisture-retaining plants, such as succulents, are left to dry for a few days or even weeks before they are put into compost. For propagation from spores see under the descriptions of ferns.

Quick method of shortening an overgrown rubber plant

Air-layering is also possible in the case of cordyline, dracaena, philodendron, scions of bromeliads and clivia.

This method is called 'air-layering' and can be used for other plants as well as for the rubber plant *(Ficus)*. Its purpose is to reduce the size of a tall plant with a bare stem. The best time to do this is June or early July.

1. Using a sharp knife, start by making an incision below the third or fourth leaf from the top, first horizontally through half the stem, then upwards, as indicated by the dotted line. As the vertical cut should remain open slightly, a match is placed in it. A rolled piece of silver paper will do equally well. The lower dotted line shows where the stem should be cut after rooting.

2. The cut and the stem are now wrapped in a good handful of moist sphagnum or bog moss, obtainable from florists. Peat fibre may be used instead, but is rather difficult to fasten round the stem. In this

case the lower edge of the polythene (see 3) should be attached first, making a small container for the damp peat.

3. The moss must be attached in such a way that the cut and the leaf-stem are approximately in the centre of this remarkable 'cracker'. Take a sheet of clear polythene and wind it a few times round the ball of sphagnum, tying it carefully at top and bottom with raffia or string. If the moss was wet when applied, it will need no further watering, provided it is closely covered. Place the plant in a warm position, but out of the sun, and note when the first roots become visible through the polythene. This usually takes about six weeks, when the polythene can be removed.

4. Potting is done as follows: When the polythene has been removed, the stem is cut below the roots that have formed (see 1). Carefully remove most of the moss and take out the match or silver paper. The part of the stem between the two cuts is discarded. The rooted cutting is now planted in a suitable pot, using pre-packed potting compost mixed with one part of sharp sand.

Twelve months of plant care

Every sensible plant lover and indoor gardener is aware of the fact that the care of houseplants is not limited to watering and feeding. Apart from these daily jobs, there are many operations to be carried out throughout the year to ensure the well-being of our plants. Some jobs obviously ask to be done, but to understand when it is time to do others we must keep a sharp eye on our plants.

Just as in the garden, the soil in large plant containers should constantly be loosened, to prevent a solid crust forming on the surface. Good potting compost will already make quite a difference, since this does not harden too easily. Nevertheless it frequently happens that, as a result of much watering, the upper layer becomes very solid, so that air cannot penetrate, leaving the roots below in distress. In such a case the compost should be loosened; this can be done quite easily with a stick, but it is of course even easier with the aid of one of those small tools sold in any good garden shop or centre.

If matters have come to such a pass that a thick layer of moss covers

Care of the compost

the surface, it will be necessary to remove the entire upper part of the compost. Replace it with fresh potting material and cover with a thin layer of sharp sand. You should in any case try to find out the cause of this moss formation, for it is certain that something was wrong with the water situation. Was the plant too wet?

A plastic frame is suitable only for not too large, one-sided plants.

Tall growing plants require support in order not to collapse. When inserting canes or circles of bent wire, one should be careful not to damage any roots. Special pots sold for soilless cultivation often have two small holes for this purpose. It is not really necessary to position the sticks or wire in advance; in fact it is better to do this as growth progresses. Wood remains the best substance for this purpose, being itself vegetal. Split bamboo is available up to a certain length; in the case of taller supports round bamboo can be used successfully. If you like, small frames can be made of this material. These usually look better than the plastic frames often used nowadays and frequently sold with the plant. Man-made materials – often in the most vulgar colours – do not go with natural plants. The same applies to the plastic clips and ties – they may be convenient, but they don't look nearly so well as a simple, natural piece of raffia. Whatever material you use for tying, the stem must never be tied close to the stick. There should always be some room for growth. If you think raffia is not strong enough, you should on no account use wire. Better to take a piece of string, which will not cut into the stalk so easily.

A stone can prevent a wall pot tipping over.

Ornamental pots for use on the window-sill, and special pots for hanging plants, are well-known accessories, which sometimes show the plant to better advantage. However, this is not the case if flowering plants are placed in pots of clashing colours or with overpowering decoration. For the plant itself the worst that can happen is water remaining in the bottom of these pots; we have already mentioned this elsewhere. Some ornamental pots have their own saucer for surplus water. White and dark brown, both neutral colours, harmonise with the colours of leaves and flowers, and are therefore to be recommended.

Often a plant imperceptibly grows top-heavy. One might come downstairs one morning to find that the magnificent flowering amaryllis has collapsed during the night, or a hanging pot may almost have tipped out of its frame. I'll spare you the other calamities which might befall you. The sketch shows how to prevent such occurrences. Most people can manage to find some good-sized stones in their neighbourhood, and when there is not time to repot immediately, they can be used as a reasonable stop gap measure.

Improving the 'microclimate' by combining several plants in a tray.

Dust

In winter so much dust falls on the plants (especially if they are not behind glass) that spraying alone is insufficient to get rid of it. Often the addition of water only results in the dust penetrating deeper into the foliage and stopping the pores. In this way breathing and assimilation are hampered. Large-leaved plants should therefore be regularly and generously rinsed with lukewarm water. The stems may be sponged as well. It is advisable to use softened water, to avoid those ugly chalk marks which may be left.
Not all plants tolerate such a shower. Cacti and succulents definitely dislike it, especially in winter. Plants with hairy leaves also should never be washed. In such cases it is better to use a soft brush to remove the dust. Small-leaved plants can be laid on their sides under the shower. It is better to give a hard-water shower than to leave pores clogged with dust. Many plants can successfully be placed outside during a light shower in the summer. If you are afraid that the container will overfill you could cover it in aluminium foil.

Washing the windows

If there are many plants on the window-sill, or hanging from the ceiling, washing the windows may be a problem. A number of plants can be combined in a trough or tray, which facilitates their removal. When you buy such a trough, look out for one with a separate drainage tray. They do exist and are, of course, better, as they prevent surplus water gathering provided you drill a number of holes in the bottom of the trough itself and cover them with crocks. The plants may be left in their individual clay pots, but it is also possible to plant them directly into the trough.

Longer window-sills may accommodate as many as three of these troughs — for instance asbestos-cement balcony boxes. A further advantage of this system is that the plants are always turned towards the light in the same way.

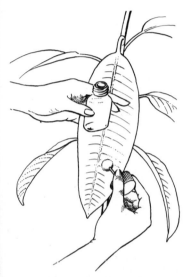

Good quality foliage-gloss products protect against dust and remove lime blotches. It is better to use a spray.

A brilliant foliage gloss has always been the pet delight of those polish-minded housewives, who like to have their plants as shiny as their parquet floors. After dusting the leaves of their rubber plant — very useful in itself — these misguided women attack the poor thing with beer, milk or even salad oil. Poor ficus — the oil irrevocably closes the last pores and moreover makes the dust adhere.

From this point of view we can only welcome the arrival of aerosols for this purpose, which do not endanger the plants while satisfying the desire for shiny leaves on the part of their owners. This high gloss effect is, of course, quite unneccessary, but these aerosols also contain a substance which dissolves the chalk marks, and this cannot fail to please.

Some products also contain anti-static elements, foliage food and even insecticides, depending on their composition.

Dried compost

When plants are properly looked after, dried out compost should never really occur. Nevertheless quite a few plants die as a result of drying out, in the first place, of course, those that are particularly sensitive on this point. Once the pot-soil is thoroughly dry, a little watering will have no effect; the water will run down between it and the pot-wall. Plunging will be more successful in such cases. By plunging we mean submerging the entire pot with its soil, and possibly part of its stems, in a bowl of water, as illustrated in the adjoining sketch. Be sure to use lukewarm water. The plant should be left in the water only as long as air bubbles rise up from the surface of the compost.

Pruning may be necessary for a variety of reasons. This is why in the section 'Plant Descriptions' we have in many cases inserted a separate heading 'pruning'. One thing that will often be necessary for instance, is to prune the roots when parts of the upper growth have been removed, otherwise the superfluous roots will rot. This often occurs in the case of pelargoniums and fuchsias.

Houseplants are rarely shaped by pruning, unless perhaps a small bay-tree requires some trimming. Small orange trees are also trimmed to improve their shape. Most other plants are pruned only to rejuvenate them. This may, for example, have a favourable effect on an old oleander or hibiscus which has grown bare in the centre. By cutting down severely to the old wood, it is possible to force these plants to sprout again at the bottom, giving us a fine dense plant the following year.

Any plant improves in appearance if lanky, dry and bare branches, growing too close together or crossing each other, are removed. It certainly does require some knowledge to be able to prune a plant in such a way that the plant will subsequently grow in the desired direction. This is an entirely different matter from random cutting. The cuts should be absolutely smooth, which can only be achieved with very sharp tools. Thin branches are always cut just *above* an eye or bud. If the cuts discharge a milky liquid, the flow can be stopped by burning with a match. In the case of cacti the juice is generally first removed by blotting; the flow is then stopped with charcoal powder. The cuts will finish drying out in the air.

Hints on pruning

Left: cutting back a pelargonium in February.
Centre: in the case of beloperone either remove the flower heads or cut back drastically to induce bushy growth
Right: Woody plant pruned drastically in late winter.

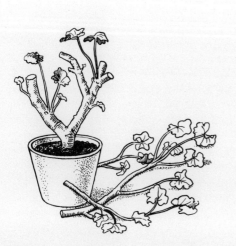

Bottle-gardens

At one time tools for bottle-gardens were not available and had to be improvised for different sized bottles.

For generations this method was practically unknown. Then quite suddenly it became the fashion: town and country were flooded by all sorts of bottles, usually green, manufactured by a rapidly growing industry and available from uninformed growers or indeed from supermarkets. Even the usually careful, serious weekly journals hailed the coming of this 'miracle bottle' from America and, again, being more or less uninformed, wrote that these bottle-gardens offered remarkable new possibilities in view of the fact that people had now less time to care for their houseplants. These apparently 'problem-free miniature hothouses' nevertheless enjoyed only a short period of success. The fashion largely caught on because when filled with a varied selection of plants, these decorative glass jars, with their wide necks, appeared virtually irresistible to the layman. Later it became clear that these bottles were not all that they should be. In other words they were not firmly sealed glass balloons, satisfying every technical and biological requirement, balloons in which – thoroughly attuned – slow-growing plants could live undisturbed in a little world of their own for up to two years. How the true bottle-garden may be achieved is shown in the drawings on the following pages. The genuine bottle-garden, as you know, dates back to the early 19th century, to the 'Wardian cases' mentioned on p. 25.

After the bottle has been planted successfully and carefully watered, an initial period begins, during which we generally find a greater or lesser degree of condensation in the interior of the bottle – a sign that more water has been given than the plants, not yet fully rooted, require.

Take note: extreme care in watering from the very beginning is one of the most important principles. If the interior of the bottle is permanently covered in condensation you have watered too freely. In this condition the bottle should not be sealed. Wait until the water situation has become balanced before you put in the cork. When the bottle shows some condensation during the morning only and then dries out the correct balance has been achieved and the cork may be inserted or, in the case of a modern bottle, a well-fitting screw-top.

A second possibility during the initial period after planting is a complete absence of condensation; this means that too little water has been given and a little should be added with extreme care, as after planting.

Bottle-garden experts manage to grow magnificent pot roses even under a Victorian glass dome.

  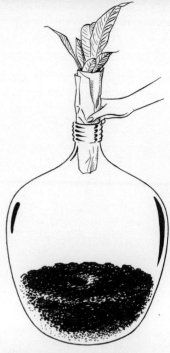

Use of a simple paper funnel to fill the bottle. First insert a layer of coarse pebbles to provide drainage; then the compost.

The drainage layer and the compost are spread evenly by means of a strip of wood narrow enough to be easily rotated. It is advisable to place a few pieces of disinfectant charcoal between the two layers.

A dangerous job: inserting the plants through the narrow neck opening. Use the improvised paper funnel and be sure to choose plants whose root system is still small and which are suitable for life in a bottle.

A bottle-garden need not necessarily be laid out in a large carboy; it can equally well be created in numerous other well-sealed objects, such as glass jars which at one time held chocolates or cotton-wool, kilner jars or a hermetically sealed aquarium. The point is that a natural cycle takes place; this happens in nature, but also in plants in pots or in soilless cultivation.

Place the bottle in a well-lit spot, but not in the sun. The plants absorb the necessary food in a watery solution. The water released in breathing is seen as condensation inside the bottle. This must happen only in the morning, when it will not be harmful.

You will remember that the plants moreover absorb carbon-dioxide and release oxygen through their pores during the day-time. If the situation inside the bottle is properly balanced, the oxygen released during the day is partly re-absorbed at night and partly used up by micro-organisms in the compost, which also take part in the production of carbon-dioxide. It will be obvious that the photosynthesis

When the plants have been successfully inserted, they are removed from the paper with the aid of a stick, tongs, or spoon and moved to their permanent position.

The same tools are used for planting in the holes, previously made with the strip of wood. Add some compost and presto!

After planting water drip by drip by syphoning through a rubber tube, best done by tying the tube to a stick. The success of the entire operation may depend on this initial frugal watering.

inside the bottle decreases, and that growing processes change. While there is not the space to go into this subject in greater detail we can nevertheless assure you that a well designed and properly sealed bottle-garden is one of the most fascinating and rewarding forms of plant cultivation provided you go about it in the right way and with patience. Obviously such a bottle garden is not cheap and cannot be bought as a 'mass product' in the shop.

Relation between available space and choice of plants

It is above all essential that the choice of plants be related to the bottle used. It goes without saying that a gigantic carboy with a 60-litre (13-gallon) content offers entirely different opportunities to e.g. a chianti wine bottle, with a much smaller capacity. In choosing our plants our starting point must be the knowledge that our very slow growing plants are suited to the height and width of the bottle.

For a small and fairly low bottle not more than two or three plants should be considered; in a 40-litre (9-gallon) bottle we can use at least 6 and in a 60-litre (13-gallon) bottle even more.

A rectangular aquarium of reasonable height undoubtedly provides excellent opportunities, but this cannot really be called a mini-garden in a bottle, even though it is hermetically sealed with a glass pane.

It is of course obvious that flowering plants are in general less ideal for the purpose than foliage plants. In England, where the 'bottle garden' and the 'fern case' were invented nearly 150 years ago, experience has shown that ferns are the most suitable. Experts go further than merely planting ferns; they even sow spores through the narrowest of bottlenecks and use various vegetative methods of propagation such as setting shoots and rhizome cuttings, or leaf cuttings of the hart's tongue *(Phyllitis scolopendrium)*. But these operations require a lot of patience; it takes months before any green appears in the bottle-garden. The following are some of the ferns which either grow very slowly, or remain very small: *Adiantum capillus-veneris; A. pedatum; A. raddianum; Asplenium adiantum-nigrum; A. obovatum; A. rutamuraria; A. trichomanes; Blechnum spicant; Pellaea* species; *Phyllitis; Polystichum;* small-growing *Pteris* species, for example *P. cretica* 'Albolineata' which has white blotches, or the aptly named 'Distinction'; in larger bottles 'Rivertoniana' and 'Wimsettii' can be used.

In the long run foliage plants are most suitable

The selaginellas occupy a very special place in the selection of bottle plants. You will find a photograph on p. 330. Finally there are the true mosses, of which more than 600 species occur in England, Scotland and Ireland. The Japanese foliaceous mosses are very popular, for instance cultivars of polytrichum, and liverwort, *Marchantia polymorpha*, which grows wild in Central Europe as well.

When designing a bottle-garden, one should really always consider the use of these 'primeval' plants, and even gnarled or leaved branches can look very pleasant. In the long run they can form a particularly attractive part of the design; in addition they are very useful for creating the balance in such a bottle-world, better than would be the case with special species of houseplants which have been inserted with difficulty. A few more of these should be mentioned; plants which grow to average height and serve as the centre point for 40–60-litre (9–13 gallon) carboys, viz. *Aglaonema commutatum* and *A. costatum*, as well as the smaller-growing *Anthurium scherzerianum*.

Plant windows in practice

What exactly is a plant window?

It might not be a bad idea to stress that an ordinary window-sill with a row of plants in pots is not a flower window. On the window-sill your plants are completely exposed to disagreeably dry air, and will have little reason for gratitude.

Height of parapet and width of container

When talking of a flower window we always refer to a container of such height and width that the pots can be entirely bedded in. As a rule, moist peat fibre is used for filling; it is best to leave the plants in their clay pots. Planting out is only advisable for relatively weak growers — such as cacti — since otherwise the roots get completely tangled up (see also p. 34). From this it will be seen that in a small flower window the minimum depth apart from the drainage system, heating cable, etc., should be about 18–20 cm (7–8 in). For larger boxes the depth should be 25–30 cm (10–12 in), while a plant box built into the floor should be no less than 50–60 cm (20–24 in) in depth. The taller the plants, the deeper the clay pots in which they grow, and to be able to contain the pots completely the above measurements are essential.

The distance between the floor and the top of the trough which might also be described as the 'height of the parapet' — should preferably not exceed 60 cm (24 in), otherwise it will be impossible to look out of the window while sitting down. As a general rule we may say that the lower the parapet, the more attractive the flower window. In modern houses the plant case is often incorporated in the floor. It will be obvious that with low windows such a plant case will receive excellent daylight and where the window extends to the floor,

The flower window is seen from different angles, depending on the level at which one sits or stands. The lower the parapet, the better the view and the more favourable the angle of the light and planting opportunities.

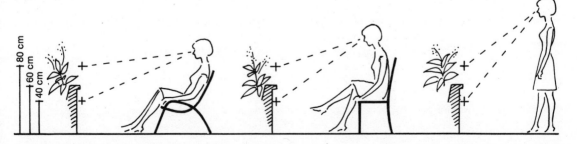

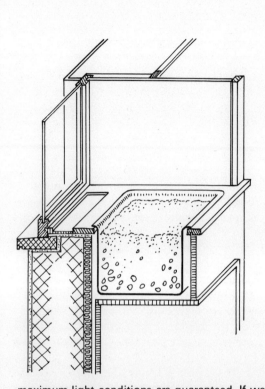 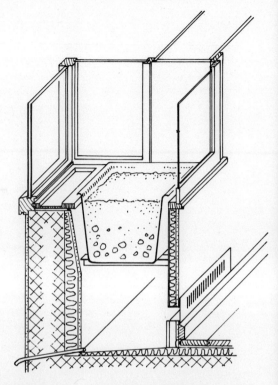

maximum light conditions are guaranteed. If we leave the flower case uncovered, the damp air rising from the soil will benefit the plants.
It is even better if a glass partition is added on the inside. In that case, atmospheric humidity can be regulated as desired, and — provided there is sufficient heat — we may speak of a 'tropical window'. I would in any case advise a minimum window height of 120 cm (4 ft) and a minimum width of 110 cm (3 ft 8 in). As regards the width of the flower case (that is, the distance between outside window and room), this is usually 40–50 cm (16–20 in). If there is sufficient space, make it 60 cm (24 in); this provides greater opportunities when planting. Of course the reach of your arms should be taken into account, for the window will sometimes have to be washed. In many cases it is possible to leave a strip of about 20 cm (8 in) between the plant trough and the window. If this is tiled, you could stand on it to clean the window and the plants would not so easily touch the glass, especially if you put the smaller plants in front.

Left: simple built-in flower window without inside partition.
Right: bay window with inside partition and heating cavity.

If the house is yet to be designed, you have the choice between a flower window built into the room, or one built out as a kind of bay. Which of these you choose depends on the design of the house, but a

Flower window built inside the room or as a bay?

bay is often best for the plants themselves. Even without a glass partition on the inside, the dry indoor air will not circulate so freely. A built-out window, moreover, receives more light. On the other hand, the light can create a problem, especially if the window faces south. In that case it is important that you should be able to screen against sunlight, for example by means of Venetian blinds. Compare also the illustrations on p. 31 and 35; both show plant windows built out into a bay.

After this explanation you will certainly have no difficulty in interpreting the section of a floor-level flower-window illustrated on this page.

Compare this also with the photograph on p. 33; the actual situation is slightly different, but the principle is very similar. You will note the unplanted strip between outside window and plants, which makes it possible to introduce electric tubular heating. Otherwise single glazing would let in too much cold. It is of course better (but rather more expensive!) to install double glazing.

Improving design with loose branches

The most eye-catching feature in a plant window — preferably a closed one — is a so-called epiphyte tree. This is a picturesque branch of an old apple-tree, or even better of a robinia (sometimes called acacia), to which all sorts of epiphytic plants are attached by various means. These branches are fairly easy to obtain and if you can handle a saw it will be no problem to cut one to size. If the branch does not have a particularly good shape, it can easily be improved by tying

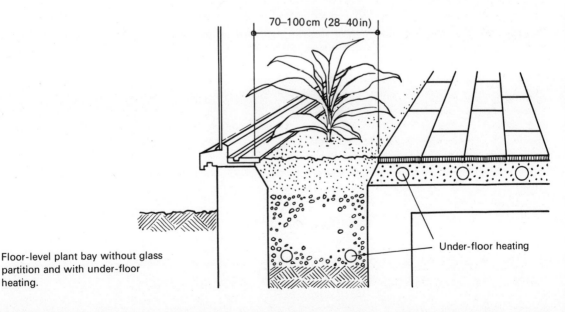

Floor-level plant bay without glass partition and with under-floor heating.

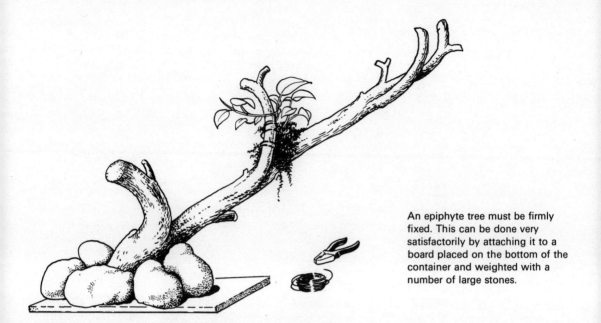

An epiphyte tree must be firmly fixed. This can be done very satisfactorily by attaching it to a board placed on the bottom of the container and weighted with a number of large stones.

loose branches to the main stem with the aid of copper wire. In this way we can also create the hollows and apparently rotting sections which are necessary for rooting the plants.

The epiphytes selected for the purpose, such as bromeliads, orchids, ferns, etc, are prepared in advance by wrapping the roots in some damp sphagnum moss.

Ordinary forest mosses are unsuitable for this purpose. Use dark brown copper wire — for instance from an old transformer — for tying. It will be practically invisible. When all the plants have been prepared in this manner, you can start to dress the tree; a measure of good taste is indispensable for this operation. Try to combine the plants so as to create a more or less natural effect. Copper wire may be used once more to tie the plants to the branch; it will be necessary to use a tiny nail here and there.

Forest moss rots too quickly

By nature epiphytes always grow on those sections of a tree which lie in the route taken by rainwater, for otherwise their roots would rapidly dry out. To save yourself a great deal of subsequent work, it

Do not water individually

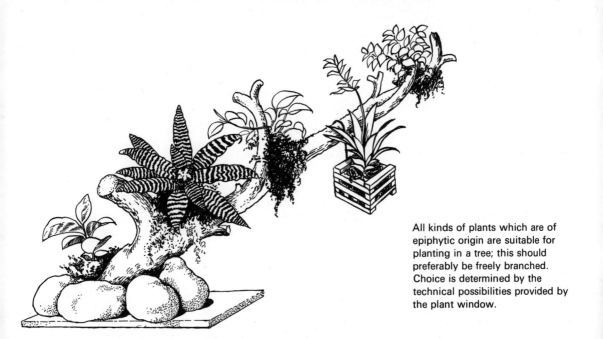

All kinds of plants which are of epiphytic origin are suitable for planting in a tree; this should preferably be freely branched. Choice is determined by the technical possibilities provided by the plant window.

makes sense to consider first how you are going to water. If you intend to fix a few sprinklers in the ceiling of the plant window, which can be operated via a rubber hose and a simple handpump, this installation should be fitted first, so that you will be able to see how the water runs down the branch. This is where you place your plants. If the reservoir is filled with rainwater, it is sufficient to pump a little every day. Some people build a cunning network of aquarium tubes along the tree, with an outlet at every plant, covering the plastic with bark, moss, etc. In this case, also, it is sufficient to move the pump a little every day. A clever handyman might make the installation automatic by means of a small electric pump switched on by an electronic hydrometer.

Choosing plants Naturally the plants selected for the epiphyte tree must be adapted to the conditions in the flower window. Trying to grow tender orchids in an unscreened window, for example, would be useless. Better to choose strong plants such as monstera, philodendron and rhaphidophora, and in cool windows possibly hedera, the common ivy.

When the plants grow, the tree may eventually become unstable. Hopefully you can insert a small eyelet screw into the ceiling, to which the tree can be invisibly attached by means of fishing line.

It need not be a whole tree

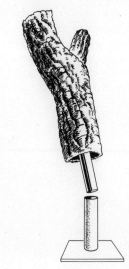

There are other ways of introducing hanging plants in a flower window in an attractive manner. From the bark of the well known cork-oak, for instance, you can easily make all sorts of 'plant carriers'. A round steel stave can be used to support the structure; the pieces of bark are attached to this with wire. The drawing shows an example with two hollow spaces suitable for plants, with or without pots, or for two small soilless containers. I constructed this example more than a year ago and planted it with a graceful hanging *Cissus striata* and a chlorophytum. It has remained in one of my flower windows, which is open on the inside, ever since.

The fact that pieces of tree-bark provide an excellent foothold for the roots of numerous epiphytes, better even than sections of tree-fern sawn into lengths, can be seen in any botanical garden. In any case plant lovers should visit such places more often.

Home-made 'tree' covered in bark of the cork-oak; space for two plants.

A permanent plant community

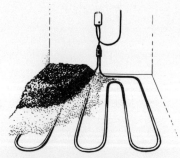

This is what a flower window can be. It is not a question of changing the plants every now and then. If the environment has been properly prepared, the plants can really grow without any problems until of course they become too large and must be reduced in size by means of cutting or division.

It is regrettable that lack of space does not allow discussion of many other aspects of the flower window, such as the best type of glazing, the glass partition on the room-side, air circulation, atmospheric humidity or heating. The wisest course is to consult a professional designer of flower windows on the possibilities available and on the question of cost. Unfortunately there are fewer experts in this field in this country than there are, for instance, in Holland and Germany. Especially in the area of lighting and climatic control, many mistakes can be made which could be fatal to your plants. If you are really at your wits' end, Rob Herwig will be pleased to give further information; you can write to him via the publisher of this book.

Heating flex (in this case thermostatically controlled) must always be laid in a 4–5 cm (1¾–2 in) layer of sand, which is then covered in peat fibre or potting compost.

Further details will be found in the chapter on Hydroponics and Artificial Light p. 43ff

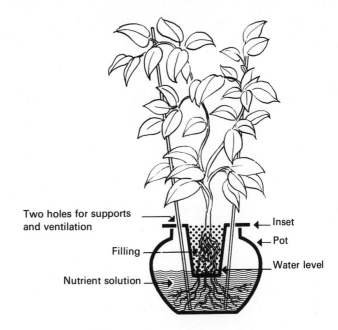

Section of soilless container for one plant.

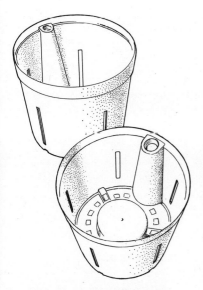

The inner container for the plant, showing the drainage holes in the base, the side openings for entry of air and the inset for the indicator.

Growing plants in hydroculture

Growing plants in water containing mineral nutrients is not a new idea. On the Continent special glass pots (see sketch) were used for fifty years or so into which the nutrient solution was placed. The plant was put into another container – in which there was a supporting material – and this was set in the neck of the nutrient-containing pot. Some of the plant's roots grew in this inset, the rest hung down into the feeding solution, the surface of which had to be level with, or just above, the base of the inset.

Until recently, these hydropots, as they were called, were the only containers used for growing indoor plants under what has come to be known as hydroculture, but during the last twenty years or so containers have been developed from more modern material such as plastics (see sketch overleaf).

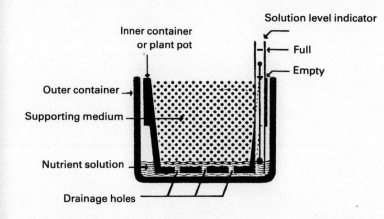

A modern hydropot of plastic, whose inner pot only is filled with the supporting medium. Note that the inner pot is nearly as large as the outer container.

Filling

The whole principle of hydroculture rests on the fact that plants can only absorb their mineral foods in solution, and today's containers ensure that they can do this with the minimum of effort on the part of the plant owner. In general, they still consist of two parts, an inner pot containing the plant, and an outer container holding the nutrient solution. The inner pot is filled with a porous air- and water-holding medium, sometimes also the outer pot, and the most suitable material for this has been found to be heat-expanded clay granules. These granules are chemically inert, and do not rot or become mouldy. When combined with the nutrient solution they do not cause undesirable reactions, and have a certain amount of capillary action. Finally, they do not change shape and can be obtained in the coarser and finer grades necessary for different kinds of plants and root systems.

Containers

The outer container should be dark in colour to keep out light and prevent the growth of algae internally, and there should be an indicator let into one side to show the level of the solution, and to provide an inlet for the nutrients. The solution should be sufficient to rise above the base of the inner container, but not so high that air is excluded from the inner space between the containers.

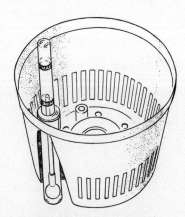

A variation on the type of inner pot; there are more air slits for a different type of plant, and the indicator is a thermometer-like bulb which rises and falls with the level of the solution.

Nutrients

As far as the plant food is concerned, it is necessary to buy special fertiliser developed for the purpose. This is for use in tap water, and has been formulated in such a way that the nutrients only become available to the plant at intervals in predetermined quantities. In general, one dose will last six months. Watering will be required about once every three weeks.

Although initially expensive hydroculture is remarkably easy to follow, and plants will be virtually free of such troubles as over- or under-watering or feeding, and leaf and flower drop. Indeed, many grow much better under this system.

A magnified transverse section through an expanded clay granule showing the degree of porosity within it.

Transference from compost to hydroculture: remove the original pot – sometimes clay pots have to be broken to avoid damaging the roots – especially when dealing with large plants such as this monstera.

Then wash off all the old compost thoroughly with warm water, cut off damaged or long roots, and pot into the hydroculture containers using a warm nutrient solution.

Changing over to hydropot cultivation

Plants can be changed over in spring, starting by putting the plant and pot in lukewarm water overnight. Next day wash off the compost, again with warm water, pull the longer roots gently through the holes in the inset, and spread the remaining roots out in the filling material in the upper container. Use lime-free water only for two or three weeks, and then give a feeding solution at half strength. Keep the plant out of the sun for the time being.

An *Echinocactus grusonii* which has been grown in water for more than ten years and has a diameter of 22 cm (9 in).

Diseases: Prevention and Cure

As far as houseplants are concerned, by no means every casualty can be attributed to insect pests, fungi, bacteria or viruses, as is the case with garden plants, where much more natural conditions of growth obtain. In the house, plants are subjected to artificial methods of cultivation, which can mean that they receive exactly what they need in the way of water, light, nutrient and humidity if their owner is an expert grower, but which, unfortunately, more often means that their needs are hardly met in any way. We must, therefore, be aware of the fact that we ourselves may be to blame for poor growth, but it is often difficult to determine, with these cultural troubles, what exactly is the cause. For instance, using an aerosol insecticide too close to a plant (the correct method is on p. 97) can make the leaves curl up and turn brown and wither, but it might be thought that a fungus disease was the culprit in the first place.

A common cultural problem is the case of green leaves losing their colour, turning yellow all over and finally falling off the plant. Rubber plants often show this trouble, starting with the lowest leaves, and the cause is overwatering. The yellowing results because nitrogen, the nutrient which helps with the green colouring of plants, is very soluble and is easily washed out of the compost. Furthermore, in a waterlogged compost lacking in air, the roots cannot function, and so

Cultural troubles and their remedies

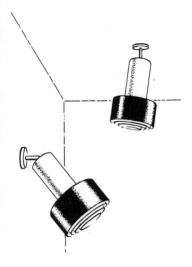

Wall lights developed by Philips-Switzerland for plant illumination (see p. 47).

do not supply food to the top growth. It is therefore unavoidable that the top growth will discolour and grow poorly.

One will often also see yellow leaves on a plant, which remain attached and keep their firm, healthy texture — the plant's growth will be slow or static. Hoyas often suffer from this, and it is usually due to watering the plant with hard water and using a lime-containing compost.

Sometimes the reverse will be seen happening, in that plants which have leaves variegated with yellow or white, become plain green all over. This is due to lack of light; such plants do not keep their variegations in shady places, and need to be in a position which is well-lit, though not sunny. Sometimes a variegated plant will produce a shoot or shoots which are plain green all over, not just one or two leaves. This will not be due to lack of light but to a tendency of the plant to 'revert', that is, produce growth typical of the parent plant from which it 'sported' or mutated in the first place. This is a genetical fault, and the only remedy is to remove the offending shoot at once, back to its point of origin, otherwise it will take over the whole plant.

Another very common type of discolouration on leaves, which is due to the way in which the plant is being grown, is browning of the leaf tips and edges. In nine cases out of ten, it is due to keeping the plant in too dry an atmosphere. Many houseplants come from areas of the world where the climate is warm and damp, if not actually hot and steamy, and moisture in the air is vital. Unfortunately it often gets forgotten, probably because it is invisible, but nonetheless it is as necessary to the plants as to the owners.

Plant table used for Luwasa soilless system; with artificial lighting.

Following on from this is another trouble very often encountered: that of dropping flowers and buds. In many cases this is also due to too dry an atmosphere — plants constantly give off moisture in the form of vapour, and the first parts to suffer are often the thin petals of the flowers, and then the buds. They can, of course, also fall for other reasons. For instance, the change of environment from nursery to florist and from florist to home will result in blossom falling almost at once, until the plant has settled down. Some plants are very temperamental, for example begonias, and will drop their flowers if their position in the room is changed; draughts can be another cause, and also not enough water at the roots: note that this does not necessarily mean a completely dry compost, simply, not enough water for that particular plant's needs.

Sometimes a plant becomes very lanky and spindly, with pale green leaves and a few wishy-washy flowers. When this happens it is because the plant is not being kept in a light enough position. In its attempts to get the light it needs, it grows towards it and becomes elongated, but even so remains pale.

Wilting of leaves and stems is often due to the obvious reason, lack of water, but surprisingly it can be because too much water has been given, drowning the roots. Too much sun is another factor, though this is more likely to result in leaf scorch, which produces pale brown, paper-like spots in the leaves. Too much warmth can be offset by spraying the plant at once with clear water, or removing it to a cooler place.

Although many houseplants are grown because they have highly ornamental leaves, more are now being grown for their flowers. If you have a flowering plant which is rather shy to bloom, the answer may be to give it more light, or to feed it with a liquid fertiliser containing a high proportion of potash, the element which helps with flowering and flower colour. Sometimes lack of flowers is because the owner has been rather too enthusiastic with pruning, and cut off all the potential flowering shoots, or has pruned at the wrong time of the year so that the plant does not have time to ripen its flower buds before the end of the season.

Lighter coloured circular blotches, on hairy leaves, are the result of drops of water left behind after watering; in a cold place these may lead to rotting of the leaf. The hairy leaves of African violets may turn white in irregular patches; this can be due to using cold water instead of water at room temperature.

Cacti are a special case, originating as they do in dry desert areas; if they appear to be 'glassy', there is too humid an atmosphere. If they are kept too cold during their growing period, or too warm and moist while resting, the stem or the neck of the roots will rot, treacherously starting inside, until one day the entire plant is found to be hollow.

In cases where there is no apparent reason for trouble, help from an expert is of great value, for instance the gardening magazines have teams who will give advice on plant care, and there are county horticultural advisers who can be contacted through your County Hall. Fellows of the Royal Horticultural Society are entitled to receive advice from the Society's plant pathologists.

Apart from the inevitable exceptional cases for which no explanation can be found, disease and insect pest control is not very difficult. Nevertheless, we should not close our eyes to the problem and simply

To prevent damage caused by cold spray, always spray from a distance of 30 cm (12 in)

Pest attack and cure

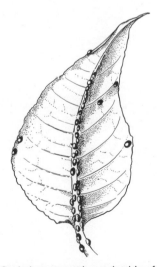

Scale insects on the underside of an aphelandra leaf.

maintain that properly cared-for houseplants cannot be attacked by pests or be subject to diseases. Proper care is, however, the best way to prevent them gaining a hold. Strong, well-fed plants will be more able to withstand infestation by parasites than specimens which have become weak as a result of wrong treatment.

Greenfly can be very troublesome, especially on cinerarias, and where plants are being grown in conditions which do not supply sufficient moisture either at the roots or in the air. In the short term, aerosols based on relatively harmless vegetable derivatives such as bioresmethrin/pyrethrum may be used, but in the long term, the growing conditions should be altered.

Red spider mite is a minute, almost invisible pest which lives on the underside of the leaves, from which it sucks the sap, causing yellow speckling to appear on the upper side. In severe attacks a web is noticeable and the leaves wither and drop off. Dry conditions and heat encourage its spread; malathion is advisable. Whitefly are tiny pests which will also be found on the underside of leaves, but which fly off as soon as disturbed; fuchsias and pelargoniums are especially popular with these pests. They are very difficult to deal with, and the use of finger and thumb to kill the adults as soon as they are seen, is vital. However, their eggs will hatch and the offspring will feed on the lower side of the leaf; they should be thoroughly sprayed with an insecticide such as permethrin or bioresmethrin. These are simple and effective; they will destroy whitefly adults within a few minutes and should be repeated every few days to destroy further adults emerging from eggs already present.

Red spider mite, larva (left) and male adult (right).
Actual size: 0.5 mm ($\frac{1}{40}$ in).

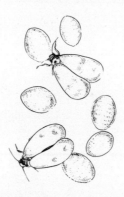

The whitefly is always found on the underside of the leaf.
Adult insect and eggs; size 1.5 mm ($\frac{1}{16}$ in).

Thrips. Left a larva; right full-grown insect. Size: 1 mm ($\frac{1}{25}$ in).

Scale insects can be mistaken for natural spotting of the leaves, or be taken for part of the bark, unless closely examined. They are small and brown, round or oval, and are another sap-feeding insect living on the undersurface, close to the main vein. Scraping gently off with a matchstick or the back of a knife, is one method of removal; another is immersion of the entire plant in a solution of insecticide.

Mealy bugs are not so often seen on houseplants; they are more often a pest encountered in the greenhouse. However, they do sometimes occur – they may have come in with the plant when bought. They are a little like small woodlice in shape, but slow moving, tending to remain feeding in one place. As protection, they produce a white wool which covers them completely, and it is important, when control is attempted, to penetrate through this to the pest beneath. Painting with a stiff brush dipped in methylated spirits is effective, but remember that they are good at hiding, and the small ones can easily be missed, so the plants should be examined several times, at intervals of a few days, until no more bugs can be seen.

If a plant is looking sick for no apparent reason, rather grey, not producing any new growth, and with a generally flabby appearance, suspect root aphis, a form of greenfly, with white fluff attached, feeding on the roots. The plant should be turned out of the pot and the soil-ball examined; the cure is to wash off all the compost, if possible, and repot, and after a few days water with a malathion solution.

Mealy bug, with white woolly secretion in the axils of a *Nerium oleander*; actual size: 1.5–2 mm ($\frac{1}{16}$–$\frac{1}{12}$ in).

Fungal diseases   Fungus diseases follow an entirely different course. Any of them will infect where a plant is weak due to poor cultivation, and can sometimes be arrested by removal of the affected tissue, dusting with flowers of sulphur and altering the cultural conditions. However, the plants may have been affected months before and the plant tissue largely destroyed before symptoms show in the form of mould, and it is often best to discard an affected plant. Grey mould *(Botrytis)*, and mildew are the two main diseases; mildew shows as powdery white patches on leaves and stems. Treatment for both can be tried with a systemic fungicide such as benomyl, or with dinocap and sulphur for mildew. Sooty mould is a minor fungus disease which grows on the sticky 'honeydew' secreted by sucking insect pests; it should be carefully wiped off so that the leaves can breathe.

Damage to a ficus leaf caused by cold.

Greenfly may completely devour a cineraria.

Yellow blotches on the leaves of an African violet – sunburn or virus disease?

3. SURVEY OF PLANT FAMILIES DISCUSSED IN THIS BOOK

Houseplants are botanically classified into families. To give you some idea of this subject, we give in the following pages a short survey of the families occurring in this book.
In the margin you will find the names of all the genera described, with the number of the page on which they occur, so that you will be able to look them up quickly. In all the book contains about 250 genera, belonging to 73 families.

Acanthaceae – Acanthus Family. Plants belonging to this family have an important place in the history of art. A number of species native to the regions around the Mediterranean and in Asia Minor served as a motif in classical sculpture. Stylised acanthus leaves were incorporated in the walls of many buildings, where they can be seen to this day.
The species mentioned in this book, which can easily be grown in the living-room or in a greenhouse, are shrublets and perennials from tropical regions of the old and new world.

Aphelandra 126
Beloperone 138
Crossandra 179
Fittonia 215
Hemigraphis 229
Hypoestes 242
Jacobinia 246
Pachystachys 279
Pseuderanthemum 310
Thunbergia 342

Adiantaceae. This is a new botanical family, split off from other fern families. *Adiantum* is the only genus; it consists of dozens of species.

Adiantum 116

Agavaceae – Agave Family. Until recently the plants belonging to this family were classified under the **Amaryllidaceae** or daffodil family. The agave itself embraces more than 300 species, most of them native of the southern United States, Mexico, Central America, the West Indies and the northern part of South America. *A. americana*, also known confusingly as Centenarian aloë, grows wild in the Mediterranean area. All the species of the agave genus flower only

Agave 119
Cordyline 175
Dracaena 193
Sansevieria 324
Yucca 350

once, when quite old, but when they do the flower stalk which is several yards high, is worth seeing. After flowering the plant dies. Other genera belonging to this family can flower several times.

Argyroderma	128
Conophytum	174
Faucaria	209
Glottiphyllum	219
Lithops	252

Aizoaceae – Mesembryanthum Family. On account of the largest genus belonging to this family, it was formerly called *Mesembryanthemaceae*. Some scholars use the name *Ficoidaceae*. This group of foliage succulents occurs especially in South Africa, and consists partly of plants of shrub-like growth, partly of stem-less plants. Some species can successfully be grown indoors. Others are popular among collectors because of their brilliant camouflage and look like living stones.

Iresine	244

Amaranthaceae – Cockscomb Family. These are grown mainly in the garden and on the balcony. Some species were cultivated centuries ago as bedding plants. Their Brazilian relation *Iresine herbstii* was used for the same purpose, until it was recently discovered in the Netherlands that it also makes an excellent pot plant.

Amaryllis	231
Clivia	166
Haemanthus	223
Hippeastrum	231
Narcissus	266
Vallota	231

Amaryllidaceae – Daffodil Family. A group of perennials, the roots of which are bulbous, related to the lily and agave families. Their natural habitat is mainly in South Africa, Central America and Australia, which indicates that they inhabit relatively dry regions. Moisture and food is stored in the bulbs and in dry periods the foliage dies. This life-cycle must be taken into account in their cultivation.

Allamanda	121
Carissa	152
Catharanthus	152
Dipladenia	191
Nerium	270
Pachypadium	278

Apocynaceae – Dogbane Family. The species which can be grown indoors originate in tropical and sub-tropical regions of the Old and the New World. A number of woody genera, such as *Rauwolfia* and *Strophanthus*, play an important part in the preparation of modern medicines, from which may be deduced that some of the related plants, cultivated as houseplants, are also to a greater or lesser degree poisonous. In practice this will present no problem.

Aglaonema	120
Anthurium	124
Caladium	144
Dieffenbachia	188
Epipremnum	314
Monstera	264
Philodendron	294
Rhaphidophora	314
Sauromatum	325
Scindapsus	328
Spathiphyllum	338
Syngonium	341
Zantedeschia	350

Araceae – Arum Family. Some people believe the name of this family is derived from the high priest Aaron, whose rod budded through God's grace. Modern philologists are of the opinion that the word is based on the Greek word *aros* (= to use).
Most members of the arum-lily family described in this book can be grown easily and successfully indoors. All forms of growth occur among the more than 50 genera of this family, spread all over the world. Some plants look like trees or shrubs without, however, ever forming real wood. Low-growing herbs also occur, while others again spread on rootstock or shoot up from perennial bulbs. Epiphytes from the jungle, lianas, rare dry-flowering plants, marsh water-plants

complete the picture. Many houseplant species, however, rarely show the characteristic inflorescence.

Araliaceae – Ivy Family. A relatively small family, although it contains no fewer than 65 genera. Most of the species are woody and thorny shrubs or trees, growing in the tropical zones of America, Asia and Australia, but also – for example the ivy – in more temperate climates. They all, without exception, occur in shady forests. It is advisable to remember this fact when growing them indoors.

Dizygotheca 192
× *Fatshedera* 207
Fatsia 208
Hedera 226
Polyscias 306
Schefflera 326

Araucariaceae – Araucaria Family. This family contains really only one genus namely *Araucaria*, one species of which, *A. heterophylla*, is fairly often seen.

Araucaria 127

Asclepiadaceae – Milkweed Family. This family's botanical name is undoubtedly derived from Asclepios, son of Apollo and god of healing. It is further connected with the main genus: *Asclepias* or milkweed, which at one time was grown in Central Europe to produce 'vegetarian' silk. The modern method of manufacturing artificial silk is rather more smelly, but apparently more effective. Climbers and unusual succulents form this family's contributions to house and greenhouse vegetation. For the garden there are a number of fine flowering perennials.

Ceropegia 156
Hoya 236
Stapelia 338
Stephanotis 339

Aspidiaceae. A group of ferns originally belonging to the large fern family *Polypodiaceae*. The family was subsequently split, since this appeared to be more logical.

Cyrtomium 186
Didymochlaena 188
Polystichum 306

Aspleniaceae. This is another fern family that has been reclassified. It contains only 3 genera. *Asplenium* itself is sub-divided into 18 species, among which several interesting jungle plants are found.

Asplenium 130
Phyllitis 299

Balsaminaceae – Balsam Family. There are now only two genera belonging to this family. In the first place, of course, the annual or perennial balsam, which can be grown in the garden as well as indoors. Who does not know the famous 'Busy Lizzie'? A tropical waterplant also belongs to the group, but this can only be grown in a heated greenhouse.

Impatiens 243

Begoniaceae – Begonia Family. By far the most important genus of this family is the *Begonia*, of which there are thousands of species and cultivars. Another genus, *Hillebrandia*, is completely unknown. The original habitats are mainly in the tropical rain forests of America, Asia and Africa. For Europe the begonias were discovered by the

Begonia 132

French botanist Charles Plumier, who named them after Michel Bégon, at that time governor of the French West Indies.

The name *Begonia* remained in use, even when it was found that the same plants had been used centuries earlier in China as a motif on painted porcelain.

Jacaranda	245

Bignoniaceae – Bignonia Family. Of this family we grow mainly garden plants, for example the trumpet-flowered *Campsis*, a beautiful climber and the fine, impressive tree, the *Catalpa*. The so-called garden gloxinia *Incarvillea*, also belongs to the family. The tropical *Jacaranda*, a decorative tree, is sometimes cultivated as a house plant, but since it never flowers here, the fern-like plant is not much valued.

Blechnum	140

Blechnaceae. Another fern family newly classified as a result of revised opinions. Indoors we grow a number of species and strains of the genus *Blechnum*, which form very strong and decorative plants.

Aechmea	117
Ananas	123
Billbergia	139
Cryptanthus	180
Guzmania	221
Neoregelia	267
Tillandsia	342
Vriesea	348

Bromeliaceae – Bromelia Family. These became known shortly after the discovery of America as a result of the importation of the first pineapples. Many genera belonging to this family were brought from Central and South America by the French botanist Plumier around 1700. The family, of which approximately 1600 species are known, is named after his friend and colleague, the Swedish physician Olaf Bromel. Like tree-orchids they are nearly all epiphytes or pseudo-parasites, growing in the crowns of trees without penetrating the wood, as do true parasites. Belgian growers in 1835 started the systematic cultivation of bromeliad plants and seeds.

Aporocactus	126
Astrophytum	131
Cephalocereus	154
Cereus	155
Chamaecereus	157
Cleistocactus	164
Coryphantha	176
Echinocactus	198
Echinocereus	199
Echinopsis	199
Epiphyllum	201
Espostoa	203
Ferocactus	209
Gymnocalycium	222
Lobivia	252
Lophophora	253
Mammillaria	254
Melocactus	259
Neoporteria	267
Notocactus	272
Opuntia	276

Cactaceae – Cactus Family. The approximately 5000 species belonging to this family form the largest group of plants, collectively known as 'succulents'. They all have strong roots, usually lying close below the surface, in order to be able to take in as much water as possible in the rainy season.

However, there are also a number of cacti with roots like tap-roots. Apart from foliage cacti which grow as epiphytes in the tropical rain forest, all cacti possess thick, fleshy tissue, in which water is stored for the dry season.

Their thick, leathery epidermis, often provided with prickles – as a defence against enemies – or other protuberances, is covered in a layer of wax which serves to restrict evaporation to an absolute minimum. The form of the plants (globe- or column-shaped) also provides the smallest possible surface. For instance, a giant cactus from the deserts of Arizona and Mexico, areas particularly rich in cacti, can store 3000 l (656 gal) of water in its 'column'. In dry periods it can

lose 50% of this water without coming to harm and in the next tropical rainstorm the loss can be made good by the pumping action of the roots.
There is, therefore, some sense in rest and drought!

Oroya	277
Parodia	284
Rebutia	312
Rhipsalidopsis	316
Selenicereus	330
Trichocereus	345
Zygocactus	352

Campanulaceae – Bell Flower Family. These occur more often growing wild than on the window-sill. Of the species which are not winter-hardy, usually originating in Central Europe, only *Campanula fragilis* and *C. isophylla* are suitable for growing in pots.

Campanula 149

Cannaceae – Canna Family. This family consists of only one genus, *Canna*, with around 50 species, growing in tropical regions of America.

Canna 150

Celastraceae – Spindle Tree Family. The deciduous and evergreen species of *Euonymus* are hardy shrubs, *E. europaeus* being the well-known common spindle of the hedge rows. Like their cultivars, the decorative evergreen species of the *Euonymus* are good plants for pots and tubs.

Euonymus 203

Commelinaceae – Spiderwort Family. For the sake of convenience these are frequently all called *Tradescantia*. Certainly a number of genera are difficult to distinguish. Our small selection shows that marked differences occur only between the pendant forms and the erect growing *Rhoeo*. They all originate in tropical and sub-tropical areas of America.

Callisia	146
Rhoeo	319
Setcreasea	333
Siderasis	333
Tradescantia	344
Zebrina	352

Compositae – Daisy Family. Together with orchids, these belong to the largest of all plant families. There are around a thousand genera embracing nearly 20 000 species and innumerable cultivars; they are spread all over the world. They include greatly varying forms of growth: from annual and perennial weeds to vegetables, herbs, medicinal plants and ornamental plants. Even pyrethrum, a greatly valued insecticide, is produced from a chrysanthemum species. On the window-sill plants belonging to the family of composites are fairly rare; this is doubtless due to their abhorrence of tropical rain forests, where they will be sought in vain.

Chrysanthemum	160
Gynura	222
Senecio	331

Crassula	177
Echeveria	196
Graptopetalum	220
Kalanchoë	247
Kalanchoë (Bryophyllum)	249
Pachyphytum	278
× *Pachyveria*	280
Rochea	321
Sedum	329

Crassulaceae – Stonecrop Family. Lay people tend to count these among cacti. 35 genera, with around 1500 species, are known. In recent years changes in botanical naming have often created confusion. For instance *Bryophyllum*, the 'brood-leaf' made famous by Goethe, is no longer a separate genus; it now belongs to the genus *Kalanchoë*; *Kalanchoë blossfeldiana* is undoubtedly the best known pot-plant of the entire family.

Carex	151
Cyperus	183
Scirpus	328

Cyperaceae – Sedge Family. Sedges are distinguished from grasses by their nearly always sharply triangular stems without internodes, and their proportionally narrow leaves. They inhabit marshes and acid meadows in all parts of the world.

Erica	202
Rhododendron	317

Ericaceae – Heath Family. This family, consisting of over 80 genera and over 2500 species spread all over the world, is of great importance in commercial horticulture. Those belonging to the European selection of indoor plants unfortunately frequently lead to disappointment. On the one hand, their marked sensitivity to calciferous water, is ignored; on the other hand, heathers are regarded as not very thirsty, whereas one should remember their ancient characteristics as moisture-loving peat-bog plants.

Acalypha	115
Codiaeum	168
Euphorbia	204
Pedilanthus	287

Euphorbiaceae – Spurge Family. This is the fourth largest plant family, whose area of distribution is chiefly in tropical America and South Africa, but which – like several wild and garden forms in this country – also occur in areas with a sub-tropical or temperate climate. Typical of all spurges is the more or less poisonous milky liquid they exude, the medicinal properties of which are said to have been discovered – as far as European civilisation is concerned – by the Roman physician Euphorbus half a century B.C. Among indoor plants a number of fine inhabitants of heated and temperate greenhouses are popular, mainly representatives from dry tropical regions. These are undemanding succulent species, resembling cacti to such an extent that their identity can only be discovered by scratching them with a pin, when the milky liquid will appear.

Exacum	206

Gentianaceae – Gentian Family. Alpine plants are definitely not the only plants included in this family. *Exacum affine*, usually cultivated as an annual, has become a very attractive pot plant in recent years.

Geraniaceae – Geranium Family. As far as house plants and plants which overwinter in the cold frame, are concerned, they originate in the Cape Province of South Africa. They are all descendants of the race imported into England in 1710 and constantly changed in cultivation.

Pelargonium 288

Gesneriaceae – Gloxinia Family. They owe their name to the Swiss physician and botanist Konrad Gesner, who died of the plague in 1565. He was the first to develop a botanical classification. All members of this family grown indoors originate in tropical and sub-tropical areas, where the same conditions obtain: a dislike of direct sunlight and calciferous water (pH 4·5 desirable!). Among them are evergreen plants, herbaceous plants which die in winter, and also shrubs; the pendant species often grow as epiphytes. The cultivated forms of *Aeschynanthus* and *Columnea* are probably the best known.

Achimenes 115
Aeschynanthus 118
Columnea 173
Episcia 202
Hypocyrta 242
Rechsteineria 314
Saintpaulia 323
Sinningia 334
Smithiantha 335
Streptocarpus 340

Gramineae – Grass Family. With their more than 700 genera and 8000 species, these form the largest group of consumer plants in the world. On the other hand, ornamental forms for indoor cultivation are of rarity value only; hence the fact that we describe only one.

Oplismenus 276

Iridaceae – Iris Family. On the one hand they are closely related to the daffodil family, on the other hand to the lilies, so it is not surprising that only the crocus is mentioned.

Crocus 179

Labiatae – Mint Family. A very well-known family, not only occurring wild, but of which many representatives are found in the herb and vegetable garden, as well as among ornamental plants.

Coleus 171
Plectranthus 304

Leguminosae – Pea Family. The third largest plant family, embracing 600 genera and around 13 000 species. However, few are suitable for use as houseplants.

Cytisus 186
Mimosa 263

Liliaceae – Lily Family. This is another large family, including 220 genera and over 3500 species, occurring in all parts of the world, even in polar regions. A considerable number of plants are suitable for use indoors and able to thrive even in primitive conditions. These include widely differing species: obvious succulents such as *Aloë* or *Gasteria*

Aloë 122
Asparagus 129
Aspidistra 130
Bowiea 142
Chlorophytum 159
Colchicum 171

Convallaria	174
Gasteria	218
Gloriosa	218
Haworthia	224
Hyacinthus	238
Lilium	250
Liriope	251
Tulipa	346
Veltheimia	348

as well as familiar houseplants like *Asparagus, Aspidistra* and *Chlorophytum*. The family also includes all sorts of bulbous plants, which for many centuries have made the lily known, and in fact famous as a symbol of high moral values and of religious and aesthetic glorification. It must be admitted that, in comparison to the biological quantity and the world-wide distribution of this family, the selection of true indoor plants might be larger. For even the finest hyacinths and the most carefully cultivated tulips cannot pretend to be anything but exceptions, as is the case with the crocus among the *Iridaceae*.

Abutilon	114
Hibiscus	229
Pavonia	287

Malvaceae – Mallow Family. Even in this era of man-made fibres, one of the members of this family remains indispensable, namely the cotton plant. Occasionally seed is available from seed merchants, but as a rule attempts to grow this plant indoors are fruitless. You will have better results by confining yourself to our selection!

Calathea	145
Ctenanthe	181
Maranta	256

Marantaceae. These are all perennial herbaceous plants requiring the warm and damp climate of the tropical forests. It is really unnatural for these plants to be grown in the living-room, although examples of lasting adaptation become known from time to time.

Bertolonia	138
Medinilla	258
Sonerila	336

Melastomataceae. The plants in this family are found especially in tropical regions of America; they are therefore typical hot-house plants and will survive in the living room only in damp conditions. The best known are *Medinilla, Sonerila* and *Bertolonia*.

Ficus	210

Moraceae – Mulberry Family. By themselves these form a considerable family, with 61 genera and around 1500 species. Lovers of house plants as a rule only know one, namely *Ficus*, or the rubber plant, but it is a very mixed family, varying from hemp and hops to the bread-tree of southern islands, the much-lauded fig-tree of Mediterranean countries, the mulberry tree once indispensable for the making of silk and occurring also in our latitude, and the rubber-trees which supply the latex for rubber. The plants we grow in pots and tubs are minute versions compared to the giants of the jungle which grow to 40 m (132 ft). The houseplants never flower or produce fruit. Apart from the straight-growing large-leaved species there are elegant forms with smaller foliage, such as the often flowering *F. deltoidea* as well as creeping and climbing species. All of them are evergreen. The real fig-tree (*F. carica*), which will survive our winters in a sheltered position, is deciduous.

Myrsinaceae. They are small, evergreen shrubs and trees growing in eastern Asia. The sole representative on the window-sill is *Ardisia*, a plant in which interesting botanical processes take place.

Ardisia 128

Myrtaceae – Myrtle Family. Like the mulberry family, this is a remarkable relationship, among which must be counted giant trees such as the Australian eucalyptus, which grows to 90 m (300 ft) and is being planted in many tropical countries. Among the more than 100 genera embracing about 3000 species, including small woody plants and dwarf shrubs, there were at one time many fine ornamental plants, used in the garden. The callistemon has enjoyed a real 'comeback' in this country, while in the United States a large number of other beautiful species are available to plant lovers.

Callistemon 147
Myrtus 265

Nyctaginaceae. A family belonging mainly to the tropics and subtropics of Central and South America. There are 30 genera with around 3000 species. Most of them are woody plants; among the herbaceous strains is the annual garden plant *Mirabilis jalapa* with its delightful flowers.

Bougainvillea 141
Pisonia 301

Oleaceae – Olive Family. Found more often growing out of doors (elder, forsythia, ash and privet) than on the window sill. Only a few tropical species have found their way into our heated and unheated greenhouses.

Jasminum 247

Oleandraceae. This is not one of the larger plant families. It contains only two species: *Nephrolepis* and *Oleandra*.

Nephrolepis 269

Onagraceae – Willow Herb Family. In the garden they are much-loved annuals, for instance *Clarkia*, *Godetia* and the evening primrose *Oenothera*. *Ludwigia* species grow in aquariums. Only the fuchsia became world-famous for use indoors and on the balcony.

Fuchsia 216

Orchidaceae – Orchid Family. With its more than 20 000 species and in addition innumerable, constantly increasing cultivars, this is regarded as the largest and most varied plant family. There are two main groups: the ground-rooting terrestrial orchids and the epiphytic, pseudo-parasite tree orchids of the jungle. In between there are a number of marginal cases, which can grow both in the ground and as

Brassia 142
Calanthe 145
Cattleya 153
Coelogyne 169
Cymbidium 183
Dendrobium 187
Epidendrum 200

Lycaste	254
Miltonia	262
Odontoglossum	274
Oncidium	275
Paphiopedilum	282
Phalaenopsis	293
Vanda	347

epiphytes; an example is the vanilla plant. Ground orchids occur in the tropics, in temperate and in cold climates. In central Europe alone there are still about 60 wild species, including some regarded as belonging to Alpine flora. Orchids are found even in the far north of Canada and Siberia. The large majority, which are also the finest, are tree orchids growing in the damp, hot jungles of the old and the New World. The Asian orchids were discovered in 1689 by the German explorer Kaempfer; in 1732 the first orchids from Central America arrived in England. The subject 'orchids' is so extensive that it cannot be discussed in depth in a general book on houseplants. In this respect specialised literature and membership of national orchid societies must come to your aid.

Chamaedorea	158
Howeia	234
Microcoelum	260
Phoenix	298

Palmae – Palm Family. With their 236 genera and over 3400 species they are among the most important crop and ornamental plants of many tropical and sub-tropical countries. The sole European species is *Chamaerops humilis*, the dwarf palm. For the rest one need only mention such names as date-palm, coco-palm, sago-palm, rattan-palm (which supplies cane for chairs etc.) or the raffia-palm, and everyone will know what the palm family consists of. Towards the end of the 19th century ornamental forms were very fashionable for growing in pots or tubs, but since then they have – unjustifiably – been neglected. At present a few are again for sale and it is quite possible that they will regain their popularity in the foreseeable future.

Pandanus	280

Pandanaceae. These tropical evergreens produce only a few species of the genus *Pandanus* which are suitable for growing in a hothouse.

Passiflora	284

Passifloraceae – Passion Flower Family. This family includes 12 genera with about 600 species. The passion flower itself occurs in about 400 species, chiefly perennial herbaceous plants and climbers, mostly originating in tropical areas of America. Green-leaved and variegated species are available for the heated or unheated greenhouse. The most striking specimen is undoubtedly *Passiflora quadrangularis*, which produces flowers 10–12 cm (4–5 in) in diameter, and edible fruits. This plant grows too wild and too rapidly for use indoors.

Piperaceae – Pepper Family. Needless to say they originate in pepper-producing countries; especially in the jungles of tropical America and Malaysia. There are two groups: species with green and with variegated foliage, and those of creeping and pendant habit.

Peperomia 291
Piper 301

Pittosporaceae – Pittosporum Family. A small family, occurring mainly in Australia. They are all evergreen shrubs or shrublets.

Pittosporum 302

Plumbaginaceae – Sea Lavender Family. These chiefly inhabit coastal areas and steppes in the Old World. The family also includes garden plants such as *Armeria* and *Limonium*.

Plumbago 305

Polypodiaceae – Oak-fern. Even after the creation of many small families as a result of new rules concerning nomenclature, there still remain sufficient genera to make this the largest fern family. One can hardly keep track of the many forms occurring in the more than 150 genera and about 700 species. They are, moreover, cosmopolitans in the truest sense of the word.

Phlebodium 297
Platycerium 303

Primulaceae – Primrose Family. This family is so well known that it need hardly be discussed, but the fact that the cyclamen belongs to this family for some reason continues to amaze us.

Cyclamen 182
Primula 307

Proteaceae – Protea Family. To this family belong 62 genera with over a thousand species. Nevertheless there is only one representative in the living-room, namely the evergreen *Grevillea robusta*, a native of Australia. Looking at this plant one would never suspect that the South African proteas flower so magnificently. They are imported into this country only as cut flowers.

Grevillea 220

Pteridaceae. At one time a genus counted among oak-ferns. As a rule these ferns are easily grown.

Pteris 310

Punicaceae – Pomegranate Family. This family has a classical history: Pliny, the famous Roman historian and writer already mentioned a 'Punic apple', *Malus punicus,* said to have been cultivated by the Carthaginians in North Africa. Its original habitat was the area between the eastern part of the Mediterranean and the Himalayas.

Punica 312

Duchesnea	196
Rosa	322

Rosaceae – Rose Family. It is difficult to realise that this is a large family, consisting of over 100 genera and 3000 species, among which are crop plants as well as ornamental plants. They occur in all parts of the world. In this country roses and strawberries grow side by side.

Coffea	170
Gardenia	217
Ixora	244
Manettia	256
Nertera	272

Rubiaceae – Madder Family. With around 500 genera and an estimated 6000 species, this family easily exceeds the *Rosaceae* in size. The coffee plant and *Cinchona*, which produces quinine, head the excellent family of commodity plants. It occurs especially in the tropics. Most of the plants are woody, but there are also indigenous herbaceous plants such as bedstraw and sweet-scented woodruff, as well as *Rubia tinctorum*, once important as a source of a valuable red pigment.

Citrus	163
Skimmia	335

Rutaceae – Rue Family. These are rich in aromatic oils. Transparent patches in the leaves indicate the presence of oil which, however, also occurs in all other parts of the plants and even in the flowers. This oil is used for scientific purposes and in medicines. *Dictamnus* and rue have been growing in our gardens since time immemorial.

Hydrangea	240
Saxifraga	325
Tolmiea	343

Saxifragaceae – Saxifrage Family (Stone-break). They are found all over the world, usually in the form of hardy herbaceous plants or shrubs. Plants occurring in high mountains are in fact capable of breaking stone with their roots. Many representatives of this family are found in the garden, for example the ornamental shrubs *Deutzia* and *Philadelphus*, as well as red currant and gooseberry, and various perennials such as *Astilbe* and *Tiarella*. On some occasions they are found in the tropics.

Calceolaria	146
Hebe	225

Scrophulariaceae – Figwort Family. These, too, are found more often in the garden than in the greenhouse. The 200 genera and around 3000 species of this family are spread all over the world.

Selaginella	330

Selaginellaceae. This family is represented by only one family, with more than 700 species. The tropical rain forests of Asia, Africa, America and Australia are considered to be their native habitats.

Pellaea	290

Sinopteridaceae. A new fern family, apparently not yet described in England. A number of oak fern genera are now classified as a separate family under this name. (The prefix *Sino* means Chinese.)

Solanaceae – Nightshade Family. This is one of the most important plant families. Among the crop plants we find potato, tobacco, tomatoes and paprika. In addition there are many beautiful garden plants such as petunia and, as well, poisonous and medicinal plants, for instance *Datura* and bitter-sweet (*Solanus dulcamara*). There are 85 genera with about 2300 species, growing mainly in tropical countries, although many have adapted themselves to the western European climate and have been growing here for centuries.

Browallia 143
Brunfelsia 143
Capsicum 150
Solanum 336

Theaceae – Tea Family. These form an exclusive group, the trees and shrubs of which prefer the cool mountain forests of the tropics and sub-tropics in eastern Asia. They are plants of distinction, admired and glorified by many a plant lover.

Camellia 148
Cleyera 165

Tiliaceae – Lime Tree Family. These are much more tropical plants than our own lime tree would make us suspect. The South African *Sparmannia* is only one example among thousands.

Sparmannia 337

Urticaceae – Nettle Family. The large stinging nettle *Urtica dioica* belongs to this family; it is a widespread 'weed'. Nettles thrive everywhere, especially in the tropics, where many of them, instead of painful stinging hairs, produce very useful fibres.

Pilea 299

Verbenaceae – Verbena Family. The extremely useful teak is one of the members of this family. Who would have thought that our small garden verbenas had such a distinguished relation!

Clerodendrum 164

Vitaceae – Vine Family. This family consists of 12 genera and around 700 species, mainly climbing shrubs or lianas. A very old plant, already cultivated in primeval eras, is the vine, *Vitis vinifera*; the Virginia creeper, *Parthenocissus*, is an important ornamental plant; in addition *Cissus* and *Ampelopsis* are grown indoors. As a matter of interest it should be mentioned that this family also embraces trees, shrubs and succulent forms.

Ampelopsis 122
Cissus 160
Rhoicissus 321
Tetrastigma 341

4. HOUSEPLANTS IN COLOUR

Once the final selection has been made when writing a book on houseplants, one is inevitably faced with the problem of how to classify the plants in a way which will enable the reader to refer to them easily. A classification in which the plants are grouped according to situation or habit, or to their suitability for certain uses, would not befit the purpose of this book, especially as many of the plants discussed are marginal cases which would have to be classified under more than one heading. Nor would classification by family be convenient in this respect.

In this book the alphabetical order, using internationally known botanical names, has seemed to be the most suitable. As few readers appreciate the use of symbols these have been omitted, resulting in an inevitable repetition of the rules for looking after plants with the same requirements. However, this is not a great disadvantage, since we often learn a great deal by repetition. The use of scientific names avoids confusion, because in many cases a plant has no English name, or has one that varies from one region to another.

Abutilon

Shrubby plants originating in the tropics; suitable for indoor use. They are easy to look after and in summer will even grow in a sheltered position on the balcony or in the garden. In summer they should have a half-shady situation, in winter as much light as possible. The plant can reach a height of two metres (6 ft).

Abutilon belongs to the *Malvaceae* family and is an evergreen shrub with slender, arching stems. The leaves are usually heart-shaped. If properly cared for, the plant will grow rapidly and, if given sufficient light, will flower freely. Abutilon-hybrids is the botanical name for the group of strains now chiefly marketed.

They all have plain green foliage and from spring until autumn untiringly produce bell-shaped flowers, veined or striped yellow, orange or reddish-purple in colour.

Of the approximately 150 species brought to Europe in the last 120 years, mainly from Central and South America, two plants of interest to the lover of houseplants have survived:

A. striatum, from which two highly valued varieties have been developed, namely 'Thompsonii', which has yellow-blotched leaves and orange-red flowers, and 'Souvenir de Bonn', with white variegated foliage and salmon-pink, red-veined flowers. Both have an erect and bushy habit. To maintain this shape, it is necessary to prune the plants rigorously at the end of February, before shoots appear. *A. megapotamicum* has drooping stems, an abundance of small green leaves and of small, orange-yellow lantern-shaped flowers with a wine-red calyx. The 'Aureum' variety has yellow-blotched leaves.

Both are popular as hanging plants and are also offered as erect growing plants. You can cultivate them in this form yourself in two ways:

a. Remove all side shoots as soon as they appear, until the plant has reached the desired height. It can reach 1·20 m (4 ft). If the tip is then removed, a crown will form.

b. The commercial grower follows the same method, but uses a fast-growing hybrid on which to graft *A. megapotamicum*. Don't forget to support the stem with a cane reaching to the crown, to prevent the graft becoming detached.

Abutilon striatum 'Thompsonii', fam. *Malvaceae*

Propagation: all abutilon species can be grown from tip-cuttings in March–April. Hybrids can also be sown. If sown in spring, they will flower within six months.

General care: during the growing period they should be watered freely: in hot weather twice a day. From March till August they should be fed weekly. Resting period is from September until February in a moderately heated room, to 15°C (58°F); do not feed, water sparingly. Repot in April in not too large pots.

Abutilon megapotamicum, fam. *Malvaceae*

Acalypha

The name is of Greek origin and means 'nettle', but the approximately 500 species of this plant, occurring in all tropical regions, have nothing in common with nettles. *Acalypha* species belong to the *Euphorbiaceae* family. Suitable for use indoors and in a moderately heated greenhouse.

Acalypha hispida, fam. *Euphorbiaceae*

For indoor cultivation the main species is *A. hispida*, the Chenille plant which has green foliage and a long, trailing inflorescence resembling deep red chenille. All other acalypha species have insignificant dioecious flowers and are therefore cultivated chiefly for their magnificent variegated foliage. The finest foliage plants are found among the *A. wilkesiana* hybrids. The colours vary from green to brown and dark red on white-edged leaves. There are also forms without white margins, the leaves being blotched with brown, green and pale red.

If too little light is provided, the leaves revert to dark green. Like the variegated strains, the most generally grown *A. hispida* needs plenty of light (though not burning sunlight) and warm conditions throughout the year. In winter the temperature should never fall below 16–18°C (60–65°F). A high level of atmospheric humidity should be provided, even during the dormant period between September and late January. If the air is too dry, the leaves will curl up and aphids and red spider will appear.

General care: during the flowering period from May to the end of September the soil should be kept moist, but care should be taken that the plant does not stand with its feet in water. From April onwards feed every fortnight. Remove the sideshoots on young plants during the first year. In the second year the plant should be topped to induce a good shape. Dead flowers must be removed.

Once the plant has produced a fair number of shoots and is growing well, it should be repotted in porous compost in a fairly large pot.

Propagation from cuttings in spring is not easy and will be successful only if a great deal of top heat can be provided.

Achimenes hybrid 'Rose', fam. *Gesneriaceae*

Achimenes – Hotwater plant

The original plant, a native of Central and South America, is small and modest in comparison with modern *Achimenes* hybrids, which are much larger and more colourful. They are magnificent pot plants, capable of being grown indoors in summer, although they prefer a greenhouse.

This plant is a member of the *Gesneriaceae* family. During the growing period, from spring until autumn, it requires a warm and light situation, but not in direct sunlight. After flowering, and when the foliage has died, the dry stem must be cut level with the soil, the tuber-shaped rootstock removed and kept in dry sand in a warm place.

Like gloxinias, the rootstock is potted in lime-free compost in February–March. Water only with softened, tepid water. When buds become visible, spray frequently.

Adenium Obesum

Among plants which have been transformed into amazing flowering succulents is the *Adenium obesum*. At one time this rare plant was found only in botanical gardens.

Origin: the name is thought to be derived from the fact that it was discovered close to Aden. It also occurs wild in many dry regions of Africa and on the island of Socotra, famous for its plant life.

Development: a house plant of international repute was developed from the botanical rarity when the German botanist Reinhard Schäfer succeeded in grafting the succulent on the closely related *Oleander*. Both plants are poisonous. New name: *Adenium obesum* 'Multiflorum', a cultivar which embraces the best traits of both plants. Flowers in spring and autumn. Easy to grow.

Adenium obesum 'Multiflorum', fam. Apocynaceae

Adiantum – Maidenhair fern

Although this graceful fern is indispensable in flower arrangements, we should not be too anxious to use it as a houseplant. Since its native habitat is in the damp atmosphere of the tropical rain forests of South America, it feels more at home in a heated greenhouse, where the atmosphere corresponds to that of its place of origin.

This means that adiantum requires a great deal of warmth and a high level of atmospheric humidity throughout the year, but particularly during its growing period from March until August. It must have a shady and draught-proof position – requirements difficult to fulfill in the living-room. Always give tepid, softened water (pH 4·5) and spray frequently.

Connoisseurs of this plant say that if the soil-ball is dry the plant will rapidly roll up its leaves and within a few hours will look like a bundle of hay. The danger of this happening is the greater the more the root system has developed, since in that case the soil cannot retain much water. In addition the correct soil mixture used for the adiantum contains so much peat that, when the soil-ball is dry, the water will drain away rapidly. The best method is to soak the plant in lukewarm water until the soil-ball is saturated.

Shrivelled up parts should be cut down to soil level. Possibly the plant may put forth new shoots at that place. In healthy plants old brown leaves are removed to make room for new shoots. As always, the exception proves the rule. There are cases where adiantum has given pleasure for many years, growing to enormous plants. It should then be given a larger pot every year.

If you are unable to buy a ready-made fern mixture, you can produce it yourself by mixing 2 parts pre-packed potting compost with 1 part peat and 1 part sharp sand.

From March until August give the plant an occasional dose of thoroughly diluted manure. In winter the temperature should not fall below 18°C (65°F); some soil heat is desirable, for adiantum dislikes cold feet. *A. raddianum* is often marketed under the old name *A. cuneatum*. 'Decorum', 'Brilliant-Else' and 'Fritz Luethli' are the best cultivars for growing in pots.

Adiantum raddianum 'Decorum', fam. Adiantaceae

Aechmea

The genus *Aechmea* belongs to the extensive and multi-form family of *Bromeliaceae*, and originates in Central and South America. A few dozen species have been in cultivation for about 150 years and many interesting cultivars have been developed in that time. After flowering the parent plant dies.

All aechmeas sold as pot plants are epiphytes. This is how they grow in the trees of the jungle, where their roots serve as support and to help them to cling to the branches rather than to absorb nourishment. For the latter purpose they have little suckers at the bottom of the rosettes. The leaves are curved like gutters, catching the water. Because of this it was given the name 'rainbutt plant' in Germany. The leaves of the funnel-like rosettes often bear thorns.

Aechmea is very suitable for use as a houseplant, since it is accustomed to varying humidity. The most important representative of the genus, and nowadays very popular, is *A. rhodocyanea*. The leaves may grow to 60 cm (2 ft); they are wide and lightly dentate. The silver-grey cross-banding of the leaves forms a magnificent contrast to the pink inflorescence above. This lasts for months and consists mainly of bracts, for the small blue flowers are insignificant and soon disappear. All bromeliad species, including this aechmea, flower only once, in the centre of the rosette, after which young plants appear at the base. From these new plants are grown, while the parent plant dies.

General care: in the living-room in a warm, light, but not too sunny spot. Can be grown fairly easily indoors, but is also suitable for growing as an epiphyte in the plant window (see p. 88). The air should not be too moist. Spray only on very hot days. Water and spray only with tepid, softened water (*p*H 4–4·5). In summer water may be poured into the funnel, but this should on no account be done in the resting period (between October and February, depending on the flowering time). Whether the funnel should be emptied of water at the beginning of the dormant period, generally in October, depends on the temperature in which it has to live during winter. In a warm room 22–24°C (72–75°F), a little water in the funnel may be an advantage, but in moderate warmth 16–18°C (60–65°F), water remaining in the funnel may lead to rotting. Temperatures below 15°C (58°F) are tolerated only by the hardiest aechmea species.

Aechmea rhodocyanea, fam. Bromeliaceae

Below 12°C (54°F) even these will rapidly suffer damage from cold and will die after some time. Regardless of temperature and growth, the soil-ball is kept moderately moist throughout the year.

Feeding: a weak solution of fertiliser, for example half the strength recommended in the instructions. During the growing period a solution can be poured into the funnel every fortnight; use only soft water. It is advisable to rinse the funnel from time to time with clear water.

Repotting: unnecessary, for as said above, new rosettes are formed and the old plant dies. This may take six to eight months.

Propagation: this is very simple; the young rosettes are removed. The new plant may flower the following year. The small rosettes must be half the length of the parent plant before being removed. Start pouring water into the small funnels two weeks in advance. When prising off a small plant, leave a well-rooted section of the parent plant attached to it. Plant in adequately sized pots, using special bromeliad compost or pre-packed potting compost mixed with plenty of peat fibre.

Other fine aechmea species are: *A. chantinii*, with variegated foliage; as easy to care for as *A. rhodocyanea*. *A. fulgens* and *A. miniata* are a little more difficult.

Aeschynanthus

A decorative member of the *Gesneriaceae* family, somewhat resembling *Columnea*, to which it is related. In the right conditions it will grow, as an epiphyte, into a magnificent, long-stemmed and freely flowering hanging plant. The best place for the aeschynanthus is a hothouse.

During its growing period (March to August) the main requirements are heat, a humid atmosphere and a half-shady situation. During the dormant period (October to January) the temperature should initially be 18–20°C (65–70°F), but it creates further difficulties in the last four weeks since for its bud-formation it requires a temperature of 12–15°C (54–55°F). To be honest, I have never followed all these rules, but nevertheless my *A. speciosus* flowered this year from spring until far into summer.

Water and spray only with lukewarm fully softened water (pH 4–4.5). Feed from March until August. Repot in February in bromeliad soil, although I have found that it is satisfied with ordinary prepacked potting compost as well . . .

Aeschynanthus speciosus, fam. Gesneriaceae

Agave

The agave is not a cactus, even though there are many prickly species originating in desert areas.

Formerly the agave belonged to the *Amaryllidaceae* family, but an international commission on plant nomenclature has classified it, together with a number of related plants, as a separate family.

The agave develops rosettes of fleshy (succulent) leaves, often ending in very sharp thorns. For that reason it is not advisable to place large specimens indoors or in tubs on the terrace when there are children about.

The largest of the agaves is *A. americana*; its leaves may grow up to 1·75 m (6 ft) long and its flower stem to 5–8 m (16–26 ft). Only small species are suitable for indoor use. Of these dwarf forms the finest is undoubtedly *A. victoriae-reginae,* a native of Mexico, whose rosettes in the course of time may achieve a diameter of 50–70 cm (20–28 in). Its most striking feature is its unusual white-edged leaf. Other species that can be recommended are: *Agave filifera*, also from Mexico, with leaves 3 cm (1¼ in) wide and up to 25 cm (10 in) long; *A. parrasana*, from the southern part of the United States, which has unusually wide bluish leaves up to 30 cm (12 in) long; *A. schidigera*, from Mexico, whose narrow 1½-cm (¾-in) leaves give it a globular shape. It may take ten to twenty years before the agave will flower indoors. You may have seen it in flower in a botanical garden.

Propagation: it is possible to grow it from seed. A comprehensive catalogue may include a seed mixture or even named species. They should be grown in the same way as cacti (see p. 71). A simpler method of propagation is by taking sideshoots, which in most species are developed by quite young plants. Using a sharp knife, cut the young shoots close to the parent plant and let them dry out for a few hours before planting them in a sandy mixture.

The dwarf species *A. schidigera* and *A. victoriae-reginae* can be grown only from seed.

General care: place in good light in summer; the larger species can be put out of doors. Hard water (*p*H 7–8) may be used for watering. Frequent feeding encourages growth; in poor soil the plants will remain small. Repot in spring if the pot has become too small. Large specimens can spend the winter in a cool greenhouse, provided the temperature is 4–6°C (40–43°F). Smaller plants can join cacti in a light, cool position. Do not feed during this period (from October until January) and water very sparingly.

Agave victoriae-reginae, fam. *Agavaceae*

Aglaonema – Chinese evergreen

This plant belongs to the arum family and one has to decide whether one should be satisfied with the magnificent foliage or grow it for its flowers, which occur only rarely. Moreover the plant is only suitable for a heated greenhouse or a tropical plant-window — more so even than the *Dieffenbachia* which it resembles.

Origin: the native habitat of aglaonema is in India and areas beyond, and in the Indian Archipelago, where it grows in tropical forests in very damp air and dim light. About 50 species are known, of which only six are cultivated over here; in addition there are numerous strains. In the United States the plant is seen more often; American plant-lovers have a choice of about 30 species.

Appearance: in broad outline its habit resembles that of dieffenbachia species; some grow almost in the shape of trees, others are low-growing and shrubby; the leaves are variegated. The typical arum-lily shaped inflorescence is modest in comparison to the magnificent foliage; a number of species develop colourful berries.

Species: *A. commutatum* (Java, Philippines) has a 50-cm (20-in) trunk, branching at the top and green foliage, with grey spots between the veins. There are beautiful strains, such as 'Pseudobracteatum' which has yellow-white leaf-stalks. This flowers relatively easily; the calyx is white, the berries red. *A. costatum* (Malaysia) is a low shrub with dense foliage; the leaves are up to 30 cm (1 ft) long, emerald green with irregular white blotches. *A. modestum* is the hardiest species and the most suitable for simple indoor cultivation. In this plant the leaves are 25 cm (10 in) long, but here are plain green. *A. pictum* (Sumatra, Borneo) has a main trunk bearing numerous leaf-shoots. The leaves, up to 20 cm (8 in) long and 5 cm (2 in) across, have grey-green and silvery white marking on a dark green background.

Position: the plants grow best if given plenty of room in a tropical plant case, where they are never turned or moved. The pot should be wide and low, since the root-system is flat. Soil must be light and rich in humus, mixed with sphagnum and pebbles.

Watering and spraying: water only with softened tepid water (pH 4–4·5), freely in the growing period, sparingly in winter. Do not spray directly onto the plant, to avoid spotting the leaves of the weaker species.

Feeding: during the growing period (April to August), the plant should be given a lime-free feed every two weeks or so.

Dormant period: between October and February give little water and do not feed. Temperature should not fall below 15–18°C (58–65°F) (air- and soil-temperature).

Propagation: usually by cuttings, in a heated soil mixture in the greenhouse. *A. commutatum* can also be grown from seed in a soil temperature of about 26°C (80°F).

Diseases: if treated incorrectly, mealy bugs may occur in the axillae of the leaves. Red spider may be the result of too much light.

Aglaonema, hybrid 'Silverking', fam. *Araceae*

Allamanda cathartica

This tropical plant, closely related to the oleander, is so beautiful that one must regret its limited suitability as a houseplant. It is too vigorous a climber for the living-room, and few people have a greenhouse large enough to grow it.

Origin: with a few exceptions all the 12 species known are natives of Brazil, including *A. cathartica*.
Appearance: climber growing to 6 m (19½ ft). Size rather like *Passiflora quadrangularis*. Long oval leaves, pointed at each end, 10–14 cm long and growing in groups of three or four. Flowers funnel-shaped, golden yellow, white at the base, 5–7 cm (2–2¾ in) in diameter; fragrant in some species. Propagation from young terminal shoots.
Species: cultivars are seen more often than the original species, for instance 'Grandiflora', which has even larger, lemon-yellow flowers; 'Nobilis', with leaves hairy on the underside, also with very large flowers 10–12 cm (4–5 in); 'Schottii', even more vigorous, hairy twigs, very large flowers, yellow, with darker yellow and brown stripes on the inside of the funnel.
Flowering from June to October.
Position: light to semi-shaded throughout the year. Wherever possible, the shoots are separately tied to threads, so that the buds and flowers can turn towards the sun. If grown in pots, the more vigorous hybrids are trained round bent wire; shoots that grow too long should be cut back. Less vigorous kinds should have their long shoots pruned to encourage bushier growth.
Watering and spraying: water freely during the growing period, from April to September or October. In winter water sparingly and do not spray. The dormant period is from November until late February.
Feeding: because of its vigorous growth and prolonged flowering season, the plant requires generous feeding. Give liquid feed weekly from April until August.
Pruning: when the plant has finished growing in November, prune drastically to encourage new shoots.
Repotting: if necessary in February or March (but certainly before the plant produces new shoots). Do not prune at this time.
Propagation: from cuttings in heated compost. Growing from seed is also possible.
Diseases: avoid draughts, dryness and overbright light, which may encourage mealy bug, scale insects, red spider mite and whitefly.

Allamanda cathartica, fam. *Apocynaceae*

Aloë

At one time the juice of this plant was used to cure gall-bladder disease and to heal wounds. In some regions it is still called 'wound-cactus'.

The aloë is a native of tropical parts of Africa, South Africa and Madagascar. There are approximately 250 succulent species, mainly tree- or shrub-shaped and generally with beautifully marked foliage grown in a rosette. Smaller, stemless rosette plants, with fine, crenate fleshy leaves are used in the garden. By placing them in a temperature of 4–6°C (40–43°F) in winter you will protect them from aphids and rotting. *A. arborescens* is one of the taller, stemmed forms. The most suitable species for indoor use is *A. variegata*, which tolerates the dry atmosphere of the living-room provided it is watered sparingly (never in the rosette).

Aloë variegata, fam. *Liliaceae*

Ampelopsis

As indicated by the family name *Vitaceae*, this is a vinous plant. At the florists' any plant belonging to this family, if not known as *Cissus* is called *Vitis*. They are trailing or climbing plants. Except for the greenhouse species they are not very sensitive to the dry, warm atmosphere of our living-rooms in winter.

The most beautiful species is the variegated *Ampelopsis brevipedunculata*, a native of Eastern Asia, sometimes marketed under the name *Vitis heterophylla* 'Variegata'. Its red stems trail gracefully over its supports. The leaves are undivided or five-pointed, dark green with white and pink to clear red streaks. The inflorescence resembles that of the garden vine. This is a beautiful plant to trail round pillars and other room dividers of this nature in rooms that are kept cool in winter. The plant is not completely hardy, but nevertheless requires a cool position in winter. It is suitable for use on the balcony, provided it is protected from severe frost, but the best situation is a cool greenhouse. The small-leaved *Ampelopsis brevipedunculata* 'Elegans' loses its green-white-pink marbled foliage in the autumn.

Ampelopsis brevipedunculata 'Elegans', fam. *Vitaceae*

Ananas comosus — Pineapple

The chief representative of the bromeliads, *A. comosus* has been internationally known for about 500 years. The first descriptions of the fragrant, golden-yellow fruit arrived here soon after the discovery of America in 1492.

Distribution: the Spanish scientist Oviedo, in his book *Natural History of India*, published in 1535, gave Europeans further information about the plant. The fact that he gave India as the country of origin was a result of the mistake Columbus made in thinking that he had landed in the West Indies. Regardless of the many legends about the plant, it remains a fact that the ananas grew wild everywhere in tropical regions of America. It was soon brought under cultivation, which resulted in changes in its appearance and practical applications. In the course of the 17th and 18th centuries, this valuable fruit also began to be cultivated in other tropical countries, in Africa and Asia. Since 1830 it has become possible to grow pineapples with enormous fruit in European hothouses. Change comes with the years . . .

Appearance: the original pineapple has leaves up to 1·2 m (4 ft) in length, with horny thorns; it grows in a rosette with a diameter of about 2 m (6 ft). Through cultivation we now have beautiful dwarf forms as houseplants, which even bear fruit. The initial strains, still occasionally used as decorative plants, had a diameter of 75 cm (2½ ft), although they already showed the white-edged, red-streaked or cross-banded yellow-green leaves. Modern dwarf forms are only half the size. Fruits are sometimes used in flower arrangments.

Flowering: the plant flowers at about two or three years old in a warm situation, after which small young rosettes develop while the parent plant gradually dies.

Care: the pineapple is not an epiphyte; like *Billbergia* (p. 139) it is a terrestrial bromeliad, and requires the same treatment.

Propagation: by removing the newly formed rosettes.

Ananas comosus 'Variegatus', ornamental pineapple, fam. *Bromeliaceae*

Anthurium – Flamingo plant

This beautiful member of the arum family with its decorative flowers is an outstanding example of houseplants rescued from their tropical isolation through systematic methods of cultivation. The modern form of anthurium is a reliable and vigorous plant flowering freely and for long periods.

Origin: the more than 500 species of the genus *Anthurium* are distributed over the relatively small area of tropical rain forests of Central and South America. The first representative of the genus was discovered in 1850 by the Austrian physician and botanist Dr. Karl von Scherzer and was named *A. scherzerianum* in his honour.

Appearance: *A. scherzerianum* is no longer available and has been replaced by the stronger and more freely flowering *Anthurium scherzerianum* hybrids, including an attractive dwarf form. All these hybrids have narrow, dark green, stemmed leaves. The inflorescence is of medium size; the spadix is usually proportional to the always glossy, red, pink, white, or speckled calyx. *A. andreanum*, a native of Columbia, and the hybrids developed from it, are larger in all their parts and are often used as cut flowers. The long-stemmed leaves are 30–40 cm (12–16 in) in length. The spadices are frequently spiral; the calyx is not glossy.

Apart from these two groups, mainly grown for their flowers, there is a third with exceptionally fine foliage. However, the plants of this group grow so rapidly that they are not always suitable for use in the living-room. Nevertheless *A. crystallinum* can be grown successfully in a high degree of atmospheric humidity in a large tropical window. The leaves are oval-shaped, about 50–60 cm (20–24 in), olive-green with ivory veining. The inflorescence is insignificant and slightly fragrant. A fourth group includes species with divided or undivided green foliage. This group includes the epiphyte climber *A. scandens* and its cultivar 'Violaceum', whose spadices after flowering develop violet-red berries. The leathery leaves are narrow and shiny. This plant is very popular in the United States.

Flowering season: if properly looked after, the two groups of hybrids developed from *Anthurium scherzerianum* and *Anthurium andreanum* may flower for most of the year. The foliage plants flower only in spring – if at all.

Position: the *scherzerianum* hybrids are the hardiest and therefore the most suitable for simple indoor cultivation. During the growing period, from March till August, some degree of atmospheric humidity is desirable. All other species and forms require a warm, light situation, away from direct sunlight, for instance in a heated north-east or east-facing plant window. *A. crystallinum* will grow only in a tropical greenhouse or in a tropical plant window.

Temperature: even in winter, the hybrids do not tolerate a soil temperature below 18°C (65°F) and an air temperature below 15°C (58°F). *A. crystallinum* requires 18–20°C.

Watering and spraying: only with lukewarm, fully softened water (pH 4–4·5) freely during the growing period. If you possess a rainwater butt and live in an area where the rainwater is still pure (that is, not in industrial regions!) the simplest method is to place a watering-can filled with this water on a radiator and use it in the first place for lime-haters like *Anthurium* and *Ericaceae*. Flowering plants should be sprayed indirectly; drops of water will spot the calices.

Feeding: since they are natives of tropical rain forests, they are very sensitive to salts. It is best to feed every fortnight with the nutrient solution used for soilless cultivation.

Repotting: when shoots appear, repot in porous soil mixed with peat fibre and sphagnum. Some firms produce special anthurium mixtures containing these ingredients; although this may be difficult to find, a well-stocked shop should be able to provide it. Naturally it can be used for all other plants requiring sphagnum and peat in their soil mixture.

Propagation: from offsets. Can be grown from seed, but it will take a long time for a good-sized plant to develop.

A very attractive method of increase is by means of stem cuttings. An electrically heated indoor propagator, as described elsewhere in this book, is indispensable for this method. The main stem of an old or drooping

Leaf of *Anthurium crystallinum*, fam. Araceae

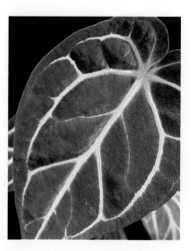

anthurium is stripped of all its leafstems, which are removed with a sharp knife, taking care not to damage the invisible eyes in the axils.
Insert the sections of the main stem upright in a mixture of 50% sharp sand and 50% peat fibre, covering them almost entirely. The temperature in the propagator should be between 25 and 35°C (77 and 95°F). After a few weeks the cuttings will develop leaves and roots.

Diseases: excessively dry air and direct sunlight make the leaves curl

Anthurium scherzerianum hybrid, very suitable for soilless cultivation

and may also lead to the appearance of red spider mite or thrips.

Aphelandra, hybrid 'Dania', fam. *Acanthaceae*

Aporocactus flagelliformis, Rat's-tail Cactus, fam. *Cactaceae*

Aphelandra

Plant-lovers who call aphelandra an ungrateful plant to grow, have generally only themselves to blame. Perhaps they forget to water in time — the soil ball will then dry out and the foliage will drop; or they neglect to syringe where the atmosphere is too dry; this will cause the leaf-edges to turn brown.

Origin: as this member of the *Acanthaceae* is a native of tropical to sub-tropical regions of America, it naturally has specialist requirements.
Appearance: nowadays it is mainly the yellow-flowering species with white-veined leaves which are available. *A. squarrosa louisae* 'Fritz Prinsler' is considered to be the best.
Position: in simple room cultivation it should be placed in as warm and light a spot as possible, but not in direct sunlight.
Flowering season: three weeks between October and January. Dead flowers should be removed together with one or two pairs of leaves.
Watering: from the time the first shoots appear until after flowering the plant should be watered very generously with tepid, softened water (pH 5·0); spray overhead at least every two to three days. Water more sparingly in winter.
Feeding: weekly during spring and summer, but not in autumn and winter.
Repotting: in March, in large pots. Nutritious compost, rich in humus. Pre-packed potting compost should be mixed with some extra peat fibre.
Propagation: cuttings are only successful in a heated greenhouse, but plants can be cut down hard after flowering and allowed to rest, then started into growth again with repotting. New shoots will be provided from stumps.
Diseases: drying of the soil-ball and lack of moisture in the atmosphere will also lead to the presence of scale insects and mealy bugs. A too sunny or too cold situation will cause the leaves to drop. Only an aphelandra with dense leaves, good-looking even as a foliage plant, is worth keeping.

Aporocactus — Rat's tail cactus

In former days a collection of favourite houseplants frequently included a rat's-tail cactus. As a rule it had joined the family as a cutting, for this modest plant was rarely marketed.

Origin: its ancestors were natives of Mexico, where *Aporocactus flagelliformis* grows as an epiphyte in rocky mountains up to an altitude of 2500 m (8000 ft). In such conditions it is not exactly coddled and for amateur cactus collectors this is therefore the best one to start with.
Appearance: the strong stems, up to 1·5 m (5 ft) in length, should hang from the pot in a close and regular pattern. The numerous buds develop in spring. Incorrect treatment and especially lack of light lead to thin and unsightly stems, and in such conditions the plant will not flower.
Flowering: it will flower only if, after a rest period in plenty of light, in a cool position without being kept completely dry, the plant is placed as early as February in a warmer, even lighter and if possible sunny spot. It flowers in spring. Vigorous specimens produce flowers up to 10 cm (4 in) in length, fully open, rose-red, resembling the flowers of the Christmas cactus.
Care: the same as for spring-flowering leaf cacti. During the growing period water freely with a nitrogen-free liquid (April to August), followed by a rest period.
Overwintering in a cool atmosphere encourages flowering. It should on no account be left in the warm living-room.

Araucaria

Araucaria, a plant with a slightly old-fashioned air, is now almost forgotten. Nobody today knows that the explorer, astronomer and cartographer Captain Cook discovered this beauty with its lacy foliage in the Philippines on his voyage round the world.

Origin: this conifer was found only in a limited area between eastern Australia and New Caledonia and as a result this region, consisting chiefly of water, has been called the 'Araucaria-province'.

Mutations: times change. The botanical name *A. excelsa* – the exalted – was changed into *A. heterophylla*, meaning 'having different kinds of foliage'. Domestic environments in western Europe also changed. In earlier days this plant decorated the handsome but cold rooms of large houses, rooms where the fire was lit only on Sundays. The more houses became centrally heated, drying out the atmosphere, the less frequently the araucaria was grown.

Summer: everything will be fine as long as there is a north-facing window or a shady place under trees in the garden. The plant needs a lot of room, not too close to others. Water with lukewarm softened water (*p*H 4·5–5). Feed from May until August with special liquid rhododendron fertiliser.

Winter: from October onwards the plant should have a cool position, 5–10°C (42–50°F) and should hardly be watered at all, since otherwise the lower leaves will drop one by one, leaving an ugly bare trunk, such as sometimes occurs in an incorrectly treated ficus, dieffenbachia or croton. No new leaves will ever grow there. However, if the transition from summer to correct winter treatment is achieved with care, many new shoots will appear at the top and form a magnificent crown.

Araucaria heterophylla, Norfolk Island Pine, fam. *Araucariaceae*

Ardisia — Spear flower

The name is derived from *ardis*, the Greek word for point. Ardisia is a graceful little tree, a native of a region stretching from Japan to southern Asia. In its original habitat this evergreen tree, usually horizontally branched, reaches heights of more than a metre (3 ft) but as a houseplant *A. crenata* (syn. *A. crispa*), which is gradually returning to favour over here, grows to about 60 cm (2 ft).

Flowers and fruit: the white flowers grow in terminal racemes from July onwards and rapidly develop into pea-sized berries, gradually changing colour to reddish purple. These last for six months or more, and constitute the real ornamental value of this modest plant. It may happen that new flowers appear at the same time, a most attractive sight.

General care: approximately the same as for *Ficus*. The most favourable winter temperature is 12–15°C (54–58°F). At a higher temperature the berries will drop sooner; if the atmosphere is dry they will shrivel. Regular spraying with lukewarm water, starting at the beginning of the flowering season, is one of the most important requirements to ensure satisfactory development and longer lasting berries. A light to sunny situation is appreciated.

Important: the dark green, leathery, elliptical leaves sometimes show bud-shaped wavy edges. These 'bacteroids' or nodules contain a certain bacillus; they occur also on the roots of beans and other plants and are a natural phenomenon, not a disease. They should not be removed.

Ardisia crenata, fam. *Myrsinaceae*

Areca — Betel Palm

Few house plants have in the course of generations had such a variable fate as the palms. While at one time they formed an essential element of all the better living-rooms, their reputation has at times dropped almost to zero. Today they enjoy a nostalgic revival; this also applies to the *Areca*.

Variations. We are of course not concerned with the true betel palm with its peculiar commercial value, but with the graceful *Areca triandra*. It is thought to have been introduced in England from India in 1810. Its elegant, narrow-leaved fronds and slow growth, as well as the fact that it is eminently suitable for hydroculture have doubtless contributed to its popularity.

Although young plants of this species are less inclined to flower than the long-established *Chamaedorea elegans*, which develops a mimosa-like inflorescence every year, the *Areca triandra* has now overtaken this as well as other rivals. Even *Chamaerops humilis*, the robust dwarf palm with its rather stiff, fan-shaped fronds can at present not keep up with it.

Suitability. In palms, which can of course not be pruned, the size which the plant may attain plays an important part. The advantage of the limited size — max. approx. 1.8 m (6 feet) — of the two other *Areca* species is balanced by the slow growth of the *Areca triandra* already mentioned. One day, however, one will inevitably outdo the other.

Occasionally the *Areca* is referred to under the difficult botanical name *Chrysalidocarpus* and cultivar name 'Golden fruit'. For care see under *Chamaedorea*, p. 158.

Areca triandra, Betel palm, fam. *Palmae*

Argyroderma

As stated in the chapter on plant families, *Aizoaceae* were formerly called *Mesembryanthemaceae* (Midday flowers). Collectors and professionals still call them 'Mesems' for short. Grown in shallow bowls or in a seed-bed they may be brought to flower every year.

A. testiculare is very popular because of its magnificent flowers, in yellow or purple, and because it is so easy to grow. In South Africa it occurs in areas where the soil has a high quartz content. The compost must contain a great deal of sand and marble chips, and must be porous and rich in lime. Cover it with a layer of gravel. The growing season occurs in summer; the plant flowers in the autumn. It is dormant in winter and spring, which facilitates treatment. Water sparingly and give very little feed, even in summer. Winter temperature 8–10°C (46–50°F); keep completely dry at this time.

Argyroderma testiculare, fam. *Aizoaceae*

Asparagus — Asparagus fern

This plant, like the vegetable species, belongs to the family *Liliaceae*. The relationship can be clearly seen, even though not all indoor forms show as much resemblance to the vegetable as do the old but still good *A. sprengeri*, now called *A. densiflorus* and *A. setaceus*.

From left to right: *Asparagus falcatus*, *A. densiflorus* 'Meyeri', *A. densiflorus* 'Sprengeri', fam. *Liliaceae*

Origin: its former specific name *sprengeri* is no longer used as such, but it reminds us that Karl Sprenger, from Mecklenburg, who for many years studied botany in Naples and who died in Corfu in 1917, introduced the plant from East Africa.
Species: apart from *A. densiflorus*,
Species. The house plant *Asparagus densiflorus* 'Sprengeri' (on the right in the photograph), which is now classified as a separate species, was at one time also popular as foliage in flower arrangements. It can also be grown in hydroculture. Unusual decorative forms are the erect-growing *A. densiflorus* 'Meyeri' (centre) and *A. falcatus* (left), which was introduced from India as early as 1792.

Asparagus densiflorus 'Myriocladus', densely branched ornamental asparagus, fam. *Liliaceae*.
This sickle-shaped asparagus – a vigorous trailing plant with narrow pseudo-leaves and occasional small single flowers – provides magnificent decoration in moderately heated rooms and a beautiful framework in flower windows. The 'feathery asparagus', still best known under its old trade name *A. plumosus*, is suitable for use in floral decorations and is distinguished by its readiness to flower. The descriptive name had to give way many years ago to the new botanical name *Asparagus setaceus* (i.e. bristle-shaped). But it

Asparagus densiflorus, 'Myriocladus', fam. *Liliaceae*

remained the same elegant climber, pleasing its owners with its numerous white flowers, later turning into inconspicuous red berries, which is also one of the advantages of the – reputedly flowerless – good old 'Sprengeri' of mature years. *A. crispus* has unusually wavy foliage; after flowering profusely this plant, too, produces an abundance of scarlet berries. Great-grandmother's *Medeola*, which now bears the botanical name *A. asparagoides*, is an undemanding treasure; it has been known in Europe since 1702. Finally the illustration shows the South African *A. densiflorus* 'Myrociadus', the 'densely branched' asparagus.
Care: in the case of *A. densiflorus*, the main species, many things can be done which would be impossible where other plants are concerned: overwintering in a warm or cold spot, no rest period; water throughout the year with moderately soft water, freely in summer; if placed in a warm situation in winter it should occasionally be fed. The only things this plant will not tolerate are a dark position and a dry soil-ball. If its situation is too dark when the growing period begins, aphids may appear.

Aspidistra elatior

This plant comes from the deep, cool shade of the Japanese mountain forests and is therefore so undemanding that at one time it was found in grocery shops and workshops. The flowers grow close to the soil, but are of no decorative value as they are dark purple and difficult to see.

The erect, evergreen dark leaves grow from a thick, fleshy rootstock. The beautiful white-edged varieties are less strong; they require a little more warmth and very little feeding; overfeeding causes the variegated leaves to revert to plain green.

Care: although it tolerates a great deal, for example lack of light, or a dry atmosphere, there are two things the plant abhors, namely: a sunny situation and excess water in the bottom of the pot. The first causes leaf-burn and the second root-rot.

Aspidistra elatior, fam. *Liliaceae*

Asplenium nidus – Bird's-nest fern

If you know our indigenous *A. ruta-muraria*, the wall rue with its graceful and plentiful foliage, you will hardly recognise the tropical bird's nest fern as its 'big brother'. Nevertheless they both belong, not only to the same genus but also since the introduction of the new fern classification, to the same family.

Origin: *A. nidus* occurs in every rain forest of Asia and Africa, and also in Australia.
Appearance and requirements: like the bromeliads in tropical regions of America, this plant grows as one of the largest epiphytes in the crowns of trees, and collects its nourishment in the deep, funnel-shaped rosette formed by the undivided, erect, shiny leaves, pale green in colour. These leaves can grow to a width of 20 cm (8 in) and a length of over a metre (3 ft). When used in plant combinations few of them live to a great age.
Care: *A. nidus* belongs in a tropical plant window or at least in a flower window with soil heating, which in winter should not drop below 16–18°C (60–65°F), while the air temperature should never be less than 15°C (58°F) at night. Water and feed as for *Aechmea* (p. 117) or *Vriesea* (p. 348).

Asplenium nidus, Bird's-nest Fern, fam. *Aspleniaceae*

Astrophytum myriostigma, Bishop's-cap, fam. *Cactaceae*

Astrophytum

Among the genus we chiefly know *A. myriostigma* (bishop's-cap), *A. asterias* (sea urchin) and *A. ornatum* (star cactus). The fact that *A. asterias* has a double among the succulent euphorbias, as mentioned on p. 206, is one of nature's little jokes.

The species most popular among cactus lovers originate in Mexico. They like a compost mainly of loam and gravel. Strains without thorns demand a situation in semi-shade. Thorny and white-haired forms should be screened from strong sunlight in spring, and in summer should have plenty of light and a warm position. They usually flower in summer. Water sparingly and feed hardly at all. From autumn onwards withhold water and gradually move the plant to a cool position 5–8°C (42–46°F).
Propagation: from seed.

Begonia

Begonias are among our most versatile and beautiful ornamental plants. Botanically they are subdivided into annuals or perennials, herbaceous plants and non-woody semi-shrubs. This extensive genus is classified into 24 sections.

The amateur generally distinguishes only three groups: 1. flowering begonias; 2. shrubby begonias; 3. foliage begonias. *Begonia rex* hybrids must be regarded as the most important representatives of the latter group. In addition there are the tuberous begonias used for bedding and on balconies. A characteristic of all species is the nearly always asymmetrical foliage.

From their native habitat it may be deduced that all begonias are shade-loving plants, growing terrestrially or as epiphytes in tropical rain forests. However, they also occur in sub-tropical regions and even in mountain forests to altitudes of 4000 m (1300 ft). The appearance of the plants and the shape and colour of the flowers have greatly changed in the course of about two centuries, during which time many hybrids have been produced. Nevertheless, certain basic qualities have been retained, so that the origin of the botanical species still provides important indications of the treatment required.

With a few exceptions, a general requirement is the desire for a slightly shaded position and a high degree of air moisture. For the magnificently marked, large-leaved *B. rex* hybrids and other representatives of this group in particular, direct sunlight is fatal. All forms also have in common a liking for friable soil, rich in humus, in which the finely branched root system can spread itself to the full. For that reason it is advisable to plant them in wide, shallow bowls.

The succulent stems, the very large leaves or very profuse flowering, together with the damp forest soil of their native habitat, indicate a great need for fairly lime-free water (pH 4·5–5·0) during the growing period. This applies particularly to flowering begonias and foliage begonias. Shrubby begonias will accept slightly harder water (pH 6·5).

Lorraine begonia 'Gloire de Lorraine', winter-flowering, fam. *Begoniaceae*

Begonia acutifolia, a shrubby begonia

1. Flowering begonias

These are generally considered to be annual herbaceous plants of three types.

a. *Begonia* 'Gloire de Lorraine'. These plants appeared on the international market in 1893 and have since conquered the world. Many hybrids of the original species, in shades varying between pale and deep pink, flower from the autumn until approximately January. A large number of these plants are bought in December.

Care: when buying a plant, keep it in a cool spot for a few days if possible, temperature 12–15°C (54–58°F); afterwards temperatures of 18–20°C (65–70°F) will be tolerated without the leaves dropping or flowering being delayed – provided that you do not place the plant in full sun, that the air is not too dry and that the plant's keeping qualities are not affected by sharp-needled pine branches or other plants placed nearby. The general rules regarding hardness of water must be adhered to. Do not allow seed formation. 'Gloire de Lorraine' is discarded after flowering.

b. *B. elatior* hybrids have been available since 1907. There are many forms, still being improved, with single and double flowers in shades of white, yellow, pink and pale to dark red. There are also fairly large-flowered hybrids. Flowering season and care as for Lorraine begonias. Discarded after flowering.

c. *B. semperflorens* hybrids, the perpetual-flowering begonias. This is actually a bedding plant for the garden, but an increasing selection is being marketed for use as houseplants; these, exceptionally, include forms which tolerate sun. Flowering season: throughout the year, depending on start of cultivation.

Diseases: a typical disease among begonias is an attack of mildew, which may affect the leaf- and flower-stems as well as the leaves. Another often occurring evil is rotting of the stems caused by fungus formation (the stems snap and drop off). Removal of affected parts as soon as seen, and treatment with flowers of sulphur or benomyl will help, together with improvement of ventilation. To avoid other plants becoming affected, the plant should be temporarily removed.

2. Shrubby begonias

These erect growing bushy species nearly all originate in tropical areas of Central and South America, where many species grow to more than a metre (3 ft) in height. The *B. corallina* hybrids with their world-famous cultivars 'Lucerna' or 'Madame Charat', which flower profusely at a very early age, may even reach a height of two metres (6 ft) and, if properly looked after, flower throughout the year. All shrubby begonias are strongly branched, but do not develop woody stems, although in older plants the stem appears to be so. The flowers, growing in pendulous racemes, are fairly large, usually red or pink in colour, sometimes fragrant. Although the leaves are not as magnificent as

▶

those of foliage begonias, they nevertheless in their shape and marking give an indication of the innate capabilities of the genus *Begonia*. As regards the erect habit, as usual there are exceptions to the rule. *B. foliosa* is a shrubby begonia with finely branched main stems, which become pendulous with age and are therefore used as very attractive hanging plants. *B. limmingheiana* (which flowers only from March to May), also has pendulous stems and pale to coral-red, densely-growing flower racemes.

General care: follow the basic rules given on p. 132.
Grow indoors throughout the year; the plant is not happy in the garden or on the balcony. If possible do not change its position, and never turn the pot. Situation: light, away from direct sunlight.

Watering: during the growing and flowering periods water freely with softened water, but avoid leaving surplus water in the bottom of the pot as much as avoiding letting the soil-ball dry out. A high degree of humidity in the atmosphere is important, especially while the plant develops shoots or flowers. Spray overhead very carefully; the flowers must never get wet.

Feeding: during the growing period (usually from February until August) feed weekly in accordance with instructions, preferably using lime-free liquid fertiliser.

Overwintering: if possible not in a heated living-room; rather in a moderately warm place, about 15°C (58°F). A too warm situation may result in discolouring of the leaves, leaf-curl and an attack of plant pests.

Pruning: shrubby begonias that have grown too tall or have become bare in the centre, must be drastically pruned in spring, before new shoots appear. This will encourage new growth at the bottom.

Repotting: when the pot no longer allows further root growth, the plants should be repotted in wide pots. The roots may be shortened a little. Pre-packed potting compost mixed with a little extra peat may be used.

Propagation: easily increased from cuttings, see chapter on Propagation, p. 70. Propagation from seed is best left to professional growers.

Further details: shrubby begonias are very sensitive to draughts, too cold, too hot, and too dark positions. Nor can they bear surplus water remaining in the pot or in the saucer, while on the other hand a dry soil-ball is detrimental as well. Every mistake in cultivation results in dropping of the leaves or delayed flowering.

Other fine shrubby begonias for house cultivation:

B. albo-picta, tall, strongly-branching; leaves narrow, asymmetrically heart-shaped, green with numerous silvery spots. Flower racemes pendulous, greeny white in colour.

B. credneri, 40–50 cm (16–20 in) in

◁ *Begonia metallica,* a typical shrubby begonia; height to 80 cm (32 in); flowers from summer to autumn

Begonia 'Cleopatra', a handsome new ▷ cultivar with dark-blotched foliage, white-haired beneath

height; branched, leaf asymmetrically heart-shaped, pointed at the top, upper side green, underside red and hairy. Pale pink flowers almost throughout the year.

B. fuchsioides, bushy habit, to 1 metre (3 ft) tall, stems lightly curved; leaves small, asymmetrically egg-shaped. Red flowers from November till April; sometimes flowers in summer.

B. maculata, resembles *B. albo-picta*, but the leaf is distinctly white-flecked, reddish underneath. Flowers white and red. Very undemanding, a most popular houseplant.

B. metallica, to 80 cm (32 in) tall, freely branching, bristly haired, leaf asymmetrically heart-shaped, ending in a sharp point, upper side dark green with a fine metallic gloss, underside red; the surface of the leaf is somewhat uneven. The fairly large flowers grow in red/white clusters. Suitable for an unheated greenhouse. In winter best kept at 12–15°C (54–58°F).

B. scharffii, erect growing, few branches. The entire plant is densely covered with bristly red hairs. Leaf asymmetrically heart-shaped, with a long sharp point, dark green on top, deep red underneath. Long racemes of large white/pink flowers from summer to autumn.

B. serratipetala was introduced from New Guinea in 1913. It is one of the most beautiful shrubby begonias we know: the leaves are deeply incised, doubly serrated along one edge, glossy green with blood-red marking. This plant is easy to grow in a warm room where the air is not too dry.

3. Foliage begonias

Without a doubt *B. rex* hybrids are the leading representatives of this group of most beautiful and unusual plants, consisting of an almost uncountable number of species and cultivars.

The original species of *B. rex* – known here for more than a century – hardly occurs any more in its pure form. It was a forest plant originating in the Indian peninsulas, with thick rootstocks and asymmetrically egg-shaped leaves, up to 30 cm (12 in) in length, showing a central, wide silver band, fanning outwards on a metallic green-black background and edged with reddish purple. The pink flowers, not very numerous, but fairly large, appear in early summer. *B. rex* has been crossed with *B. diadema*, imported from Borneo, and also rarely found in its specific form today. These crossings resulted in the *B. rex* hybrids, even more strikingly marked and much more suitable for use as houseplants; the beauty of the freely-flowering blossoms are nothing compared with the glorious foliage. These world-famous standard forms are joined by many equally attractive species, some of which were subsequently named.

General care: on the whole the rules are the same as those described for shrubby begonias, but in view of their native habitat in the damp soil of

Begonia rex, hybrid

Begonia hispida cucullifera (B. phyllomaniaca). Adventitious buds sprout from the main veins of the leaves

tropical rain forests, we should be even more cautious when selecting a situation, which should on no account be sunny. A north-facing window in a not too warm room (for stronger species 15–20°C (58–70°F) in winter) is best. The growing season is from February until the beginning of autumn. There is no actual rest period, but in the dark months of the year no new leaves develop. There is no cause for alarm if some leaves drop; the loss will soon be made up.
Watering: with lukewarm, soft water (pH 4·5–5·0). Spray overhead only; moisture on the leaves can cause spots. Beware of the soil-ball or the atmosphere becoming too dry.
Repotting: in spring when necessary, in the widest possible, fairly shallow pots with a thick layer of crocks on the bottom. Compost mixture: prepacked potting compost mixed with some peat fibre and pieces of charcoal.
Propagation: in room cultivation this is best done in summer by means of leaf cuttings (see the 'green section', p. 72).
Diseases: the same as in the two other groups. Leaf-curl, spotting of the foliage and pests are caused by excessively dry air.

Other fine foliage begonias suitable for the living-room:
B. crispula arrived from Brazil in 1950. Leaves circular to wide kidney-shaped, green and glossy, uneven on the upper surface as in *Peperomia caperata* (p. 292), and veined. Flowers white inside, pink outside, small.
B. hydrocotylifolia, resembles the Pennywort, a marsh plant; develops a creeping main stem from which grow short leaf-stems with almost cylindrical fleshy little leaves. As is the case with watercress, the stems are attached to the 'navel' of the leaves. The edges of the leaves are red and hairy. Flower-stem up to 20 cm (8 in), flowers pink. An attractive houseplant.
B. imperialis, with hybrids such as 'Hildegard Epple' and 'Marbachtaler' has creeping runners with smooth-edged, almost emerald green leaves, later turning brown. Very fine.
B. manicata and its cultivars 'Crispa' and 'Aureomaculata' belong to the permanent selection of reliable, well-established houseplants. On their upper surfaces the leaves are covered in red scales; their edges are frilled. The small pink flowers on their tall stems flower for weeks on end in December and January. In favourable conditions the plant may even develop fruits.
B. masoniana is a very beautiful and interesting novelty of recent years, but not as suitable for indoor cultivation as one might wish. The apple-green, asymmetrical shield-shaped leaves bear brown stripes which from a distance look like a cross giving it its former name of 'Iron Cross'. In spring the plant should have greenish white flowers.

Beloperone guttata — Shrimp plant

B. guttata, a delightful shrublet from Mexico, is generally underestimated. It is often bought in spring as an inexpensive plant for the balcony or for a south-facing spot in the garden, and its virtues as a freely-flowering, long lasting houseplant are ignored.

The shrimp plant is anything but a disposable plant; it is far too beautiful and durable for such practices. The bracts, overlapping like roof tiles in the ear-shaped inflorescence are browny red to violet and yellow in colour.

Care: a well-lit position, but not too sunny. In the garden the pot should be half-buried in a semi-shady position, since the plant is sensitive to cold in autumn. From the time the first shoots appear until August water freely; afterwards reduce watering, take care that the soil-ball does not dry out. Feed every ten days. To maintain the bushy shape, cut back drastically in February. In winter keep in good light at a temperature of 12–15°C (54–58°F) — on no account in a fully heated room. Propagation from cuttings which will root rapidly if given some bottom heat.

Beloperone guttata, Shrimp Plant, fam. *Acanthaceae*

Bertolonia

This delightful plant with its unusually marked, velvety leaves and pink flowers, is a native of the southern part of Brazil. With the shrublet *Sonerila* (see p. 336) and their mutual sister, the enormous *Medinilla* (see p. 258), it belongs to the family of *Melastomataceae*.

Most of these plants, including *Bertolonia maculata*, which has been known since 1850, should only be grown by experts with hothouse experience. They should have a moist and shady position in an even temperature, 22–25°C (72–78°F). Their beauty is so short-lived that a constant supply of new specimens should be maintained; they are easily grown from seed. A comprehensive catalogue will include seeds of *B. maculata* and of *B. marmorata*.

Bertolonia maculata, fam. *Melastomataceae*

Billbergia

Of all members of the bromeliad or pineapple family, *B. nutans*, with its spiked inflorescence, is the simplest and easiest flowering among the species known today. With proper care this plant can be kept on the window-sill for many years.

Origin: it originates in a region stretching from Mexico to southern Brazil. Among the more than 60 known species *B. nutans* is the only one considered suitable for room cultivation.

Appearance: unlike most other bromeliads, which grow as epiphytes in trees, billbergia is a terrestrial plant. In the course of a number of years it forms a group of small plants up to 40–60 cm (16–24 in) high. The leaves of the species are grass-like, narrow, grooved and leathery; pale to dark green; those of subsequently cultivated strains are broader and rather more grey-green.

Inflorescence: the flowers droop from the end of the curved flower-stems and consist of large red or pink bracts from which the reddish sepals and yellowish green and blue petals emerge. The flowering season varies between early summer and winter, depending on the method of cultivation.

General care: situation in summer in good light, away from direct sunlight. The plant can also live in a sheltered spot out of doors, but does not tolerate a draught. The same conditions are valid for winter; the most advantageous temperature is 16°C (60°F); nevertheless it will accept the warm air of a living-room. It is not sensitive to a dry atmosphere. Damage caused by cold occurs if the temperature drops below 12–14°C (54–57°F).

Watering: normally at the top; freely during the growing season in summer; in winter sparingly to a greater or lesser degree, depending on room temperature. If grown indoors, spray only occasionally in summer. Fairly hard water is tolerated (pH 6·5).

Feeding: from May until September, once a week.

Repotting: this is best done in June–July in a roomy, low pot. Until the plant has settled, water sparingly and do not feed. Ordinary prepacked potting compost with extra peat and sand added may be used.

Propagation: by division when repotting. The plant will flower again in a year's time.

Diseases: scale insect occurs very rarely; otherwise, it is not subject to attacks by pests.

Billbergia nutans, fam. *Bromeliaceae*

Blechnum

With the introduction of the new nomenclature, blechnum has become the progenitor and representative of a genus of more than 200 fern species. All over the world they have adapted to climatological conditions; they are found everywhere and are suitable and appreciated for many purposes.

Origin: from our point of view the most interesting is *B. gibbum*, a magnificent plant discovered only about a century ago in New Caledonia (an island in the Pacific) and since cultivated as an indoor plant. The requirements of this beautiful fern as a house plant may be deduced from its natural habitat. It originates, not in the damp, warm tropical jungle, but in regions with an almost temperate climate, freshened by winds from the sea. In summer temperatures vary between 22 and 36°C (72 and 96°F), and in the cool season between 13 and 25°C (55 and 78°F).

Appearance: older specimens may grow to a height of a metre (3 ft). The photograph shows a plant whose trunk has grown horizontally, while retaining its palm-like crown.

Care: *B. gibbum* is often described as a hot house plant, but if this implies that high temperatures and atmospheric humidity are essential, it is incorrect. It is much more a plant for the moderate greenhouse, and if carefully looked after, it will even feel at home in the living-room. It thrives in a roomy and airy open flower window without bottom heat. During the dormant period in winter (October–February), the soil temperature should not exceed 16–18°C (60–65°F).

Watering: very freely during the growing period (March to July); moderately during the rest of the year. The soil-ball must not be allowed to dry out. Do not spray directly onto the foliage, but provide some air moisture while new leaves are developing.

Feeding: during the growing season with a weak solution (1–1·5%). Sensitive to minerals.

Repotting: as described for other indoor ferns; the same compost as well (cf. *Nephrolepis, Pteris*).

Propagation: from spores; fairly easy. Warm seed-bed. See also under other ferns.

Blechnum gibbum, fam. *Blechnaceae*

Bougainvillea

This plant is a native of Brazil. It owes its name to the French admiral Louis Antoine de Bougainville (1729–1811), who is said to have discovered the plant on a world tour, first in South America, and later in an island in the Solomons which also bears his name.

Worldwide success: the bougainvillea has since become internationally famous. It is found in numerous gardens from India to the Riviera, and even in south-western England and southern Germany.
Appearance: an evergreen or deciduous climber. In its wild state it grows to 3–4 m (9–13 ft). Thorny stems with vigorous green foliage. The flowers are relatively inconspicuous; they grow in groups of 3 and are surrounded by 3 bracts varying in size, purple, scarlet or orange, or more rarely white, depending on species or strain.
Use: during the flowering season, which naturally occurs between March and July, but can be varied as a result of cultivation, the bougainvillea is one of the most attractive pot-plants. It also thrives in a slightly heated flower window where the very vigorous *B. spectabilis* can climb against the wall. Occasionally the woody main stem is trained along sticks or bent wire. The less vigorous strains such as the now most valued *B. × buttiana* 'Mrs. Butt', as a rule have a shrubby habit. They can grow to a great age.
Summer: from March onwards place in a well-lit and warm situation and provide a high degree of atmospheric humidity. As soon as the weather allows, put in a sheltered spot out of doors. If the plant is kept indoors it should be given a great deal of fresh air. It is very sensitive to cold, so be sure to bring it indoors in good time in autumn.
Winter: rest period until March. A cool position in good light at a temperature of 6–8°C (44–46°F).
Watering: freely in summer, sparingly in winter, but take care that the soil-ball does not dry out, as this will result in dropping of leaves and flowers.
Feeding: in summer weekly with normal solution.
Repotting: every year while the plant is young; later at longer intervals. Soil mixture as for a hortensia hydrangea, though ordinary potting compost may be used instead.
Propagation: only by professional growers.
Peculiarity: in cool, rainy weather the bracts will slowly fade. Be patient.

Bougainvillea glabra 'Alexandra', fam. *Nyctaginaceae*

Bowiea

This plant is a real curiosity, a native of South Africa, which is also the home of very strange dry-flowering plants, cork trees growing to six metres (43 ft), etc. The photograph shows what the bowiea looks like.

Origin: it is a lily plant, occurring only in South and East Africa; its full name is *B. volubilis*, the twining bowiea. James Bowie, after whom the plant was named, lived from 1789 to 1869 and collected plants for Kew Gardens from all over the world.

Appearance: it starts with a yellow-green bulb, at most the size of a tennis ball. In mid-winter, January–February, several twining stems grow from the bulb; this fact is reflected in the specific name. The stems are fleshy and strongly branched, and at first bear small leaves which, however, soon drop. The insignificant greenish flowers grow on other stems.

Growth and care: the parent bulb must be planted in loamy soil in a fairly large, flat bowl or pot. The bulb multiplies rapidly, so that the entire pot will soon be filled. In summer the green stems drop off; this is the beginning of the dormant period. It should then be given plenty of light and a cool situation, 6–10°C (44–50°F); withhold water. New shoots may be trained up stakes.

Bowiea volubilia, fam. *Liliaceae*

Brassia

The genus *Brassia* belongs to the group of botanical orchids, very popular at the present time, since the plants are fairly hardy and therefore suitable for orchid-lovers who do not own a tropical plant window or greenhouse. Although not very large or striking, the flowers are nevertheless graceful.

Brassia maculata, fam. *Orchidaceae*

Origin: its mother country is tropical America, a region between Mexico and southern Brazil, where around 1813 the first of a total of 38 species were found. However, the spotted *B. maculata* is thought to be a native of Jamaica.

Appearance: all brassia species have fairly large, close-growing pseudo bulbs – cylindrical stores for food and water, appearing on top of the soil. Such a pseudo bulb has no connection with flower bulbs; it is a typical phenomenon among epiphyte orchids. The pseudo bulbs of the brassia bear two leathery leaves, while the flower-stem, which grows up to 50 cm (20 in), develops from the underground bulb. The best known species is *B. verrucosa* which, like *B. maculata*, flowers in early summer.

Care: simpler than for many other houseplants. During the growing period – from autumn to spring – it should be given a half-shady position, be kept moist with fully softened water and sprayed frequently. Shrinking of the pseudo bulbs in summer indicates lack of water. Water very sparingly in winter; temperature 18–20°C (65–70°F). Feed only with a very weak, limefree solution ($\frac{1}{2}$ g 0·5 oz) to 1 l ($1\frac{3}{4}$ pt) water, fortnightly in summer.

Browallia

This tropical nightshade plant owes its name to the lutheran bishop Johan Browallius, who was not only an important dignitary of the Finnish church, but also a clever botanist, a friend of Carl Linnaeus. They were both born in 1707. He never knew the plant named after him.

Browallia speciosa, fam. *Solanaceae*

Origin: *B. speciosa*, the species most frequently cultivated nowadays, is thought to have been discovered in Peru in 1830. Nine species are known; from *B. speciosa* hybrids have been developed.

Appearance: eight of the species are annual herbaceous plants, among which the summer-flowering *B. grandiflora* and *B. viscosa* enjoy a sunny and warm position in the garden, together with other nightshade plants. *B. speciosa* is a shrubby plant, growing to 50 cm (20 in), freely branching and with dark green foliage. The violet-blue flowers with their white throats appear in large numbers among the leaves, giving a beautiful contrasting effect.

The flowering season depends on the time of sowing: if they are to flower in summer, they should be sown in the greenhouse in February–March. If pot plants for the winter season are required, the grower sows in July–August in an unheated greenhouse.

Care: browallia is not very sensitive in regard to temperature. Winter-flowering plants may be placed in the living-room, but their flowering period is not so prolonged as that of summer-flowering specimens, which may give pleasure until far into autumn without requiring much care. Do not place in full sunlight. Water and feed normally. Dead flowers should be removed immediately; that is all.

Brunfelsia

Otto Brunfels, physician, theologian, botanist and professor in Mainz and Strasbourg, lived from 1489 to 1534. He compiled the first large book of herbs, entitled *Conrafayt Kreuterbuch*. He had been dead for centuries when Linnaeus gave him the honorary title 'Father of Botany' and named brunfelsia after him.

Brunfelsia pauciflora calycina, fam. *Solanaceae*

Origin: this genus belongs to the interesting family of nightshades, or *Solanaceae*, natives of Central and South America. Their woody character indicates their durability.

Appearance: these sideways growing little shrubs do not branch freely. They are evergreen and have dark green, leathery leaves. In this country only *B. pauciflora calycina* is available, usually marketed under its former botanical name *B. calycina*.

Inflorescence: if properly cared for it will produce numerous large flowers from March until September at least. The violet-coloured flowers of the winter-flowering form 'Floribunda' have a small white eye.

Care: a half-shady position, a little lighter in winter. During the growing season, from March until September, water normally; some degree of atmospheric moisture is desirable. Feed every two weeks. During the dormant season, from October to January, water sparingly and do not feed.

Repotting: if necessary repot in shallow pots as soon as the plant begins to put forth shoots. Prepacked potting compost is suitable.

Important: very sensitive to variations in temperature. Normal room temperature is satisfactory. Dry air and sunlight frequently lead to an attack of plant lice (greenfly).

Caladium

When you see the caladium at the florists', where it is appearing with increasing frequency, you will admit that this is one of the finest foliage plants in existence. The colourful leaves are borne on stems which are often several metres (feet) high. Formerly this plant was rarely available, but the situation has now improved.

Origin: caladium originates in tropical regions in America, chiefly in the jungles of Brazil and the Amazon.

Appearance: originally there were about 15 species, of which especially *C. bicolor*, *C. picturatum* and *C. schomburgkii* have supplied many magnificent strains. The pure species have practically disappeared. We now speak of *C. bicolor* hybrids and of arrow-leaved *C. schomburgkii* hybrids. Only the small *C. humboldtii* with its green, white-flecked foliage, continues to be grown in its specific form. It is a tuberous plant.

Inflorescence: if the plant is going to flower at all, the typical arum-lily shaped inflorescence appears in April–May. Very insignificant.

Application. caladium is above all a somewhat sensitive greenhouse plant. It feels most at home in larger tropical plant windows or hot-houses. It demands a half-shady position, away from direct sunlight, but there are strains which, after careful hardening, can be grown for months on end as simple houseplants.

Care: from spring until August the plant demands plenty of water, and until July a weekly feed as well. From August onwards, gradually reduce the water supply, until the leaves shrivel up and fall. The tubers are left in the pot at a winter temperature of 18–20°C (65–70°F). In February they are taken from the pot, old compost and old roots are removed, and the tubers repotted singly or in groups of three in soil with a high humus content. They should be given a warm position, 24–26°C (76–79°F) and kept damp. Only when the leaves have fully developed should the plant be placed in a slightly cooler situation in order to harden off. While the shoots are emerging, the air must be kept very moist. Older leaves of strains suitable for indoor cultivation will tolerate a drier atmosphere.

Propagation: during the growing period small offsets are formed, which are suitable for propagation. Do not remove them too soon; wait until they put forth leaves themselves.

Caladium bicolor, hybrid, fam. *Araceae*

Calanthe

These terrestrial orchids cannot really be recommended for beginners, for they are not easy to grow. Their native habitat is in eastern Asia, Japan, the Pacific Islands, tropical regions of Africa, Central America and Mexico. Some experience is needed before you attempt to grow these orchids.

Appearance: they all have large pseudo bulbs, with strong, stemless densely veined leaves (in the photograph the fern leaf has been added as a background). Some species are evergreen, others lose their foliage every year. The latter and their hybrids are at present frequently cultivated. One of these is *C. vestita*, a native of Burma. The foliage of this plant turns yellow and drops in late summer, which does not detract from its magnificent flowering towards Christmas. The hybrid 'William Murray' in the photograph is one of the *C. vestita* hybrids which are very popular in the United States as well. The graceful umbelliferous inflorescence varies considerably both in shape and in colour: golden yellow, white, pink or multicoloured with a darker, often spotted throat.

Care: most calanthe species belong in a well-lit, tropical plant window; a few will be happy in a moderately warm greenhouse. As they are deciduous they actually require two different situations. The bare plants do not look very attractive and moreover need a dry resting period.

Calanthe vestita 'William Murray', fam. *Orchidaceae*

Calathea

Among the *Marantaceae* to which we shall refer several times, the *Calathea* genus contains the largest number of species. Its manifold beauty of foliage never ceases to attract us. Unfortunately these plants, unlike the simpler marantas, cannot grow outside a tropical plant window.

Origin: there are about 150 species, which occur in the chain of islands between India and Australia, in warm tropical rain forests. A number of species also occur in tropical areas of Africa.

Appearance: they are shrubs, with stems growing from the root; as a rule they have long and narrow leaves, marked in several shades. Small basket-shaped flowers appear in spring or early summer.

Species: *C. backemiana*, leaves pale silvery-green with dark green bands, lower side green. *C. lancifolia* (its former specific name, *C. insignis*, is still frequently used in the trade), is the finest.

C. makoyana: leaves rounder, red on the underside, olive-green with dark green streaks on top. The plants may reach a height of 30–60 cm (1–2 ft).

Care: half-shady to shady position. When new leaf-shoots appear, a high degree of atmospheric humidity is essential. Winter temperature must not drop below 16–18°C (60–65°F). The growing season is from March to September; the dormant period from October to January. Water with lukewarm, softened water. Drying out of the soil-ball is as harmful as excess water collecting in the bottom of the pot. Feed every fortnight during the growing period. Propagation by division. The new plants should be put into small pots.

Calathea lancifolia, fam. *Marantaceae*

Calceolaria — Slipper-flower

A cool and damp atmosphere is the right environment for this popular annual, which around Easter is among the most frequently bought plants. Their native country is in the mountain forests of the Andes in South America, but the original species have long since been replaced by cultivars.

Calceolaria hybrids are bushy, herbaceous plants, available in low-growing, medium and tall-growing forms. Their unusual flowers with their 'slipper'-shaped lower lip can be single- or bi-coloured, streaked or spotted. They are sold in full flower and will last for a fairly long time, provided they are kept away from sun and heat. The best situation is on a north-facing window-sill in a temperature of 10–12°C (50–54°F) but not everyone will be able to provide this. Be generous with watering and give the plant plenty of room. It is rarely saved for another season. Occasionally an attempt is made to keep cuttings from the parent plant throughout the winter, but there is little chance of success. Sowing in July, with the intention of keeping the plant through the winter, is also best left to the professional grower.

Calceolaria, hybrid, Slipper-flower, fam. *Scrophulariaceae*

Callisia

Strange to say, the botanical name *Callisia* means something like 'Lily of Beauty'. It belongs to the *Commelinaceae* family and is closely related to the *Tradescantia*. Until recently the species were called *Setcreasea*, but since the latest reclassification the latter now represent a separate genus.

C. elegans is a creeping, ground-covering plant which thrives in a tropical plant window. It is still frequently sold under its former name *Setcreasea striata*. The leaves are olive-green with fine white central nerves on the upper surface, deep reddish purple underneath. They grow in overlapping rows like roof tiles round the stem. The flowers are white. *C. repens* is another creeping plant. Cultivation is the same as for creeping forms of *Tradescantia*.

Callisia elegans, fam. *Commelinaceae*

Callistemon

One day the little boy next door asked me why I called this amusing plant growing on my terrace 'bottle-brush' and whether you could really use it for cleaning. The extraordinary flowers with the stamens arranged around a long central axis, and no conventional petals, are exactly like a bottle-brush, as can be seen in the picture.

Origin: there are about 25 species, all found in Australia or in the islands of Tasmania with their fresh climate. However, *C. citrinus* (formerly *C. lanceolatus*) is the only one cultivated in Europe.

Appearance: in its native country it is an evergreen shrub or small tree up to 3 m (9 ft). Grown in pots, and later in tubs, it may achieve a third of this height. Firm leaves, 4–5 cm (1¾–2 in) long, with clearly defined veins, they are usually lancet-shaped, hence the old name *lanceolatus*.

Flowering: it is an ever-recurring surprise to see the inflorescence. Quite young plants produce the bottle-brush shaped flowers, growing in all directions from the top of the crown. They are up to 10 cm (4 in) in length; the individual florets with their striking stamens, grow densely on a central stem. Flowering season from June until August.

Use: it is a very useful, undemanding plant for pots and tubs, once very fashionable. After having been forgotten for some time, it has once more found its way into many living-rooms, especially on Mother's Day. The larger, freely flowering specimens are rather expensive, but since they are as easy to cultivate and increase as the *Myrtus* to which it is related, prices will doubtless drop.

Care: if possible it should be given a place out of doors in summer; otherwise as light and airy a position as possible indoors. A cool situation in winter, 6–8°C (44–46°F). These are the main conditions for successful development and flowering. From March until the time when it is put out of doors, around mid-May, the plant prefers good light and a warm room; it tolerates dry air.

Watering: with softened water (*p*H 5).

Feeding: fortnightly from the time the first shoots appear until August, with lime-free feed.

Repotting: in February–March; trim a little at the same time. Use lime-free compost, rich in humus (or loam with peat-fibre); otherwise the foliage will discolour.

Callistemon citrinus, fam. *Myrtaceae*

Camellia japonica – Camellia

This may be considered a queen among plants. Its flowers are of a truly regal beauty. Whether it makes 'regal' demands as well is doubtful, for if we look after it properly, we shall find its requirements very modest and shall enjoy the camellia for many years.

Origin: belonging to the family of *Theaceae*, it is among the oldest plants in cultivation in eastern Asia. It competed at an early stage in history with another valued plant, the tea plant, which did not become known in the Occident until later. This was due largely to their similarity in appearance, expressed even in their names. Not many people know that the tea plant is called *C. sinensis*; the botanical name for the common ornamental camellia is *C. japonica*. It is therefore not surprising that it created a problem even for a man like Linnaeus. He urgently needed a genuine tea plant for certain scientific investigations. As everybody was anxious to be of assistance to the famous scholar, travellers from China brought him several tea shrubs in the middle of the 18th century. However, each of them proved to be a camellia with single red flowers. The two species owe their generic name *Camellia* to the Jesuit father Georg Josef Kamel, a missionary who died in Manilla in 1706. The tea plant is unsuitable for use as a pot plant. Our camellia originates in the cool mountain forests of Japan and is also found in northern China and in Korea. In Europe the cultivation of camellias started soon after 1800 in England, and from there spread to the Continent, first to Belgium, later to other countries as well. It is thought that in the United States the first camellia was grown as early as 1787, but its cultivation in that country did not come to full development until more than 150 years later, after 1935. America now has many 'Camellia-clubs', whose members grow the plant out of doors wherever the climate allows.

Appearance: an evergreen shrub with shiny, dark green, leathery leaves, from which we may deduce that this is a strong and undemanding plant. Two/three year-old plants, ready for the market, are 40–60 cm (16–24 in) tall. Although slow-growing, they develop into stately tub plants. All present-day camellias are cultivars.

Inflorescence: wide open flowers, single, double or semi-double, up to 6 cm (2½ in) across, depending on the strain. They resemble roses, but have no scent. Their colour varies, again depending on the strain, between white (not much in demand) soft pink and deep red; the most popular is a deep rose-red and white streaked variety such as 'Chandleri Elegans', which has been in cultivation since 1824. The normal flowering season runs from January to April, but because of the numerous strains available, it is possible to have flowering camellias from November until April.

Winter: there is one basic rule for treating this plant correctly: moderation in all things. Any change in situation and temperature, any sudden transition from moist to dry air, any irregularity in watering, will endanger the flower buds. Hence the frequent disappointments when a flowering

Camellia japonica, Camellia, fam. *Theaceae*

plant, with numerous buds, is taken from the flower shop to the living-room with its warm and dry atmosphere: buds and leaves will drop at once. The correct treatment is to keep the camellia at a temperature of 10-12°C (50-54°F) until the buds open. If after that you are unable to resist the temptation, you may transfer the plant very carefully to a slightly warmer room, where it must face the light in the same way as before. The temperature should not be above 16°C (60°F). To avoid any risks it is better to leave it in its original spot. After flowering the temperature should not exceed 6-10°C (44-50°F); this will particularly benefit the foliage.
Summer: if possible bury the pots in an airy, half-shady position in the garden between mid-May and mid- or late September. A balcony free of draughts might do. Indoors the plant will get too little fresh air in summer.
Watering: always with lukewarm, softened water, pH 4-4·5. During the winter maintain regular moisture. In summer, water normally until July, but from late July onwards keep the plant drier in order to encourage bud-formation.
In winter provide adequate atmospheric humidity until flowering begins. The leaves should be rinsed regularly with lukewarm water.
Feeding: from December until July with lime-free feed.
Propagation: in a commercial nursery, from cuttings.
Diseases: too warm a place in winter and incorrect treatment in summer lead to attacks by scale insects.

Campanula — Bell flower

It is a well-known fact that the majority of indoor plants do not tolerate hard water. Among the few exceptions to this rule are two campanula species, which for many generations have been famous among many plant lovers as hanging plants, or those which climb round pillars.

Origin: *C. fragilis* and *C. isophylla* are both natives of the chalky mountains in Italy and other Mediterranean countries.
Appearance: *C. fragilis* has curved or trailing stems with smooth, pale green, small leaves and blue flowers from June until far into the summer. *C. isophylla* is relatively densely hairy, which gives it a more grey-green to whitish appearance; this species flowers about four weeks later with larger, blue, wide open, star-shaped flowers. It is generally known by its cultivar name 'Mayi'. There is also a white form called 'Alba'. These two are often planted together in a pot and trained up a stake as 'bride and bridegroom'. All these house-campanulas have very fragile stems which break at the slightest touch and secrete a milky (non-poisonous) liquid. They are perennials and if properly looked after will flower repeatedly for many years. They may flower so profusely that the foliage is hardly visible.
C. pyramidalis, which grows to 1·5 m (5 ft) in full flower provides a wonderful spectacle, but it is biennial and too large for indoor cultivation.
Care: in summer a spot in good light in the living-room, but screened from strong sunlight. From the end of May onwards it may be placed on a sheltered balcony or in the garden. As the plant is very sensitive to frost it is absolutely essential to bring it indoors before the first night frost occurs. It should then be allowed to rest until early February in the best possible light at a temperature of 6-8°C (44-46°F). From March onwards give it a warmer position; drastically cut down old shoots and if necessary put into a larger pot. Compost should be rich and humusy, with a little sand added.
Watering: during the growing period and in the flowering season maintain a regular degree of moisture. Drying out of the soil-ball is very harmful, particularly for *C. fragilis*. Water sparingly during the dormant period.

Campanula isophylla 'Mayi', Bellflower, fam. *Campanulaceae*

Feeding: in summer every two weeks.
Propagation: in spring, from cuttings; *C. fragilis* can also be grown from seed.

Canna

In splendid isolation canna is the sole genus in the family of the same name; about 50 species occur in tropical regions of America. Why the main species, divided into numerous strains, is nevertheless called *C. indica*, can only be due to the 'West Indies' mistake of Columbus.

C. indica hybrids are actually garden plants, but there are magnificent dwarf forms, growing only to 50–60 cm (20–24 in) and therefore more suitable for use in the living-room. 'Lucifer' is a most graceful variety; in addition there are 'Alberich', salmon-pink; 'Perkeo', cherry red and 'Puck', yellow. As a result of modern growing methods it is no longer necessary for us to start the plants into growth ourselves; we can buy them as young plants, complete with growing instructions. The rootstock is treated in the same way as tuberous begonias.

Canna indica, hybrid, 'Lucifer', fam. *Cannaceae*

Capsicum – Red pepper

The well-known houseplant *C. annuum* is closely related to the Spanish pepper, much eaten in southern countries. The species of the winter cherry described on p. 336 belong to the same group: they are all nightshade plants.

Origin: the beautiful, increasingly varied cultivars for indoor cultivation have been developed from the original species occurring in Central and South America, where they grow into shrubs up to 2 m (6 ft) in height, or into annual plants up to about 50 cm (20 in). It is therefore understandable that it is the annual species which are used for cultivation in Europe.
Appearance: the pot plants available are bushy small shrubs of varying size. It appears that the dwarf form 'Gnom' is the one now most in demand; it is a trailing plant, only 20 cm (8 in) tall, but which can produce as many as 50 fruits, entirely hiding the green of the foliage.
Inflorescence and fruits: they are usually propagated by professional growers between February and April in a heated seed-bed. The first plants

Capsicum annuum, Red Pepper, fam. *Solanaceae*

appear on the market as early as late August, but the large majority of plants in full bloom generally do not come into the shops until November; they are then available until the end of the year; large numbers of these plants are sold around Christmas, when you will not see the inflorescence, usually white or pale yellow, sometimes violet in colour. The shape of the fruits varies greatly; there are globular forms, but also erect-growing or pendulous cone-shaped fruits in clear colours; the photograph at the bottom of p. 150 gives a vivid idea.

Interesting detail: you may be wondering whether the fruits of the red pepper are edible, like those you buy at the greengrocers', and whether you can serve them at dinner. There is nothing to stop you experimenting. As the plant belongs to the nightshade family, all paprika species are slightly poisonous. Children are unlikely to eat the fruits, since they are so hot, and adults will certainly not swallow the entire harvest at one go. If you want to try it, it is essential to remove the seed, as you would do with large peppers.

Care: there is not much I can tell you about this. Drought and heat rapidly cause the colourful fruit to shrivel up. A cool situation and regular watering will prolong the plant's life, but it will come to a sudden end and will then only be fit to be thrown away. As the name *C. annuum* implies, it is truly an annual plant.

Carex

There are more than a thousand species of this grass-like plant, distributed all over the world. They create wild meadows, annoy the garden lover by their tendency to spread and in the high mountains grow up to the snow line. As a houseplant carex is quite rare: only one species is suitable for this purpose.

C. brunnea occurs in southern Asia as well as in Australia. A white variegated form, 'Variegata' has been developed, usually sold under the name *C. elegantissima*. It is a graceful, yellow-green striped grass, growing to 20–30 cm (8–12 in) and very undemanding. In contrast to its close relation *Cyperus*, the carex will not grow in an aquarium and does not even tolerate surplus water in the saucer. It prefers a normally moist soil-ball. It does not feel at home in a very warm room, for it is really a plant for the unheated greenhouse and likes a winter temperature of 8–16°C (46–60°F). Apart from this it requires no special position: both sun and shade will be suitable. Repot when the plant is obviously outgrowing its container. Propagation by division; seed will produce plain green plants only.

Carex brunnea 'Variegata', fam. *Cyperaceae*

Carissa

Why is this splendid evergreen shrub so little known? Is it perhaps because, like all other members of the *Apocynaceae* family it is poisonous? But this is after all accepted as part of the bargain where the oleander and many other greenhouse and houseplants are concerned.

Origin: these arborescent evergreen shrubs, which as a rule have unusually double-thorned twigs, green foliage and large, fragrant flowers, originate in tropical parts of Africa, Asia and Australia.

Use: because of their size and their greenhouse constitution, winter temperature 12–15°C (54–58°F), they are not very suitable for indoor cultivation. One should also be warned about the extremely poisonous nature of the finest species, *C. spectabilis*; it is in fact advisable to wear gloves when handling this plant, for instance, when pruning it. But in spite of these disadvantages it gives forth a delicious scent during its flowering season from February until April and therefore deserves a place in the greenhouse. This species, moreover, has no thorns.

◁ *Carissa spectabilis*, fam. Apocynaceae

Catharanthus roseus, fam. Apocynaceae
▽

Catharanthus roseus

When you hear its former botanical name: *Vinca rosea* you will recognise this attractive plant as an old friend. *Catharanthus roseus* is in fact an exotic relation of our well-known garden plant *Vinca minor*, the periwinkle; it occurs in many tropical regions.

Commercial growers usually cultivate *Catharanthus roseus* as an annual by sowing it in a heated greenhouse from February onwards. The shrub-like plants will flower from early summer until autumn, and their splendid rose-red flowers with a background of dark foliage fill many a bare spot. Greenhouse owners appreciate this particularly. It is possible to prolong cultivation by means of cuttings, taken in February, or by seeds, fairly easily obtainable, and sown in February.

Cattleya

Undoubtedly the genus *Cattleya*, with its enormous number of species and hybrids, contains some of the most striking orchids. It is certainly an attractive thought to own such a plant, but nevertheless one should be warned not to undertake its cultivation without some experience.

Origin: all cattleya species are natives of South America. Many of them were discovered in the area between the Amazon and the Rio Grande; others originate in Venezuela, Colombia and Equador. In cultivation the climatic conditions of the native habitat should be reproduced as far as possible. Cattleya is named after William Cattley, of Barnet, who around 1820 owned one of the largest orchid collections.

Appearance: the pseudobulbs are long and narrow and cylindrical in shape; the thick leathery leaves appear singly or, very rarely, in pairs; the root develops strong runners; flowering occurs at different times, depending on the species; the inflorescence rises from the leaf-sheaths. The number of blooms varies considerably, between 2 and 12–15; the spikes of the small-flowered varieties contain an even larger number of blooms. Large single flowers may attain a diameter of 20 cm (8 in). When cut they may last from ten to fourteen days.

Care: cattleya species are more difficult to look after in winter than in summer. For this reason the tropical epiphytes — species as well as hybrids — do not belong in the living-room or in the plant window, but in a heated greenhouse, where high temperatures and a very high degree of atmospheric humidity can be provided during the growing period, and a cool and dry environment during dormancy. All this requires a profound knowledge of the subject and it is outside the scope of this book to go into details. If you are interested in knowing more, you are advised to join an orchid society and to consult specialised literature while gaining experience.

Watering: with completely demineralised water (pH 4–4·5). Because of air pollution rain water, once so highly recommended, is no longer suitable. In winter rain water is, in fact, quite unusuable because of the fumes caused by oil-fired central heating. It is advisable to plunge the pot rather than to water.

Atmospheric humidity: this should not be provided by spraying the plant, but only by means of a fully automatic humidifier, to prevent the formation of water-drops. Spotting of leaves and flowers is thus avoided.

Compost: this consists of sphagnum moss, and chopped fern roots (osmunda and polypodium), with some further additions. The most modern mixtures contain polystyrene granules as well.

Cattleya trianae, fam. *Orchidaceae*

Cephalocereus senilis – Old man cactus

'Respect for old age' presents a slight problem here. Scientists are still not agreed on whether *C. senilis*, the well-known Old Man Cactus, is the sole representative of his genus or not. The Nomenclature Commission has already decided that there are at least six other species.

Origin: the present situation is that all columnar cacti have been brought together in one genus. They all originate in Central and South America, in regions between Mexico and Brazil.

Appearance: *C. senilis*, Old Man Cactus, remains the leading species. In its native country it may grow to an imposing column up to 15 m (48 ft) in height, the top crowned by long, wavy hair. Old Man Cactus does not flower until it has attained a height of 5–6 m (15–19½ ft); it is therefore not surprising that an amateur will never see this phenomenon in his greenhouse. Even small seedlings already have soft hair reaching to the soil.

Care: Old Man Cactus requires a great amount of light and in spite of its origin in the hot rocky areas of Mexico covered in chalky slate, it needs humid warmth under glass during its growing season in summer; frequent misting is therefore essential. During winter, when it is kept dry, the temperature should not drop below 15°C (58°F). The white hair is sensitive to draughts. To prevent it getting dirty, place small stones round the column.

Watering: normally in summer. From November to February keep dry, with transitional periods in autumn and spring.

Feeding: because of the hairy surface, great care should be taken when feeding. Avoid solutions rich in nitrogen. It is best to give a special cactus fertiliser once a fortnight.

Repotting: younger plants annually; at longer intervals as the plants mature, into only slightly larger pots. As for all cacti, leave the soil-ball intact. After repotting provide temporary support.

Propagation: easy, and very interesting from seed. A good seed catalogue will contain various strains, from 'Aureus' to 'Tetetzo'.

Diseases: lack of overhead spraying in summer will lead to the appearance of red spider and mealy bug. Too much moisture in winter may cause fungus formation. For the same reason cacti occasionally rot close to the neck of the root.

Cephalocereus senilis, Old Man Cactus, fam. *Cactaceae*

Cereus – Hedge cactus

The name *Cereus* means 'wax-taper'. This simple meaning covers an enormous variety of species and forms. Recent classification has simplified matters for the amateur.

Origin: at one time cereus cacti formed a genus which was regarded as the most important of all *Cactaceae* and which, in addition to enormous hedge cacti, also contained shrubby, creeping and trailing types. As a result of re-classification, the genus now embraces only columnar cacti, including branched tree-cacti and organ-pipe cacti. Their native habitat is in South America: from the northern Argentine to Mendoza, Paraguay, Uruguay and eastern and southern Brazil.

Appearance: most cereus cacti eventually grow too large for small collections or for the window-sill. Consider also that their fascinating flowers, usually opening at night, appear only at a more advanced age. The white, sea-green or bluish wax layer which covers the epidermis of many species to reduce evaporation, gives a beautiful effect. As they are so undemanding and possess enormous vigour, columnar cacti are frequently used as stock on which smaller and weaker species are grafted – often without much expertise. It will rarely be successful.

Species: an attractive exception to the disadvantage of excessive size is *C. chalybaeus*. This species does not branch freely; its columns grow to 'only' 3 m (9 ft) in the course of time, and are covered in a bloom so deep blue that it is regarded as the bluest cactus of them all. Flowers up to 20 cm (8 in) in size may be produced by relatively young plants about 50 cm (20 in) tall. These are scarlet to rose-pink outside and white on the inside.

Cereus peruvianus 'Monstrosus', Hedge Cactus, fam. *Cactaceae*

It is a nice plant to grow from cuttings or from seed. This is also worth the effort with *C. jamacaru*, another blue cactus, but which becomes so large that it soon outgrows its suitability for being used indoors. The oldest and largest specimens grown in Europe can be found in the park facing the casino in Monte Carlo.

C. peruvianus, of which the 'Monstrosus' cultivar is well-known among cactus lovers, is slightly less vigorous (illustrated on p. 155).

Care: hedge cacti are among those plants of which it is said that the more they are neglected, the better they grow. I myself have such a 'Monstrosus'; I obtained it as a cutting, and ▶

after allowing it to dry out a little before planting, it rooted readily. This was ten years ago; since then it has been repotted twice. Every summer it decorates the terrace; it is brought indoors at the first signs of night frost and passes the winter in company with other balcony- and tubplants in good light at a temperature of 6°C (43°F). It has gradually grown so large that, to prevent capsizing, the pot is now placed in a tripod, of which the construction can be seen on p. 56.

Watering: in summer it should be given some water from time to time, and the plant should occasionally be sprayed to remove dust and open the pores.

Feeding: cereus requires feeding only very rarely. An occasional feed in summer is sufficient.

Propagation: cuttings will readily root. The plant is also easily grown from seed, but the choice of seed is so large that one should select with care.

Ceropegia

If you study the relationship between various houseplants you will discover that the graceful ceropegia belongs to a most interesting family. It is difficult to believe that this succulent hanging plant is so closely related to the good old *Hoya*.

Origin: the genus *Ceropegia* contains about 160 species, occurring from South Africa via the tropical area of Arabia to south-eastern Asia and western China. They also grow in New Guinea and in northern Australia. A few species are found in the Canary Islands and in Madagascar. Species grown as houseplants are as a rule restricted to *C. linearis* and the much better known *C. woodii*, distinguished from the former only by the shape of its leaves. Both are natives of South Africa.

Appearance: thread-like stems appear from rooted corms which may have a diameter of up to 5 cm (2 in). Pairs of heart-shaped fleshy leaves appear at the joints, marbled on their upper surface, pale green underneath. The amusing lantern-shaped little flowers grow from the axils, followed by new cormlets. Many collectors of succulents use these little corms as stock on which to graft difficult *Stapelia* species, which, like ceropegia, are related to the *Asclepiadaceae*. The species mentioned flower nearly all the year round.

Species: *C. distincta* ssp. *haygarthii* has even more remarkable spotted flowers. It prefers to climb up a post or a pole. *C. fusca*, a native of the Canary Islands, with stems the thickness of a finger ending in flower racemes, also prefers to climb. The largest green-marbled flowers topped by a 'lantern roof' are found in *C. sandersonii*. It is a pity that these plants are so rarely obtainable.

Care: my own ceropegia plants thrive in soilless cultivation and need very little care. The pots hang all the year round from the frame of a large south-facing window. They develop numerous shoots, creating a living curtain. If planted in soil they should be watered very sparingly and even less in winter. Give an occasional small amount of feed from April to August. They really should pass the winter in a cool position, but it is pos-

Ceropegia linearis woodii, fam. *Asclepiadaceae*

sible to keep them in the living-room, as withered stems will subsequently produce new shoots.

Diseases: do not occur in practice.

Chamaecereus

At one time it belonged to the collective genus *Cereus*, but, as is the case with many other plants, it is now the sole representative of its genus, *Chamaecereus*, under the name of *C. silvestrii*. It is a freely flowering individual; its family relationship has been enlarged in some measure by various hybrids.

Origin: its native habitat is in western and northern Argentina, where at a certain altitude the shrub-grown slopes provide the most favourable light conditions. Robbing the chamaecereus of light is the worst one can do to it.

Appearance: in summer each shoot, the size of a finger both in thickness and in length, bears several fine large flowers, bright scarlet in colour. If properly cared for new shoots will surround the plant, soon forming a circle of small pillars and flowering at an early stage.

Situation: I repeat: a light, but not too sunny situation is the prerequisite for favourable development. Lack of light will result in thin, lanky and pale green 'fingers'; too much sun will turn them brown; both are harmful to the plant's health. In winter it must be kept cool to very cool; in fact a night temperature of a few degrees below zero is desirable. The plant will shrivel up, but will flower all the more profusely afterwards.

Watering: water regularly only in summer.

Feeding: feed generously during the growing period, from April to August, using a nitrogen-free fertiliser.

Repotting: very difficult, as the shoots are extremely fragile.

Compost: must be very rich and friable; it may contain some humus (peat fibre).

Propagation: detached shoots will generally grow again. Allow the broken surface to dry out thoroughly before planting. Very easy to grow from seed. A good seed catalogue will present a choice of seed of various hybrids.

Germinates in light; the seed should be covered very thinly with fine sand.

Diseases: in summer lack of light and air may lead to an attack by red spider. In winter the plant may die as a result of excess heat or moisture.

Chamaecereus silvestrii, fam. Cactaceae

Chamaedorea

As we said under *Areca*, the Betel palm, p. 128: indoor palms have had an unexpected revival, doubtless, due to a feeling of nostalgia for the past. Without doubt, the *Chamaedorea* has always stayed in the running – probably because of its readiness to flower.

Origin: the genus embraces about 120 species, all natives of Central and South America. *Chamaedorea elegans* originates in Mexico.

Appearance: they are all thornless small plants with a tubular stem and usually pinnate foliage. Occasionally the pointed ends of the leaves are split. Hybrids of *C. elegans* are available, for instance the American 'Bella', a delightful mini-palm, which flowers as readily as the species. Among chamaedoreas with non-pinnate leaves we should mention the interesting *C. tenella*, also a native of Mexico. This plant rarely grows higher than 70–80 cm (28–32 in) and the 7-cm (3-in) wide leaves are 10–12 cm (4–5 in) long.

Inflorescence: always dioecious; the flowers appear on single or irregularly branched stems growing below or among the leaves. The yellow inflorescence resembles that of the mimosa. After flowering, berry-like fruits, the size of peas, may develop. Even quite small plants in late winter provide the unusual, but annually recurring spectacle of 'flowering palms'.

Situation: *C. elegans* is one of the most hardy dwarf palms and should be given a light, but not sunny, position. During the growing period, from spring until July, and on into autumn, any normal room temperature is acceptable. In the resting period from October until February the temperature should not exceed 12–14°C (54–57°F) especially at night. The plant tolerates a fairly dry atmosphere, but in a very warm room dry air will affect growth.

Watering: freely in summer; in winter only the top layer of soil should be kept a little moist. The water should not be too hard; pH 5–5.5 is correct.

Spraying: this will encourage leaf growth in summer. Spray only on warm days.

Feeding: a weekly dose of highly diluted liquid fertiliser from March until July is best. Do not feed at other times.

Repotting: in spring, if it is necessary. The older the plant, the less frequently it should be repotted. Use tall palm-pots.

Compost: well-draining, with humus; it will do no harm to mix it with some extra sand.

Propagation: from seed only; this germinates slowly at a high degree of bottom heat, 24–26°C (76–79°F). Sow the seeds in threes in small pots; prick out the seedlings with extreme care, as the fine roots are very fragile. Screen young plants against direct sunlight; keep warm, shaded and moist for the first few years. Flowering usually occurs in the third year.

Diseases: incorrect care will lead to attacks by scale insects and red spider.

Chamaedorea elegans, fam. *Palmae*

Chlorophytum — Spider plant

Among the more than a hundred species known to professionals, only *C. comosum* is cultivated here as a houseplant. It has in fact already been described in a little book dating from 1809 as a 'pot-grown ornamental plant', and was first mentioned in 1794.

Origin: the species occur in tropical areas of the Old and the New World. The chlorophytum we know under many different names originates in South Africa, where it grows only as a plain green plant.

Appearance: the 'Variegatum' forms, with a white centre stripe, have long been cultivated indoors; they have 20–40 cm (8–16 in) long, narrow leaves arising from the soil. From spring until summer the plant develops thin flower-stems of tremendous vigour; they may become several metres (feet) long. The weight of the flowers appearing at the end, followed by small leaf rosettes, makes the stems curve, and because one always has the chlorophytum in this form, as a bundle of flower-stems ending in young plants, it gives the impression of a hanging plant.

Care: chlorophytum is justly regarded as one of the hardiest and least demanding of indoor plants. It accepts any and all conditions: it adapts immediately to warm as well as to cool rooms; tolerates little as well as a great deal of sunlight; cares nothing about dry air, although it benefits by spraying in the growing period in summer. Very suitable for soilless cultivation.

Watering: in the growing period, from February to September, normally or freely. Reduce watering during the dormant period from October to January, depending on whether it is kept in a warm or a cool position.

Feeding: weekly from March till September.

Repotting: as soon as the fleshy roots push the plant out of the pot. It can be done at any time except in winter. Use a wide, fairly shallow pot.

Soil mixture: not too heavy, somewhat loamy.

Propagation: by removing the young plants, or by lay division of the parent.

Diseases: badly cared-for plants will be attacked by greenfly.

Important: if the compost dries out the tips of the leaves may turn brown. Curling leaves are due to lack of air or to feeding in winter; this makes the plant limp.

Chlorophytum comosum 'Variegatum', fam. *Liliaceae*

Chrysanthemum

This plant's ancestors came from China and Japan; its full name is *Chrysanthemum indicum*. With its enormous progeny of cultivars — *indicum* hybrids or garden chrysanthemums — it is among the most popular autumn-flowering plants.

They are grown in pots, in bowls, on balconies; as bedding plants and on graves, so they are not simply houseplants. Take care not to allow the soilball to dry out. Dead flowers should be removed immediately. The flowering period will be more prolonged in a cool situation than in a warm room. After flowering the question arises as to whether it is worth while to keep the plant, cutting it down to 5 cm (2 in), the temperature should then be 4–6°C (40–44°F), but it is not really advisable.

Chrysanthemum indicum, hybrid, fam. *Compositae*

Cissus

What would a book on houseplants be without the cissus! Such a book would hardly be worth buying, for it would lack one of our most versatile and faithful green friends. In this book it is fortunately given the attention it deserves; an extensive assortment, representing more than 350 species, is mentioned.

Origin: there are a few species which do not depend on a tropical climate, but by origin all representatives of the genus *Cissus*, of which the first was introduced into Europe to be cultivated in 1790, are tropical plants. The native country of each plant will be mentioned separately, as they are spread all over the world.

Appearance: because the plant has adapted itself to climatic conditions everywhere in order to maintain itself as a climbing plant, it varies greatly in outward appearance. Something that is not generally known is that, apart from vinous, trailing lianas, there are succulent representatives of this group as well, greatly resembling cactaceous spurges. Some climb, others do not, but instead form a thick, succulent stem, 2–4 m (6–12 ft) in height. These remarkable oddities can certainly be used as pot plants, so a collector of unusual plants could include them.

The species generally used as house plants can live to a ripe old age.

Species for simple room cultivation

Species: first we should, of course, mention *C. antarctica*, a native of Australia. Little need be said about the treatment of this plant. There are three things which it definitely does not tolerate: 1. Too bright light. This leads to a non-parasitical foliage disease. In serious cases the glassy, transparent blotches on the leaf first turn yellow, then brown, and finally the leaf drops. 2. A compost containing peat — but this is easy to avoid. 3. Surplus water in the bottom of the pot (a foot-bath); this can be destructive, especially in the resting period in winter. However, everyone knows that the amount of watering in winter

Cissus antarctica, fam. *Vitaceae*

depends on the temperature. Further details of treatment required will be found under the general instructions on care.

Secondly there is *C. rhombifolia*, often marketed under its old name *Rhoicissus rhomboidea* and not to be confused with *R. capensis* described on p. 321. Not that such confusion would lead to disappointment, for the latter plant also is suitable for simple room cultivation.

C. rhombifolia was not introduced in this country until about twenty-five years ago. In Sweden, where green foliage plants are very popular, it was known earlier; in the U.S.A. a more erect and compact growing hybrid has meanwhile been cultivated, namely 'Mandaiana'. The stem and the underside of the leaves are red; the upper side is green; all its parts are slightly hairy. The central leaf is always a little larger than the two somewhat asymmetrically placed outer leaves, as can be clearly seen in the illustration of *C. rhombifolia*.

Whereas even after several years of successful indoor cultivation we hardly ever see flowers on *C. antarctica*, *C. rhombifolia* will in the course of time produce insignificant little flowers, half hidden by the foliage; this happens in summer. Just as in the case of the wild vine, these flowers can develop into hard, inedible berries. Cultivation is the same as for *C. antarctica*.

A plant which is unfortunately difficult to obtain is *C. striata*, a native of Brazil: a small, weakly trailing climbing shrub with slightly hairy stems and leathery green leaves. It is very suitable for indoor cultivation, and if it were marketed on a larger scale, this plant would find a warm welcome from people with little space at their disposal.

Species for plant- and tropical-windows

In this group the magnificent *C. discolor* is of course the leader. It is a plant from Java, with pointed, heart-shaped leaves which grow up to 15 cm (6 in) in length and 8 cm (3 in) across; they are velvety, with violet-red veins and edges; the upper side is marbled metallic olive-green, the lower surface red. This magnificent foliage appears on strong stems, erect rather than pendant and eventually becoming woody. While the plant is still young, the stems, too, are red. If the plant is correctly cared for, at a high temperature and atmospheric humidity, flower-stems bearing small yellow flowers will, at a later stage, grow from the leaf axils. In this instance the plant's vigour should be taken into account; it is a tropical climber of indomitable strength.

This applies even more to *C. gongylodes,* which is sometimes allowed to grow carelessly along the ceiling of a tropical window, where its long red aerial roots, appearing from square branches, form a living curtain. The leaves may grow up to 30 cm (1 ft) in length and the adhesive suckers at the ends of the trailing stems attach themselves just where you don't want them. Something which in a large greenhouse is overpoweringly beautiful, in the small space available to private owners becomes all-intrusive. I speak from experience . . .

Succulent species – very rare

For the sake of completeness we should mention a succulent species. *C. quadrangularis* lives in the dry jungle and steppes of East Africa, where it climbs high into the tops of trees, from where, like a kind of degenerate 'Queen of the Night' it lets its stems droop down.

C. juttae, *C. bainesii* and *C. crameriana*, at one time belonging to the *Cissus* genus, have now been reclassified as belonging to the *Cyphostemma* genus.

Care: we have already said something about *C. antarctica* when describing the assortment. It should be added that, although it is often seen in a south-facing window, this is in fact not the right place. In this position it

▶

Cissus rhombifolia, fam. *Vitaceae*

Cissus discolor, fam. *Vitaceae*

Cissus striata, hybrid from the U.S.A., fam. *Vitaceae*

will be prone to the 'non-parasitical foliage disease' manifested by transparent leaves, described above. The correct position in simple room cultivation is a well-lit, but not sunny spot. The word 'room' should be stressed, for all cissus species belong indoors throughout the year. As a rule it does not accept a situation on a sheltered balcony. This generally results in dropping of the leaves at the onset of autumn, and since before winter no new leaves can grow, the plant would have to be cut back vigorously, which does not improve its appearance. All cissus species suitable for room cultivation tolerate both a warm and a slightly cool position in winter. The correct temperature lies somewhere between 22–24°C (72–75°F) and 12–15°C (54–58°F).

Watering: sparingly in summer; in winter the amount depends on whether the plant has a warm or cool position. Degree of acidity: pH 6–6.5. A high degree of air moisture is not essential.

Feeding: every week from the time when the first shoots appear until mid-August; then every two to three weeks.

Repotting: young plants every year; older specimens at longer intervals in a slightly larger pot. When you remove a cissus from its pot you will find that the root system is fairly restricted. Like those of a vine, the roots can penetrate very deeply to obtain an adequate water supply; this is impossible in a pot, and the root-clump therefore remains relatively small. The plant will not appreciate too much fresh compost when being repotted.

The best mixture for *Vitaceae* is one based on garden compost. Unlike ordinary prepacked potting compost, which is based on extremely acid garden peat, this compost contains a fair amount of rotted organic elements of all kinds, which together provide an adequate lime reserve. The roots appreciate this greatly; they also like a very good layer of crocks for drainage. Never put a cissus into an ornamental pot without drainage.

Propagation: if adequate heat can be provided, 25–30°C (78–85°F), this is best done in an indoor propagator. Propagation from cuttings in spring or early summer is fairly simple. The cuttings should be 3–5 cm (1¼–2 in) in length and should be inserted in small pots containing a special compost sold for this purpose.

Diseases: the greatest risk is that of excess water at the bottom of the pot, especially in winter: this causes mildew, brown blotches and other fungus diseases.

Hothouse species

Treatment as described in the instructions for the tropical window.

Citrus

There are two ways in which you can raise enthusiasm for citrus plants among plant lovers. The first is to grow plants from seed which you have gathered yourself, and to give these away. The second is to buy for a friend one of those magnificent new strains laden with golden-yellow fruit.

Origin: their native habitat is in eastern Asia. From there citrus plants spread over the countries of the ancient world to the coasts of the Mediterranean. The first specimens appeared on the estates of the nobility in Western Europe at the beginning of the Renaissance. In summer they were looked after in the beautifully planned gardens; in winter they were brought into the orangeries in their tubs. *C. sinensis,* the orange tree, and *C. limon,* the lemon tree, were the chief species cultivated in gardens.

Appearance: in their wild state they are fruit trees not unlike plum trees, with evergreen leathery leaves. As the fruit takes more than a year to develop, even the 'mini-trees' serving as indoor plants provide the spectacle of bitter-sweet fragrant flowers appearing at the same time as the ripening fruit. Plants grown from the pips of oranges, mandarins, lemons or grapefruit are a different matter. These wildings grow into bushy, sometimes thorny, shrubs, hardly flowering and practically never bearing fruit. Like any apple- or pear-tree they must first be improved by grafting, provided that a suitable stock is available. If this is the case, it is also possible to grow small standard trees. However, in spite of their slow growth, these become too large in the long run and the amateur would therefore do better to restrict him- or herself to the cultivated, smaller forms such as *C. aurantium myrtifolia,* a Chinese dwarf orange, or *C. mitis* 'Calamondin', a native of the Philippines and cultivated on quite a large scale in Germany. Another attractive Chinese kind is *C. taitensis* 'Otaheite', which has ripe fruits around Christmas.

Care: this plant poses a problem: all citrus species require to overwinter in good light and a cool position at a temperature of not more than 4–6°C (43°F). In summer they should go out of doors if possible, though they are satisfied with a balcony or a permanently open window. Water freely until mid-August, regardless of whether the water is hard or soft. Keep the soil-ball moist during the rest period in winter. Feed frequently in summer, but not afterwards. In early spring the plant may be cut back if absolutely necessary; it will involve the loss of many a flower-bud. Repotting and other measures as described under *Nerium oleander.*

Important: warmth and too moist soil in winter cause the leaves to drop.

Citrus micrantha microcarpa, fam. *Rutaceae*

Cleistocactus

Formerly cleistocacti formed part of the large collective genus *Cereus*. Since the latest classification by the Nomenclature Commission they form a separate genus. One of their typical traits is that the plants do not tolerate being kept dry in winter; they demand moisture throughout the year, otherwise they will rapidly shrivel up.

Origin: they originate mainly in western Argentina, Uruguay, Paraguay, Central Bolivia and Peru, where — paradoxical as it may sound — they grow in rocky places with a high degree of atmospheric humidity.

Appearance: slender columnar cacti, branching from the base and partly growing erect up to heights of 2 m (6 ft), partly remaining squat and supine. The tubular flowers grow like fingers from the supine sections, constantly throughout the summer. The species *C. smaragdiflorus* untiringly produces its pale red fruits in the process. Another species, *Cleistocactus strausii*, has dense, snowy white spines and carmine-red flowers, while the cultivar 'Fricii' in addition has white hairs, 3–5 cm (1¼–2 in) in length. *C. wendlandiorum* is recognised by its unusually drooping yellow and red flowers.

Care: in summer place in a warm and sunny position in rich soil with plenty of room for the roots. Provided they are given sufficient water and nitrogen-free feed, they are very vigorous and flower readily. In winter they should have a light and cool situation and should not be kept entirely dry.

Propagation: from cuttings, but also grown easily from seed.

Cleistocactus wendlandiorum, fam. Cactaceae

Clerodendrum

An approximate translation of the Greek botanical name is 'tree of fate' or 'tree of chance'. Why the plant got this name, or what it has to do with fate, is no longer known. Until the beginning of this century it was called 'innocent love' in Germany.

Origin: the genus *Clerodendrum* contains about 400 species, most of which occur in Asia and Africa. Only three or four have come into cultivation over here and have become magnificent, eye-catching plants in exhibition greenhouses. Its application as a houseplant is limited.

Appearance: the best known species is undoubtedly the *Clerodendrum thomsonae*, a climber growing up to 4 m (12 ft), originating in the Congo. The twining stems bear long heart-shaped leaves. The rather short-lived scarlet flowers appear in dense racemes in April. They are surrounded by inflated, creamy white calices, which remain on the plant for a long time and, as with several other plants, constitute its chief ornamental value. Another native of Africa is *C. splendens*, also a twining plant, bearing bright red flower clusters from December to May. An erect growing species is *C. philippinum* (formerly *C. fragrans*), which comes from eastern Asia. The flowers resemble those of the hortensia hydrangea, they are white shading into pink. The flowering season varies.

Care: *C. thomsonae* requires a resting period at 10–12°C (50–54°F) from October to February, during which time it usually loses its foliage. In spring old branches must be cut back in order to make room for new flowering shoots. It is possible to trim them in the shape of a standard tree for indoor use.

Clerodendrum thomsonae, fam. Verbenaceae

Cleyera

The photograph of this plant hardly does justice to the elegance and decorative value of the cleyera. An investigation into its keeping qualities will show that it will remain attractive for a long period without causing much trouble; it is therefore regrettable that it is so rarely marketed.

Origin: Andrew Cleyer was a 17th-century Dutch physician and botanist. The genus *Cleyera* which was named after him two centuries later (formerly it was called *Eurya*) includes a large number of species occurring in Central America. However, these trees and shrubs proved to be unsuitable for cultivation. The only species cultivated is called *C. japonica* (syn. *C. ochnacea*), whose native habitat is in those countries where the tea plant (*Camellia sinensis*) and the camellia (*C. japonica*) also grow.

Appearance: small tree or shrub with evergreen foliage of varying shape, usually 6–8 cm (2½–3½ in) long and 4 cm (2¼ in) wide. It is from this species that *C. japonica* 'Tricolor', the only kind now in cultivation, was developed. Pale to dark green marbled leaves, edged with creamy white markings, in youth often shading into red. This truly magnificent pot plant is now cultivated especially by Belgian growers. Because of its extremely slow growth it can decorate your living-room for many years.

Inflorescence: it is unlikely that the variegated cultivar 'Tricolor' will flower. The all-green species has fragrant white flowers and produces round red fruits.

Care: the fact that it does not flower simplifies treatment; this sometimes causes problems with the closely related camellia. No need to worry about failure of winter flowering. The cleyera is a robust plant, which may be placed in a half-shady position out of doors in summer. It does, however, like to spend the winter in a cool spot at 10–12°C (50–54°F).

Watering: the very finely branched root system rapidly dries out and care should be taken to prevent this happening. The compost should be kept moist in winter as well. Occasionally spray or rinse the foliage to remove dust. The water must be soft (*p*H 4·5–5).

Repotting: in spring if necessary. Normal prepacked potting compost. Press down thoroughly.

Feeding: from December to July give an absolutely lime-free liquid fertiliser.

Propagation: the grower cultivates new plants from cuttings.

Diseases: too high a temperature in winter and incorrect treatment in summer lead to attacks by all sorts of pests.

Cleyera japonica 'Tricolor', fam. *Theaceae*

Clivia

The name calls up visions of festivals in ancient Rome and of mild summer nights, but this is wide of the mark. At one time the plant was called *Himantophyllum*, but when in ± 1866 a new classification was introduced, it was renamed after Lady Clive, duchess of Northumberland.

Origin: the genus *Clivia* embraces three species, all natives of South Africa. The only one brought into cultivation is *C. miniata*, originally found in the mountain valleys of Natal, where the soil is loamy, with a porous subsoil and a rich layer of humus. On occasion we may see *C. nobilis*, a native of the Cape, more modest in all details and as undemanding as the imposing *C. miniata*.

Appearance: the clivia is not a true tuberous plant, but its rootstem is built up in layers such as is found in onion plants; these form the centre of the fleshy roots. The leaves are strap-shaped and evergreen; they grow directly from the soil. The plant develops new leaves in pairs, so that after a number of years a strong stem consisting of layered leaves is formed. It may achieve a height of 60 cm (2 ft).

Inflorescence: the firm flower-stem arising from the foliage is slightly flattened and ends in a cluster of 10–12 fine, orange-red flowers, each growing on a separate stem of their own. In early spring the flower-stem starts to grow close to or between the leaves, but never exactly in the centre. The question often raised, whether after flowering the clivia should be discarded, has no biological basis whatsoever. In fact, if well cared for, the plant may flower for a second time in autumn. And if you do not remove the young plants which appear after flowering, you will in the course of time possess a magnificent family with a dozen flowers at a time. The photograph shows an approximately ten-year-old plant.

Use: the clivia has been recognised for more than a century as an undemanding plant which looks attractive even when not in flower. As it is a very tolerant plant, satisfied with a modicum of light and with no special requirements as regards atmosphere, it has become a kind of standard plant in restaurants and hotels.

Care: one essential is to leave the plant in the same position year after year, summer and winter. Do not turn it; mark the pot and its situation. A well-lit window where the morning sun enters, is the best place. If possible do not let the temperature in winter exceed 12–15°C (54–58°F). This appears to be in contradiction to the position it used to occupy between curtains on the window-sill, but as there was no heating radiator underneath the sill in those days, it worked. The plant is not sensitive to dry air.

Watering: this involves one of the main causes of clivias dying: the roots do not tolerate stagnant water in the bottom of the pot. Compare the description of the soil condition in its native habitat. Although, therefore, you should water freely in summer, you should avoid a footbath, and withhold water until the surface of the soil-ball has become dry. During the resting period from September to January, the plant should be kept as dry as possible, depending on temperature. In a cool position you need hardly water at all. Spray when the foliage starts to grow and a new flower-stem is produced. Occasionally rinse to remove dust. The pH of the water should preferably be about 6·5–7·0.

Feeding: fortnightly during the growing period from February to July only. From late July onwards, when leaf growth has ceased, prepare for the dormant period by feeding only once every four to six weeks.

Repotting: the older the clivia, the less frequently it should be repotted. Young plants every year, afterwards at increasingly longer intervals. Carefully remove the compost from between the roots with a small stick. For large specimens add a slow-acting fertiliser to the compost, as for tub plants. Prepacked potting compost may be used for the clivia, but better still is a mixture of rotted turves, coarse leafmould and sharp sand, to which a fertiliser in the form of dried blood, hoof and horn or bonemeal may be added. If you are unable to obtain all these ingredients, try at any rate to mix some clay or loam into the compost. The time to repot is just after flowering.

Propagation: wait until after the second year before carefully removing the sideshoots, the new little plants formed at the base of the plant; they must have at least 4–5 leaves and be growing well; the roots should first be detached (see chapter Propagation, p. 70). Insert the small plants in small pots in special propagation compost: 1 part sand, 1 part peat. As soon as a small root-ball has formed, transfer to a larger pot containing potting compost. However, it remains doubtful whether it is advisable to move the young plants. This depends largely on the space available for a spreading clivia, which may in the long run achieve huge dimensions.

Diseases: too warm a position in winter frequently leads to an attack by scale insects. This is most unpleasant, since it can be obstinate. The pests lodge themselves in the axils of the leaves and increase at an alarming rate; they can only be got rid of by consistently applied measures. At the same time the plant's general

condition should be considered in order to improve its health. When using a pesticide, following the instructions, do not neglect the leaf-axils. In the case of a severe attack, repeat the treatment at intervals of eight to ten days, but above all remedy the cause of the attack, for instance a wrong situation. Yellowing of the leaves may be caused by too much sun as well as by overwatering, especially with cold water. This excessive moisture is also frequently the cause of the well-known cork disease (oedema) on the foliage. This starts with pale green spots, later turning yellow or red and finally forming the brownish cork spots. Keeping the plant drier will not get rid of the cork spots already present, but will prevent the trouble spreading.

Important: usually when clivias refuse to flower it is because they have not been given the necessary resting period in winter. Briefly this means that flower-stems 'nipped in the bud' are the result of lack of warmth in late winter; this is necessary for their development. After flowering, orange-yellow seed capsules may develop. There is little point in leaving them; they absorb too much energy. These seeds will not as a rule germinate. On the other hand it is possible to grow plants from suitable seed, but it is a lengthy process. With maximum care it will still take at least three years from sowing to flowering; possibly even longer.

Clivia miniata, fam. *Amaryllidaceae*

Codiaeum — Croton

With this plant the exceptions prove the rule. Its use is certainly not confined to owners of greenhouses or plant windows. Fortunately fine, flowering specimens with shiny foliage are seen increasingly often in living-rooms.

Origin: this tropical member of the spurge family originates in Indonesia and Polynesia, where the croton occurs in 14 species of variegated shrubs and trees, growing up to several metres (feet) in height. Probably the magnificent shrub was discovered there in 1860 by the famous English botanist John Gould Veitch. It was he who brought it to Europe as a novelty; the decorative plant was at that time already highly valued in its native regions and has since become so in other tropical countries as well. In Europe the species have so often been crossed that it is almost impossible to bring order into the names; for that reason we generally refer to them merely as *C. variegatum pictum*.

Appearance: evergreen shrubs with leaves of varying shape, narrow or wide, sometimes spiralled. The majority are three- or four-coloured in clear shades of green, red, white, pink, yellow and brown, spotted, marbled, veined or blotched.

Inflorescence: once the plant has settled down satisfactorily, it will produce insignificant flowers in late winter or early spring.

New methods of cultivation: until quite recently the croton was regarded exclusively as a greenhouse plant. Young plants used in bowls and troughs for their decorative value rarely survived the first winter in the living-room. As a result of rigid selection and propagation of the hardiest species a remarkable change has occurred: crotons have become tougher.

Situation: some shade and no direct sunlight both in winter and in summer. Lack of light, however, causes the variegated foliage to fade or to turn plain green.

Temperature: an even air- and soil-temperature, 16–18°C (60–65°F) in winter; lower temperatures cause frost damage and dropping of the leaves; higher temperatures rob the plant of its resting period, continuing its growth and making it produce small, less finely marked leaves, which spoil its appearance.

Watering and spraying: keep fairly moist from March to August, water moderately from autumn onwards. Throughout the year beware of dry air and drying out of the soil ball; this applies particularly to simple room cultivation without a plant window. Dry air caused by central heating results in the appearance of mildew and makes the lower stem become bare.

Feeding: when the plant begins to put forth shoots and does not drop leaves, feed every week from March to August.

Propagation: only in a heated greenhouse.

Diseases: faulty treatment leads to obstinate attacks of scale insects. Red spider mite and thrips occur, especially in dry air in winter.

Codiaeum variegatum pictum, Croton, fam. *Euphorbiaceae* (see also p. 39)

Coelogyne

Many people shake their heads and refuse to believe that it is perfectly possible to grow orchids in the living-room. Nevertheless the group of delightful flowering coelogyne species contain some of the most important plants, much loved by orchid collectors and beginners.

Origin: the chief species, C. cristata, was found around 1830 in Nepal in the Himalayas, where it grows on damp rocks and in dead trees at an altitude of up to 2000 m (6000 ft) and more. Other coelogyne species occur in Ceylon and in the island of Samoa. C. fimbriata, a dwarf species also suitable for room cultivation, was introduced into Europe from southern China in 1825. C. dayana and C. massangeana are natives of Sumatra, Java and Borneo.

Appearance: in all there are about 120 species, some of which are terrestrial orchids, while others grow as epiphytes or tree-orchids.
C. cristata has egg-shaped to globular pseudobulbs, about 6 cm ($2\frac{1}{4}$ in) long, each with two leaves up to 30 cm (1 ft) in length. C. dayana and C. massangeana are capable of developing pseudobulbs of up to 12 cm ($4\frac{3}{4}$ in) and leaves of up to 50 cm (20 in). The larger and stronger the pale green pseudobulbs, the healthier and more readily flowering the plants will be.

Inflorescence: C. cristata produces single flowers of up to 10 cm (4 in), hanging in groups of three or four. The flowers are creamy white with yellow-crested lips, and always open simultaneously; older plants, which may have dozens of flowers, are a magnificent sight. The same applies to the graceful single flowers of the two other species, they are up to 5 cm (2 in) in size, growing in racemes from stems up to a metre (3 ft) long; these flowers are white also, but have yellow-edged and brown-marked lips.

Unfortunately none of the coelogyne flowers are suitable for cutting, but on the plant they last for an incredibly long time. The flowering period of C. cristata is January to April, that of the two other species from late spring to high summer.

Care: as the stems are strongly pendant, they are best grown in special orchid baskets. These are filled with the usual compost, consisting of fern roots and sphagnum, to which some chopped turves may be added. If you would rather use pots, these should be filled up to at least a third with crocks. The plant should be watered with demineralised and lime-free water; you should be particularly careful with spraying, for the flowers, especially those of C. cristata, are easily stained by moisture or damp air. In summer the plants are kept in a shady greenhouse; C. cristata can also stand in an east-facing window indoors. However, in winter the temperature should be lowered to 14–16°C (57–60°F), while watering is reduced to a degree at which the pseudobulbs just fail to shrivel up.
Repotting should take place after flowering, in other words usually in spring. At the same time the plants may be divided to obtain further specimens.

Coelogyne cristata, fam. *Orchidaceae*

Coffea arabica — Coffee plant

The coffee plant belongs to the not inconsiderable group of houseplants which in a sense owe their existence to the interest of amateur growers. Although it is simple to cultivate, it is rarely obtainable at the florists'. Occasionally a good catalogue includes seeds of the plant.

Origin: everyone knows that coffee, and therefore the coffee plant as well, comes from Arabia. From there the coffee was introduced for purposes of consumption and delight to the tropical plantations of the Old World, later to appear in other continents as well. The genus *Coffea* includes about 50 species of shrubs and small trees, mainly originating in tropical regions of Africa. Since time immemorial the most important species has been *C. arabica*, of which there is a dwarf form, 'Nana' which is suitable for simple room cultivation.

Appearance: in cultivation on tropical plantations the coffee plant grows into a somewhat bushy plant up to 3–5 m (9–15 ft) tall. In a greenhouse in a botanical garden it will reach 3 m (9 ft); properly cared for old plants can certainly achieve 2 m (6 ft) in the course of time. The normal size of an indoor plant, initially one-stemmed, is 60–80 cm (24–32 in). The horizontal side-branches bear dark evergreen, slightly wavy, glossy leaves in pairs opposite each other.

Inflorescence: the small, star-shaped flowers, snow-white and fragrant, appear from the leaf axils in the third or fourth year, sooner in the dwarf form. The flowering season is in the summer months.

Care: it remains in the living-room throughout the year, in half shade, not in bright sunlight. If possible give it a slightly lighter situation in winter. The growing and flowering period is from March/April to August/September. In this period the plant requires slightly moister air, so it should be sprayed frquently. Resting period from October to February at a moderate temperature 15–20°C (58–71°F); a low bottom temperature below 16–18°C (60–65°F) is more harmful than too low an air temperature.

Watering: in summer, while the plants are growing, water freely, but be careful that no surplus water remains in the bottom of the pot. Use soft water, pH 4·5–5. In winter, depending on the room temperature, water sparingly, but do not allow the soil-ball to dry out.

Feeding: from April to August give a lime-free feed every week.

Repotting: in case of necessity, at the beginning of the growing season in roomy pots. Compost should be rich, humusy and porous.

Propagation: the best results, for room cultivation as well, are achieved with plants grown at home from fresh seed. The red berry-like fruits imported by seed merchants are guaranteed viable; so are the ripe coffee beans of plants from a heated greenhouse. Home-grown plants as a rule do not produce viable seed. In spring the beans are placed in groups of two or three in damp sand at a temperature of 24–26°C (75–79°F) (indoor propagator). Seedlings should first be planted out in ordinary potting compost, to be repotted in summer in the mixture described above.

Diseases: in winter the plant may be infested by scale insects.

Coffea arabica, Coffee Plant, fam. *Rubiaceae*

Colchicum autumnale – Meadow saffron

They often grow wild in damp meadows and because of their leafless stems are called 'naked boys'. Apart from the wild species there are at least 50 others from which beautiful hybrids have been cultivated, which have found their way to the living-room as dry-flowering plants as well.

Origin: the species which are particularly suitable for indoor cultivation originate in North Africa and western and central Asia. Magnificent strains have been developed.

Appearance: the egg-shaped bulbs, whose size depends on the strain, are encased in a brown skin to above the 'neck'. Colchicum has been cultivated in western Europe since 1581, but has played a part as a popular remedy against all kinds of complaints for much longer. It contains a poisonous alkaloid, colchicine, which in diluted form has a beneficial action and is still incorporated in medicines against gout, rheumatism, etc. The seeds are particularly poisonous, but in the cultivated forms these are no longer produced.

Use: the bulbs are now sold not only to professional growers, but also to nurseries and in all sorts of shops. This also applies to *Sauromatum venosum*, described on p. 325; this is often sold under its old name *Arum venosum*. They are then treated as dry-flowering plants, placed on a saucer with damp moss, or even placed directly on the window-sill in a warm room. Within a few weeks shoots with flowers – very attractive in many strains – will appear. After flowering the bulbs are planted in the garden to allow them to develop their roots in late autumn. In spring they will put forth leaves, which must not be removed until they turn yellow. Without the ripening of the foliage and the simultaneous development of embryo flowers, they will not flower for a second time out of doors.

Species: for dry flowering the most suitable forms are the large-flowered garden hybrids such as 'Autumn Queen', 'Lilac Wonder', 'Princess Astrid', 'The Giant' and 'Water-lily'. The large flowers of the cultivated meadow saffron are also very suitable to be used for cutting.

Colchicum autumnale, Meadow Saffron, fam. *Liliaceae*

Coleus – Flame nettle

It is almost impossible to describe the beauty of the multi-coloured foliage of coleus. You can buy it quite cheaply. Three errors in cultivation may lead to disappointment: 1. lack of light makes the foliage become pale or revert to green; 2. calciferous water and soil damage the plant; 3. drying out of the soil-ball is fatal.

Origin: there are about 200 species of herbaceous and shrubby plants, originating in Asia and Africa. Many of them have been in cultivation for over a century.

Appearance: there are two distinct groups:
1. Strains cultivated solely for their magnificent multicoloured foliage; these are collectively called *C. blumei* hybrids. Depending on the strain, these grow to 30–60 cm (1–2 ft); they have angular stems, turning woody at the base, and elongated oval dentate leaves, ending in a point, varying in size and colour range. Insignificant flowers in summer; these are best pinched out immediately. Although the plants are perennial, it is not advisable to keep them through the winter in this country.
2. Annual and perennial species which in winter produce fine flowers; these are nearly all plants for the heated greenhouse. An example is *C. fredericii*, a native of Angola, which grows to a metre (3 ft) in height, freely branched, with pale green leaves. In December it produces long racemes of deep blue flowers. It is an annual, but self-sowing. *C. thyrsoideus*, another native of Africa, also grows up to a metre (3 ft); it is erect-

▶

Coleus blumei, hybrid, Flame Nettle, fam. *Labiatae*

growing and non-branching and produces stemmed terminal clusters of pale blue flowers from December to February. Perennial. Between these two groups we find *C. pumilus* (formerly *C. rehneltianus*), the hanging coleus. This is a native of the Philippines; it has slender, pendant branches and graceful multi-coloured leaves and from November to spring produces magnificent sky-blue flower racemes, 10–20 cm (4–8 in) long. Greatly recommended as a durable hanging plant.

Application: *C. blumei* hybrids are bought at the florists' in spring as vigorously growing bushy plants in splendid colours. They are very sensitive to frost and should therefore not be put in the garden or on the balcony until after mid-May.

Care: maximum colouring of the foliage occurs only in full light; the plant should therefore be given the lightest position in a window-sill, but near a south-facing window it should be slightly screened around noon. Lack of light makes the foliage fade or revert to green. If you wish to keep the plant through the winter, it should also have as warm and light a position as possible; however, this may lead to an attack by plant lice – a difficult problem. In any case such plants will be lank and bare at the bottom in the second year, so there is really little point in keeping it. Better to discard the plant in autumn and buy a new one in spring. The growth rhythm of winter-flowering species entails a different kind of treatment. Watering, spraying and feeding of these plants is concentrated in the months from November to April; the plant is allowed to rest in summer and kept somewhat drier, but it should not dry out. This applies particularly to *C. pumilus*, which must in addition be cut back slightly to encourage vigorous new shoots.

Watering: very sensitive to hard water. The pH must not exceed 4–4.5. *C. blumei* hybrids must be watered freely in summer; spray frequently on very warm days. For winter-flowering species see under Care.

Feeding: during the growing period give liquid fertiliser containing little or no lime.

Repotting: when repotting annual or greenhouse species, be sure to use lime-free compost. One which is based on peat is very suitable, also the John Innes potting compost (acid). Take extra care not to let the plant dry out.

Propagation: this is rarely done, but fairly simple. Succulent tip shoots will root within a few days in small pots containing special cuttings compost – equal parts of sharp sand and peat fibre. They will also root easily in water in a dark bottle. Afterwards they can be planted in a lime-free soil mixture. The best time to take cuttings is from spring until mid-summer. The young plants are repotted once or twice to encourage bushy growth and are stopped once. These young plants can successfully overwinter. Sowing coleus in February–March is a pleasant job for the amateur. Many kinds of seed are available.

Important: if you look after the plants in accordance with instructions, you will obtain incredibly fine colours. In foliage plants the flowers should always be pinched out.

Columnea

Just as the poinsettia was at one time difficult to keep, so *Columnea* used to cause us worry. In the atmosphere of the living-room, and sometimes even in a tropical plant-window, the branches became bare and flowering was long delayed. New strains have brought about changes for the better.

Origin: the plant owes its name to the botanist Fabio Colonna (1567–1640). There are 160 species, originating in the tropical rain forests of Central America, Costa Rica, Panama, in Mexico and in Haiti and Puerto Rico. Most of the cultivated species were unknown before 1900.
Appearance: they are epiphytes, growing in the form of shrubs, shrub-like plants, and climbing, hanging or creeping herbaceous plants high in the trees of the jungle. The species in cultivation are plants with long, drooping shoots and opposite evergreen, marked leaves.
Inflorescence: as a rule the flowers grow singly on short stems from the axils of the leaves. Some species produce their flowers in clusters. The colours are red or orange-red, often with a yellow throat. The flowering season is in autumn or late winter; it frequently flowers twice a year.
Species: of old the main species are the most beautiful and least demanding such as *C. gloriosa* with its hybrid 'Purpurea', and *C. hirta* with its very large flowers and long flowering season from late winter into spring.
C. microphylla is an attractive species with tiny leaves; *C. schiedeana* is the most vigorous – it is adorned from March until May with brownish flowers, yellow inside with a network of veins. All these are outshone by the greatly praised, almost erect-growing *C. kewensis* and also by the 'Stavanger' strain with its long stems and small leaves and by the new group of columnea hybrids, which do away with many of the problems of cultivation.
Care: hang them in good light in semi-shade, away from direct sunlight. Winter temperature 15–18°C (58–65°F). The plant requires a constantly high degree of atmospheric humidity without direct spraying. Water only with lukewarm, soft water. Feed weekly in the growing period, using a lime-free solution.
Repotting: where necessary, in spring, in larger, fairly shallow pots.

Diseases: draught may cause the leaves to drop.

Columnea, hybrid, fam. *Gesneriaceae*

Conophytum – Living stones

Among the *Aizoaceae* family with its rare forms and often even more remarkable habits, the genus *Conophytum* contains the largest number of species. Some botanical dictionaries mention only 33 species of these 'living stones', but the genus includes more than 300.

C. ficiforme plays us a special trick: its violet-pink flowers open only at night. Another peculiarity of this species, which, because of its reversed timing, demands extra attention from its owner, is that its resting period is from April to August. During this time it should be kept completely dry so that the 'pebbles' wrinkle up. The epidermis will then split and the arrival of the new plants is accompanied by the appearance of the flowers; this is the beginning of the growing period, which lasts from autumn to spring. During this period it should be kept in good light and well-ventilated, and remain moderately moist. Being sparing with water and occasionally syringeing are the secrets of its treatment. Compare also *Argyroderma*, p. 128 and the other *Aizoaceae* on pp. 209, 219 and 252.

Conophytum ficiforme, Living Stones, fam. *Aizoaceae*

Convallaria – Lily of the Valley

In books on houseplants containing a separate section on the method of starting bulbs into growth, the lily of the valley is often included. This indicates that the plant is a border case in two senses of the word. In the first place houseplants do not really include bulbs that are to be started into growth, and in the second place the lily of the valley is not a bulbous plant.

Origin: Linnaeus refers to it as *Lilium convallium*, that is, lily of the valley, as we still call it. In regions of Europe, Asia and America with a temperate climate, the plant often grows wild in chalky soil in not too densely wooded valleys. We can therefore not really regard the completely hardy plant as a houseplant. In catalogues it is offered as the 'genuine, large-flowered, very fragrant garden lily-of-the-valley', often with the added statement that 'stronger plants' are available for indoor cultivation.

Appearance: no need to tell you what the upper part of the plant looks like. Below ground they are among those plants which put out strong offshoots and are therefore popular as ground cover to crowd out troublesome weeds.

Use: in order to have flowering lilies of the valley at Christmas one must buy crowns retarded in refrigeration in the previous winter — at the latest in March. They are potted in damp moss at the correct time and marketed in this form (in November or December) — all you have to do is to provide an even moisture and a moderate temperature for the plastic or cardboard pots. Alternatively you dig up prepared crowns at an earlier point and, after pruning the longest roots, place them in groups of ten or twelve in flat bowls filled with a mixture consisting of equal parts of sieved loam and peat fibre. The tops should only just be visible and must be covered with a layer of damp moss the thickness of a finger. Put the bowls in a warm and damp spot until the crowns put forth shoots. Now remove most of the moss, leaving a thin layer, and place the bowls in a light, cool position. The best temperature for lilies of the valley in bud or in flower is about 12°C (54°F). In a normally heated room they will flower for only a few days. By planting bowls at intervals and placing them in a cool spot, it is possible to organise matters so that you have flowering convallaria indoors until well into March.

Lillipots: at Christmas and Easter, lilies of the valley are now available in 'lillipots', which contain ready planted crowns. These are watered when received and left in the dark for seven to ten days, and are then brought into the light and kept at about 12°C (54°F). They are supplied mail order.

Important: occasionally lilies of the valley produce pretty red berries. Like the entire plant these are poisonous.

Prepared convallaria may be planted in the garden after flowering, but as a rule will not flower again.

Convallaria majalis, Lily of the Valley, fam. *Liliaceae*

Cordyline

Formerly this plant belonged to the lily family, but since the most recent reorganisation by the Nomenclature Commission it has been classified among the agaves. Needless to say this botanical reclassification does not detract from the cordyline's excellent qualities as a houseplant.

Origin: there are about 20 species, of which the majority originate in south-eastern Asia and Australia.

Appearance: the genus includes tropical and sub-tropical trees, shrubs and shrub-like plants, some of which grow to 12–15 m (39–45 ft) in their native habitat, while other, stemless, species develop into shrubs as tall as a man. A further distinction may be made between species with stemless leaves, often pointed like a sword, and others whose foliage is broader and more distinctly marked. Older species produce more or less insignificant flowers.

Confusion: as a result of their close relationship with *Dracaena*, the botanical names and long since obsolete trade names are often confused. Nevertheless the two genera are easily distinguished: all *Cordyline* species have white roots ending in club-shaped tubers; the roots of all

▶

175

Cordyline terminalis 'Tricolor', fam. *Agavaceae*

Dracaena species prove to be yellow when cut.
Species: *C. australis* (syn. *C. congesta*) and *C. stricta* (syn. *Cordyline stricta*) with sword-pointed leaves are plants for a cool greenhouse; they have green or strangely variegated foliage. Older specimens of *C. australis* grow so large that they are suitable only for large, high rooms, for instance in office buildings. *C. stricta* remains smaller. *C. terminalis* with its many strains is strictly a plant for the heated greenhouse, with the exception of the red-leaved cultivars which may be used indoors. For care see under *Dracaena*, p. 193.

Coryphantha

The genus *Coryphantha* plays a very special role in cactus collections. Many people believe that they are very difficult to care for, while others are enthusiastic about their readiness to grow, their unusual spines and their uninterrupted flowering. The correct care is most important for success.

Origin: it is not generally known that there are two groups, each with entirely different requirements. Although the majority of coryphantha species originate in the south-western region of the United States and in Mexico, one group is among the inhabitants of barren deserts, and in fact is difficult to look after, while the other group embraces the more 'normal' prairie plants.
Appearance: the globular cacti, resembling *Mammillaria* grow to 8–12 cm ($3\frac{1}{4}$–$4\frac{3}{4}$ in). *C. elephantidens*, an easy species to grow, may attain as much as 18 cm ($7\frac{1}{4}$ in). They are all covered in warts, from the centre of which grow firm spines, producing a colourful contrast to the globe. The silky, glossy lemon-yellow or pink flowers grow in a circle.
Care: the desert group (recognisable by its stronger spine formation) requires poor, stony soil without humus. Cover the neck of the root with gravel; water very sparingly. The prairie plants, which are fleshier and have fewer spines, like a little humus in the compost and more moisture during the growing period. In winter all coryphantha species should be kept completely dry in a light and cool position.

Coryphantha cornifera, fam. *Cactaceae*

Crassula

The many names alone of these delightful succulents provide a great deal of material to write about. The plant described here as the most important representative of crassula (suitable for room cultivation) is *C. arborescens*, an arborescent succulent; the lower surface of the foliage has a scarlet margin.

The branched crown gives the plant the appearance of a gnarled miniature tree — it looks a little like a very small oak-tree. This plant has several nicknames, but since the botanical names of the various species have been altered several times it is possible that nicknames have become confused to some extent. For instance, *C. argentea* ('silvery succulent') is now called *C. obliqua*. The green foliage of this plant has a metallic gloss, which is supposed to look like silver. The leaves are circular and about the size of a 5p piece, and in some countries the plant is called 'money tree'. *C. obliqua* has white flowers, but, as with *C. arborescens*, they will rarely appear in room cultivation.

At one time both these species appeared under the name *C. portulacea*. This name is now obsolete, but at least you will know where you are if you meet these species under their old name. The plants originated in dry regions of South Africa and we may conclude from their origin that they are undemanding, satisfied with very simple treatment.

Growing period: from March to August. Sunny situation, as long as the sun is not shining too brightly. It is advisable to bury the pot in a sheltered spot out of doors in May. Water sparingly; feeding once a month is adequate. Spray only if the leaves are dusty. When in autumn the night temperature drops below 8°C (46°F) the crassula must be brought indoors again.

Resting period: this lasts until approximately the end of February.

Crassula arborescens, fam. *Crassulaceae*

Finding a suitable position with the correct conditions for this resting period is frequently a problem in modern homes for, like many cacti and other succulents, crassula definitely does not tolerate the high temperature of a centrally heated house at that time; on the contrary it requires good light and a cool position in a temperature of at most 6–10°C (43–50°F). A higher temperature at this time leads to lank growth, greenfly and loss of foliage. If the air is too moist, mildew may occur as well. ▶

Crassula lycopodioides, fam. *Crassulaceae*

You should not of course feed the plant in this period, and only water very sparingly.

Repotting: when the plant begins to put forth shoots in early March, it is best repotted in a fairly shallow, not too large, pot or bowl. Make sure of drainage by means of a layer of crocks on the bottom. Use a sandy, humusy mixture.

Inflorescence: if you are successful in bringing it into flower, this will occur in May–June. There are no fixed rules for obtaining flowering *C. arborescens* and related species. There are specimens which flower after three years; others will make you wait ten, fifteen or even twenty years. We do better to regard them as attractive foliage plants.

Propagation: by cuttings of shoot tips or leaves.

In the growing period the cuttings will speedily root in a mixture of equal parts of peat fibre and sand. It is advisable to allow the cuttings to dry for a few hours before inserting them in the compost.

Other *Crassulaceae* for indoor cultivation. The choice of succulents for use in the living-room is much greater than is generally realised. Those who are interested in succulents and would perhaps like to establish a small and select collection, will discover that this can be as fascinating as collecting cacti.

Crassula falcata, fam. *Crassulaceae*

The crassula family is so large that it is impossible to discuss all the distantly related species. We shall therefore confine ourselves to those members of the family which bear the name crassula.

C. falcata is still frequently sold under its now botanically obsolete name *Rochea falcata.* It is a native of the eastern part of the Cape, where it grows as a small shrub up to 90 cm (3 ft) in height. As a houseplant it is valuable not only because of its silver-grey, succulent, sickle-shaped curved leaves, 7–10 cm (2¾–4 in) in length, but certainly also because of the splendid flower panicles. In older plants these may consist of hundreds of clear red florets and achieve a diameter of 20 cm (8 in); in the course of flowering the colour will gradually fade. In its native country the plant has the nick-name 'South African alpine rose'.

C. lycopodioides is a remarkable plant. The layered leaves entirely surround the erect growing stalk, almost like a coat of mail. This succulent also originates in South Africa, where it grows to 30 cm (1 ft) on stony, dry slopes and forms fine ground cover. The greeny yellow flowers growing from the leaf axils are insignificant.

Attractive hanging plants:

C. rupestris, whose creeping, winding stems turn a clear red in bright sunlight.

C. socialis, which has pale green rosettes and small white flowers in winter.

C. teres, a very rare plant, which dies after flowering.

Care: as for *C. arborescens.*

Propagation: can be grown from seed. When ordering, be sure to mention both names.

Crocus

The crocus is of very ancient, noble descent; its native habitat is in the Mediterranean area of Asia Minor. It has corms, not bulbs, and has been used for indoor cultivation for many years. Nevertheless it is not among the first harbingers of spring on the window-sill.

This is due to the fact that it is usually brought into the heated room too soon, with the result that the flower-bud will not develop and the plant will only have leaves. The correct method is to place groups of six to eight corms in flat bowls or special crocus-pots about mid-September. To force them they must be placed in a cool and dark position. Water only once and do not bring them into the light until mid-January; still in a cool position. Water sparingly, for too much water may cause root-rot. Once a third part of the flowerbud is visible they may be placed on the window-sill.

Crocus hybrids, Crocus, fam. *Iridaceae*

Crossandra

Crossandra infundibuliformis (syn. *C. undulifolia*) was not brought to Europe from India until 1817. It soon fell into oblivion and was 'rediscovered' in 1950. Like many other *Acanthaceae* this plant is more suitable for a heated greenhouse, but there are now cultivars available which will thrive on the window-sill.

In its native habitat, in Ceylon and India, this herbaceous, shrub-like plant reaches a height of about a metre (3 ft) and produces fine, soft pink flowers throughout the year. It is the flowers which have given the plant the name *infundibuliformis*, which means 'funnel-shaped'. This refers to the proportionally long tubular crown to which the fairly flat petals are attached in the shape of a funnel. The flowers appear in clusters on a long stem which grows from the leaf-axil and is covered in scrubby bracts. One advantage of this plant as a pot plant is its prolonged flowering season, from spring until autumn; a disadvantage of the original species is their sensitivity to temperatures below 20°C (70°F) and to a dry atmosphere; the leaves will curl and the plant will cease flowering.

There is now a Swedish hybrid available: *C. i.* 'Mona Wallhed', which has so many advantages that it has achieved fame as a houseplant everywhere. It is a more vigorous form than the more compact species. In the United States, also, this plant is very popular as a house plant.

Situation: in semi-shade in summer, plenty of light in winter, but not in bright sunlight. It must be kept indoors throughout the year; in winter the temperature must not fall below 18°C (65°F).

Watering and spraying: rainwater at room temperature is best. Water freely in the growing period, moderately while the plant is dormant (from around October to February). It likes moist air, so spray often; in a too dry atmosphere the leaves curl up and the plant will not fully come into flower. Take care, however, that no water falls on the flowers.

Feeding: as soon as crossandra begins to put forth new shoots (usually in February), feed once a week, diluting the fertiliser with water and

Crossandra infundibuliformis, fam. *Acanthaceae*

▶

making sure that on those occasions the soil-ball is thoroughly soaked. Stop feeding in August.
Repotting: the best time is before the flower buds appear, that is, in spring. The compost must be light and rich in humus.
Propagation: *C. infundibuliformis* can be grown from the seed produced by the plant. New plants can also be grown satisfactorily from cuttings in May/June, provided there is some bottom heat, 20°C (70°F). The modern hybrids can be increased only from cuttings.
Various: to obtain a fine bushy plant it is advisable to pinch out the tops of the young plants occasionally. Older plants can be cut back.
Diseases: lack of moisture in the atmosphere may lead to infestation by red spider mite and whitefly.

Cryptanthus

Cryptanthus is an interesting representative of the large bromeliad family. In its wild state it sometimes grows on tree trunks, a magnificent sight, but as a rule this species is terrestrial and for this reason it does very well as a houseplant in a shallow bowl or trough. The foliage is beautifully marked.

All the species which are suitable for the greenhouse or the living-room — about a thousand in all — are natives of South America. Hybrids with beautifully formed, variegated leaves have been developed. Cryptanthus does not originate in tropical rain forests, but is a ground-covering plant in drier regions where the trees grow low, for instance in eastern Brazil. It is therefore quite a tough and adaptable plant which is not too difficult to look after. Do not obtain this plant for the sake of its inflorescence, for the flower remains hidden between the leaves and is very small.
General care: when grown in a bowl or pot, it needs a position in good light, but not in bright sunlight. Give it a friable compost, rich in humus, and if possible add some sphagnum moss. If you grow the plant as an epiphyte, the care is the same as for *Aechmea* and *Tillandsia*.
Winter temperature for the variegated species should be 20–22°C (70–72°F). Non-variegated forms tolerate slightly lower temperatures. Water very sparingly and do not feed. During the growing period water normally. The flowering period falls between March and August. Be careful not to let the soil-ball dry out. Feeding is best done by occasionally spraying with a liquid fertiliser diluted with water. Various plant sprays are obtainable in the shops. This is also the best method for an epiphytic cryptanthus.
Propagation: from the small plants formed in the axils of the rosettes. When they are large enough to be potted they will detach themselves. Use shallow bowls.
The finest: *C. beuckeri, C. bivittatus, C. bromelioides* (especially 'Tricolor'), *C. fosterianus* and *C. zonatus*.

Cryptanthus zonatus zebrinus. Zebra-striped. Fam. *Bromeliaceae*

Ctenanthe oppenheimiana

The generic name *Ctenanthe* is derived from the Greek and means roughly 'comb-flower'; the plant was given this name because of its asymmetrical shape. Unfortunately nothing is known about the Mr. Oppenheim who inspired the plant, but a little light is thrown on this Greek beauty when we hear that it is a member of the *Marantaceae* family.

Nearly all the plants belonging to this family are sold under the name *Maranta*, but as we have already told you on p. 145, under *Calathea*, there are distinct differences among these fine foliage plants which flower extremely rarely; when they do, the flowers are insignificant. Who among plant lovers knows *Phrynium* and *Stromanthe*? Or *Thalia*, which was not named after the flighty Greek muse, but after the 16th century German botanist Johann Thal? The entire *Calathea* and *Maranta* genera with their numerous species originate in the tropical rain forests of South America. *Ctenanthe*, which includes about 20 species, belongs in Brazil. Barely half a dozen are found in botanical collections. Only the most hardy, but at the same time the largest and most beautiful, reach the amateur via the trade.

If you should happen to see a so-called 'Maranta' about a metre (3 ft) in height, with dark green leaves, silver-striped along the veins and reddish purple underneath, then it will be a *Ctenanthe oppenheimiana*. The cultivated form 'Variegata' has even denser growing leaves, 30–40 cm (12–16 in) long and in even brighter colours.

Care: this beauty will grow only in a tropical environment. Temperature and atmospheric humidity must be carefully controlled in accordance with the plant's seasonal demands.

Repotting: ctenanthe likes to grow in a wide and shallow pot filled with a mixture of coarse lumps of leafmould, equally coarse peat and matured cow manure. Provide good drainage and do not disturb the plant again for a couple of years.

Watering and spraying: only with soft water (*p*H 4·5), simultaneously with the other inhabitants of the greenhouse. Feed fortnightly, but only during the growing period.

Propagation: by division of overgrown offshoots in spring.

Ctenanthe oppenheimiana 'Variegata', fam. *Marantaceae*

Cyclamen persicum – Cyclamen

Nowadays this plant is not only the most popular houseplant; it is also the one most often thrown away. The way in which it is cultivated, as well as the centrally heated homes where it ends up, make it almost impossible to care for it correctly and to provide the conditions which allow it to survive for subsequent seasons.

But once you have succeeded in getting the plant to flower for a second time, the cyclamen will, if treated consistently, reward you faithfully every year. The elegantly shaped blooms may decrease in size and approach more closely to the original form *C. persicum,* a plant which itself, after having been cultivated and crossed for more than a century in order to create the 'most important pot plant in Europe', has long since been lost.

General: the cyclamen comes into your house in late autumn in full flower. If looked after properly, it will continue flowering until late winter.
Situation: cool, in winter 10–12°C (50–54°F), some shade. The ideal situation used to be between two north-facing windows. In the sun in a warm room the plants will soon wilt.
Watering: during the flowering season, the cyclamen requires a good deal of water, even in a cool place. The water must be soft (*p*H 5–6) and lukewarm.
The corms and the young buds should be kept dry. If the corms are planted deeply, it is better to pour the water into the saucer. Excess water should be removed after half an hour. It is advisable to water plants with many flowers and a mass of dense foliage two or three times a day.
Feeding: plant food can be given throughout the year; once a fortnight during the flowering period; once a week while the corm and foliage are growing.
Very important: dead flowers and yellowing leaves should immediately be cut close to the corm. Prevent seed-formation.
Keeping: when the leaves start to yellow, some weeks after flowering has finished, gradually withhold water until the corm is completely dormant, and put the pot on its side in a dark, cool place until late May. Then plunge the pot outdoors in a lightly shaded place until the corm begins to shoot new leaves again in late July. Repot in an acid, humusy compost, leaving half the corm above the surface. Keep slightly shaded and in a temperature of 8–10°C (46–50°F) until the buds appear, then increase it a little. Start feeding as the leaves develop, and bring indoors before night frosts start.
Propagation: this is only possible by growing from seed. Diseased plants rarely recover.

Cyclamen persicum, fam. *Primulaceae*

Cymbidium

The botanical name indicates the elongated lip of this orchid. Most of the *Cymbidium* species now for sale are terrestrial orchids, suitable for the unheated greenhouse. Since growers have succeeded in developing the cymbidium mini-hybrids, orchids can now also be grown in the living-room.

The large forms such as *C. giganteum* and *C. hookerianum*, whose unbridled vigour makes them outgrow their pots and which produce stems of over a metre (3 ft) in length, are not very popular among amateurs. The cymbidium 'Galiodin' in the photograph is one of the newest and finest miniature forms; these flower more readily than their ancestors and do not stop flowering if they are repotted.

General care: a very light and airy situation on a 'cold foot'; in other words, on no account on a windowsill with a central heating radiator underneath. The plant needs to have plenty of room and not have too many other plants around; this is best achieved by placing its bowl on an upside down flower pot, which in turn is put into a bowl filled with pebbles. Temperature in winter 7–12°C (45–54°F). In summer, from May onwards, cymbidium appreciates a half-shady position under trees in the garden. When the night temperature falls below 8°C (46°F) the plant must be brought indoors once more.

Flowering season: usually in spring, sometimes in autumn; the hybrids produce 6–8 or more magnificently coloured flowers to a stem. Also very attractive and long-lasting as cut flowers.

Watering and spraying: during the growing period in summer water freely, especially in hot, sunny weather. Use only water free from lime and chlorine (pH 4–4.5) at room temperature. After the foliage has died down in autumn, gradually reduce watering and in winter give only just enough water to prevent the pseudobulbs, where the water is stored, from shrivelling up.

Feeding: feed very little; it is very easy to give too much. During the growing period in summer a fortnightly dose of a highly diluted nutrient is quite adequate.

Repotting: only when the youngest shoots have reached the rim of the pot. For terrestrial orchids use medium size, ordinary clay pots. The compost should not be rich; it should consist of equal parts of fibrous loam, peat and

Cymbidium mini-hybrid 'Galiodin', fam. *Orchidaceae*

sand, or use pure peat fibre, mixed with sphagnum moss. There should be plenty of drainage material at the bottom of the pot.

Cyperus – Umbrella plant

This is a plant which occurs everywhere, feels at home everywhere and is popular everywhere: cyperus, a member of the *Cyperaceae* or sedge family. Its pedigree goes back to the papyrus of the ancient Egyptians. It is a welcome plant on the window-sill, but also grows willingly in the aquarium.

C. alternifolius, the umbrella plant, is one of about 600 species occurring all over the world as marsh and shore plants in tropical, subtropical and temperate regions, and particularly in the morasses of Madagascar. It is remarkable that a plant originating in tropical and subtropical regions adapts itself so well to the conditions we are able to provide in the living-room. It grows as well in a pot as in an aquarium, and in summer appreciates a semi-shaded spot in the garden as much as a place in a trough in the pond. It is also one of the standard plants used in soilless cultivation and as such easily grows into magnificent specimens up to 80 cm (32 in) in ▶

Cyperus alternifolius, Umbrella Plant, fam. *Cyperaceae*

height, occasionally flowering. This amusing brown inflorescence appears in the centre of the crown on slender stems, often rising far above the pointed, sword-shaped leaves, and is the reason why the cyperus is called umbrella plant in many countries. Occasionally a variegated cultivar, 'Variegatus', is available at the florists'; it resembles the green botanical form, but because of its striped foliage is even more attractive. Unfortunately newly developed leaves gradually turn green once more, especially if the plant is well cared for! *C. argenteostriatus* (syn. *C. diffusus*) is a particularly attractive dwarf umbrella, a native of Australia. It resembles its larger relation in all details, but does not grow beyond 30 cm (1 ft).

General care: the most important rule is that, in contrast to other houseplants, *C. alternifolius* must always have its roots in water; in other words, it is a plant that likes wet feet and its pot should therefore always be placed in a water-filled bowl or saucer. If it is put in an aquarium or in a trough on the edge of a pond, it is advisable to put some soil round the root-ball and cover this with pebbles. The water may reach to the top of the soil-ball, but the stems must remain dry. If you grow the plant in this way in the garden during the summer, it will be a simple matter to remove it to a frost-free, cool situation in autumn without damaging the soil-ball.

In winter the most desirable temperature is 10–12°C (50–54°F); a fairly well-lit situation. Do not forget to keep the saucer filled with water. However, the plant is not very sensitive to temperature, for it will tolerate as much as 25°C (78°F) throughout the year and can therefore overwinter in the warm living-room. Nevertheless it prefers to be kept in a fairly low temperature during the dormant season from October to February.

Flowering season: from spring until early summer.

Feeding: pot plants should be fed every week from April to August. If the plant is grown in an aquarium or pond the rules for feeding waterplants apply and ordinary fertilisers should on no account be used.

Repotting: because cyperus grows so rapidly it requires a great deal of nourishment. Normally it should be repotted once a year, but if a large number of new shoots are produced it may be necessary to do it twice a year. The most suitable potting compost consists of a mixture of clay or loam and leafmould, to which some rotted manure or blood-meal is added. If you do not want to go to all this trouble, remember that cyperus will grow almost equally well in the ordinary pre-packed potting compost which you can buy in any garden shop or store. If you want to grow the umbrella plant in a glass tank, you are advised to obtain the large, strong species *C. alternifolius*, and start by potting it in adequate soil in a clay or plastic pot. The pot should fit in the tank with room to spare. Next, you fill the tank with marble grit until the clay or plastic pot is hidden from view. The plant now appears to 'float' among the marble grit and the stems emerge from the marble as if no soil has been used at all. Finally the tank is filled with water to approximately halfway up the pot. When it is time to repot, the marble grit can be washed at the same time to remove the green deposit. This is easily done in water to which a little hydrochloric acid has been added, but do not forget to rinse thoroughly, for hydrochloric acid is not exactly beneficial to plants.

Propagation: the umbrella plant is very simple to increase; this can be done in four different ways. The easiest method is by division, which can be done at the time of repotting. Propagation from cuttings is another satisfactory method. Choose young shoot tips and remove these with about 5 cm (2 in) of stem. Shorten the leaves to half their length with a pair of scissors. This operation, often carried out when growing cuttings, serves to reduce the evaporating area to the minimum in order to minimise the risk of wilting. Push the little bundle deep into damp sand, so that only the tips of the leaves are visible. The cuttings will soon root, especially if you keep the sand at a temperature of 20°C (70°F). Do not pot the little plants until they have formed an adequate root system. And finally it is possible to raise cyperus from (imported) seed, but germination takes rather a long time.

Diseases: too warm a situation and excessively dry air in winter may lead to greenfly. Remove the pests and improve conditions. The tips of the plant turn brown if the soil is too firm or has become sour. The *p*H of the water can be 5·5–6·5.

Two other species worth growing are: *C. diffusus*, with slightly broader leaves and fairly low-growing. This species requires to be treated like other pot plants; in other words, no 'wet feet' and a warmer position than the species described above, since otherwise the foliage will turn yellow and growth will be restricted. *C. haspan* has a very graceful habit, but needs more care during the winter than the species above, as it requires a great deal of light and warmth.

Cyperus argenteostriatus, fam. *Cyperaceae*

Cyrtomium falcatum – Holly fern

This is an indoor fern whose native habitat is in Japan, China, India and Ceylon. It has now been classified among the small *Aspidiaceae* family. The cultivar *C. falcatum* 'Rochfordianum' is more beautiful than the species; its foliage is very pointedly pinnate. The leaves are of a fine glossy green.

This is an easy and long lasting plant. In the right position it can be a very decorative acquisition for the living-room. It will grow even in fairly dark spots.
General care: the same as for *Pteris*, that is, a warm and shady situation in summer. In winter it tolerates fairly low temperatures, 10–12°C (50–54°F). Water generously and during the growing season from April to August spray frequently to increase the atmospheric humidity. Resting period from October to February. Fertilisers must be given in a very weak solution. Removal of old leaves will encourage growth.
Propagation: by division or from spores sown in damp, warm peat. This is a very meticulous job.

◁ *Cyrtomium falcatum,* fam. *Aspidiaceae*

Cytisus – Broom

In the garden, broom is a well-known shrub. As a houseplant *C. × racemosus* is a freely flowering, undemanding representative of the non-hardy forms, but it likes to overwinter in a cool spot, at 4–8°C (40–46°F). If you cannot provide this it will usually not flower quite so profusely.

C. × racemosus has clear yellow flowers which resemble butterflies. One advantage of this plant is that it likes lime both in the soil and in the water, so that you do not have the problem of water-softening.
Cytisus reacts to acid soil and lime-free water by producing yellow foliage and poor growth.
Flowering period: March–April. In summer the plant likes to be outside in a well-lit, sunny spot.

Cytisus × racemosus, fam. *Leguminosae*

Dendrobium

The genus *Dendrobium*, with its many species, is definitely among the orchids most valued by amateurs. It includes both easy and demanding plants. One will be satisfied with a warm spot in the living-room, while another demands a hothouse with a high degree of humidity.

The origin of these magnificently flowering plants explains the variation in requirements. They occur from northern Japan to the southern part of New Guinea. Some of the finest species are natives of humid, hot regions in the Himalayas and in other parts of eastern Asia. *D. nobile* is one of the best loved species, for indoor cultivation as well. It does, however, require a higher temperature during the growing period, 24°C (75°F) and very good light, preferably near a south-facing window. To induce the plant to flower it must be kept in a cool spot, 8–10°C (46–50°F) during the resting period. Flowering will then take place between March and June. *D. pierardii* has entirely different requirements: it is a tropical epiphyte and does not have a dormant period. It needs a great deal of light and warmth throughout the year and a high degree of atmospheric humidity while growing. Its bulbs point downwards and a slanting orchid basket is indispensable for this plant.
D. phalaenopsis is much appreciated for its large and colourful flowers which last well when cut; as it originates in southern tropical regions (Queensland, southern New Guinea), it definitely requires hothouse conditions. The illustration gives an idea of its beauty.
Nevertheless amateurs, and particularly beginners, would do better to confine themselves to less demanding and hardier strains.
D. chrysanthum, cultivation similar to that of *D. nobile*; golden yellow flowers in August–September. *D. densiflorum*; just possible for indoor cultivation. It flowers in March–April with a pendant raceme. *D. wardianum*, treatment as for *D. nobile*, single flowers, 8–10 cm (3–4 in) in size, in January–March.
Important: all dendrobium species like a small pot and prefer to be re-potted twice a year. They are best grown in sphagnum moss without further additives except a very occasional dose of liquid fertiliser, preferably when new growths are starting. Apart from the correct temperature and moisture, light and fresh air are of great importance for growth.

Dendrobium phalaenopsis, fam. Orchidaceae

Didymochlaena truncatula

Do not be put off by the difficult, almost unpronounceable name of this plant. Freely translated, the combination of Greek and Latin words mean something like 'shortened double cloak'. It is a magnificent fern with few requirements and will be a valuable acquisition in your houseplant collection.

Modern growers usually place the young plants in a moderately warm greenhouse. This means that they will do well in the living-room also, but in winter the plant needs a cooler environment, with a temperature of 14–16°C (58–60°F).

Didymochlaena occurs in tropical areas all over the world. It easily adapts its appearance to its environment and is known by different names in various countries.

The species marketed here has long, somewhat brownish green, double-pinnate leaves with a leathery appearance, which grow close together. *D. truncatula* is the only representative of its genus. It formerly belonged to the larger family of *Polypodiaceae*, but it has now been classified among the *Aspidiaceae*.

General care: a shady position throughout the year, warm and damp in summer. It is advisable to spray frequently. In early March the plant begins to grow once more; from then onwards until the beginning of the resting period it should be watered freely with soft water (*p*H 5).

From April until August new leaves will unfold, which like to be sprayed twice a day with tepid water (syringe). Feed sparingly; give only an occasional dose of highly diluted liquid fertiliser. The resting period is from October to February; during this time a temperature of between 14 and 16°C (57 and 60°F) is desirable. The plant should be kept drier; do not feed or spray.

Repotting: when the soil-ball is fully taken up by the root system, repot in a humusy mixture, consisting for instance of loam and leafmould. Unsightly old leaves must be cut off.

Propagation: *D. truncatula* is one of the ferns which can be grown in the living-room from spores. The spores are as fine as specks of dust and are found in small clusters on the underside of old, fully grown leaves. When they are ripe they drop and germinate like seeds in the damp soil. The simplest way to catch them is to place the pot in a bowl of damp peat fibre into which they will fall. Keep warm, 24–26°C (75–79°F). Spores can also be obtained from a well-stocked seedsman.

Didymochlaena truncatula, fam. *Aspidiaceae*

Dieffenbachia

In general dieffenbachia may be said to head the list of popular foliage plants. For more than a century about twenty species have been in cultivation from which new, constantly improved strains have been produced. Occasionally this graceful plant, a native of America, creates problems.

The fact is that, even more than the rubber plant, it has the disagreeable habit of losing its lower leaves. Any book on houseplants will tell you that the thick green trunk should be covered in a regular pattern of leaves. If the plant is constantly kept in the living-room, lacking the climate of a heated greenhouse, its trunk will grow bare in the course of time. Nevertheless in large apartments we often see specimens with several stems where the bare trunks are not obvious because of the dense crown of leaves above. Such older plants, which may grow up to two metres (6 ft) in height, occasionally even produce flowers. The inflorescence is arum lily-shaped, as in *Philodendron*. It appears at the top of the crown in ▶

A very large specimen of *Dieffenbachia* ▷ *bowmannii*, fam. *Araceae*

Dieffenbachia 'Exotica' ('Arvida'), fam. *Araceae*

Dieffenbachia maculata. Young plant. Fam. *Araceae*

spring. *D. seguine* is the species most likely to produce this exceptional phenomenon. I must confess that I have never seen a flowering dieffenbachia in private ownership, only in botanical gardens.

Dieffenbachia is a most decorative plant which now occupies an important place in many interiors. It is often seen in plant combinations, but it is best potted separately.

To maintain the plant's natural appearance you will need a heated greenhouse; in warmth and a high degree of humidity the foliage will also develop more favourably.

General care: a slightly shady situation, not in direct sunlight. From March onwards, when the plant starts into growth, and during the summer, provide a warm and humid atmosphere. Keep the soil constantly moist by watering generously with tepid softened water (pH 4·5–5). Drying out of the compost makes the edges of the leaves turn brown. If the plant is kept in the living-room it must be sprayed frequently. From March until August feed every week; the fertiliser must be lime-free. Diluted liquid cow manure is also suitable. In winter the temperature must on no account fall below 15–18°C (58–65°F) and the plant must not be in a draught. The resting period lasts from September to February; during this time do not feed and water sparingly. In this way you will prevent the formation of new foliage which in this dark season will not develop properly. However, if the plant is kept in the living-room, some degree of atmospheric humidity must be maintained by occasional spraying. It is also advisable to rinse the leaves regularly with lukewarm water, otherwise the pores will become clogged with dust.

Repotting: both young and old plants must be repotted every year in larger pots.

The compost must be rich, coarse and crumbly. You can mix it yourself by combining half-rotted leaves with garden compost and peat fragments. A few pieces of charcoal may be added.

The plant will grow in pre-packed potting compost as well.

Propagation and rejuvenation: a dieffenbachia which has lost much of its lower foliage may be rejuvenated by cutting back the stem to about 10 cm (4 in) above the soil. If the plant is healthy the stump will put forth shoots, but some patience is required. The crown of leaves which has been cut off can be used for leaf cuttings; a stem with 2–4 leaves will suffice. The plant can also be increased from approximately 8-cm ($3\frac{1}{4}$-in) long sections of stem with a bud visible in a former leaf axil. These sections are placed horizontally in a pot and covered thinly with soil.

All these methods of propagation will be successful only in adequate bottom heat, 24–26°C (75–79°F) and constant humidity of the soil. It can be done quite easily in an indoor propagator with heating flex.

Toxicity: very few books on plants mention the fact that the juice of leaves and stems is fairly poisonous, for it contains strychnine which was formerly used to kill rats and mice. Do not worry if a little juice touches your hands, for the poison is not absorbed by the skin. But beware if you have a little wound on your hand, or even worse, if some of the juice accidentally gets into your mouth.

Species: *D. picta* and *D. seguine* with their numerous cultivars are most easily obtainable. A species seen with increasing frequency is *D.* × *bausei*, which has variegated foliage, yellowish green, dark green and white. Strains of *D. seguine* are occasionally confused with forms of *D. picta*, for instance the beautiful cultivars 'Janet Weidner', 'Mary Weidner' and 'Rudolph Roehrs'.

The largest species, whose leaves are dark green flecked with pale green and up to 75 cm (30 in) long, is *D. bowmannii*. The weakest grower is the magnificent *D. picta* 'Jenmanii', with white bands between the side veins.

Dipladenia

Dipladenia, like the oleander (*Nerium*), belongs to the family of *Apocynaceae*. It has twining stems, like a liana. In recent years this plant has been fairly successful, but we should warn you that, because of its great need of moist air, this climbing plant sometimes presents problems.

For about half a century dipladenia was forgotten, but lately it has once more aroused interest. As a young plant its growth is bushy; the stems start to twine only at a more advanced age. The foliage is of a fine glossy green.

The best known species are: *D. boliviensis*, which bears white flowers with a yellowish throat and *D. sanderi*, with slightly larger, pink flowers, also with a soft yellow throat. At present the strains developed from these two are in greater demand than the species. These cultivars are more beautiful, flower more profusely and have more distinctive colours. To mention but a few: 'Amoena', clear pink with a darker centre and a yellow throat; × 'Rosacea', soft pink with a darker margin and yellow throat. × 'Rubinia', deep pink with a diameter of 8 cm (3¼ in). The latest hybrid of *Dipladenia sanderi* is 'Rosea'; the flowers are larger than those of the other cultivars and of a wonderful salmon-pink shade. The throat as usual is yellow; all dipladenia flowers have this in common.

As already mentioned, this is a climbing plant. Smooth stems with glossy dark green leaves give it an attractive appearance, heightened by the inflorescence. If properly looked after, dipladenia will reward you from May to November with its uninterrupted sequence of magnificent flowers which often last two weeks or more.

General care: the original habitat of dipladenia is in the tropical rain forests of South America. In order to simulate this environment we give the plant a light but half-shady position, without direct sunlight.

From March onwards, when the plant starts into growth once more, it requires a higher degree of atmospheric humidity to stimulate bud formation. From May until November it is satisfied with normal room temperature, but the foliage should be syringed every day. In January and February it has a short resting period in a temperature of 12–15°C (54–58°F). During this period you should water very sparingly. During the growing and flowering season water normally, but never leave water in the saucer. From April to August feed every week, but never feed later than August.

Repotting: in early spring in a slightly larger pot with good drainage. Excessively dry air and too much light lead to leaf curl.

Dipladenia sanderi 'Rosea', fam. *Apocynaceae*

Dizygotheca — Finger aralia

People who mention a finger aralia are in fact nearly always talking about *Dizygotheca elegantissima*, for other species of the genus are little known here. This species is certainly the most graceful, but in other countries *D. veitchii*, and especially the cultivar 'Gracillima' are also very popular.

The appearance of these two plants is extremely different. *D. veitchii* has graceful, much smaller leaves; the main vein is white. Both species require the same treatment, so if you have been successful with *elegantissima*, you will succeed with 'Gracillima' as well. With good care these plants will give you pleasure for a long time. The photograph below left provides a good example of what may be achieved with the slow growing plant in twelve years. As you see, this tropical plant, a native of an island in the Pacific, will in the right conditions develop a vigour of which it might not have been considered capable.

Its tropical origin should, however, always be taken into account, something which is not always possible in the living-room. It will feel most at home in company with other plants, for instance in a trough; this will keep the air more humid.

General care: plenty of light without direct sunlight. A high temperature and a high degree of atmospheric humidity. Even in the dormant period from October to February the temperature should never fall below 15°C (58°F). An even more important requirement is that the soil temperature should be constant at about 18–20°C (65–70°F), for many a healthy young finger plant has become the victim of cold and wet feet. Older plants, which have stood in the living-room for some time and have acclimatised themselves, are less sensitive in this respect.

Watering: during the growing period water moderately with slightly softened water (*p*H 5–6). Younger plants in particular should frequently be sprayed. During the resting period water sparingly, but take care that the soil-ball does not dry out. If the surroundings are on the cool side, spray less frequently. During this dormant period it is advisable to pinch out new shoots.

Dizygotheca elegantissima, fam. *Araliaceae*

Dizygotheca veitchii 'Gracillima', fam. *Araliaceae*

Feeding: feed moderately during the growing season. The plant is slow-growing by nature and therefore does not require a great deal of nourishment. Do not feed at all in winter.

Repotting: in early spring, before the plant puts forth new shoots. Young specimens should be given larger pots every year, older plants every two or three years. Ordinary prepacked potting compost is best.

Propagation: at one time this was done only from cuttings in a heated greenhouse. Nowadays dizygotheca can also be raised from seed. A good seed catalogue will include imported seed and give instructions for cultivation.

In the first year the plant will grow to 20–30 cm (8–12 in). As in *Philodendron*, plants grown from seed are recognised by the small leaves at the base of the stem.

Diseases: direct sunlight and dry air often lead to an attack by red spider mite and usually result also in an infection by scale insect. Rather than spraying with various poisonous insecticides, you should try to change the conditions, which will encourage these unpleasant pests to disappear.

Dracaena

In a magazine from 1901 I read that a plant lover was asked which foliage plant she regarded as most suitable for the living-room. The reply was: 'I would unhesitatingly choose the dracaena. It is an undemanding plant, graceful in appearance and never subject to pests. My *D. fragrans* is now seventeen years old. . . .'

Origin: the genus *Dracaena* includes about 20 species, growing in the tropics and sub-tropics, from the Canary Islands across Africa and Asia to the Australian archipelago.

Appearance: they are tree-like plants which in their natural habitat grow to many metres (feet) in height; they do not produce offshoots. The leaves are lancet-shaped, erect growing or tending to curve, depending on the species, green or variegated; hybrids with variegated foliage have been cultivated as well. There are also species with 'normal' foliage on densely branched shrubs. Dracaena is often confused with the closely related *Cordyline*, but the latter's knobbly white roots are easily distinguished from the orange-yellow dracaena roots.

Inflorescence: they vary considerably in their readiness to flower; as a rule flowering occurs only at an advanced age. The flowers are produced in loose umbels or racemes; these are greenish white or creamy white, sometimes fragrant. The flower stem appears close to the growing centre; this is why the plant does not die off after flowering. Flowering season: spring and summer. See also the sketch of a flowering dracaena on p. 62.

Species: *D. deremensis* is a native of Derema in the Usambara region of East Africa. It is now rarely found in cultivation, but its variegated strains 'Bausei' and 'Warneckei' (illustration p. 16) are much in demand. A very popular plant is *D. draco*, whose origin lies in the Canary Islands. This plant has become fashionable since tourists took seeds home by way of souvenirs. I have raised the plants from seed myself; they have by now grown into tub-plants. *D. draco* is the hardiest species, insensitive to heat and dry air; damage will only occur if the temperature in winter drops below 10°C (50°F). *D. fragrans* is a trunk-forming plant for the heated greenhouse, but if well looked after will stand living indoors. The broad to very broad leaves, 50–70 cm (20–28 in) long, curve limply downwards in a wide arc. Deliciously scented flowers when older. The plant considered to be the best of all indoor dracaena forms is its strain 'Rothiana', a modest plant which will last for dozens of years. The leaves are 70–80 cm (28–32 in) long, about 7 cm (3 in) wide, thick and leathery, dark green in colour with a white margin. Produces fine flowers at irregular intervals. Other strains for the heated greenhouse or large tropical plant windows: 'Lindenii', 'Massangeana' and 'Victoriae'; unfortunately the latter rapidly turns green.

A native of the Congo is *D. godseffiana*, freely branched; the leaves are pointed oval in shape, streaked with white. Yellowish white fragrant flowers in spring, followed by beautiful red berries. There are two strains: 'Florida Beauty', even more strikingly marked, and 'Kelleri' with marbled

Dracaena fragrans 'Victoriae', fam. *Agavaceae*

Two examples of the now so popular Ti-plant, *Dracaena*. These plants have started into growth on dormant eyes of trunk sections. They readily develop into magnificent decorative plants. Left: *Dracaena fragrans* 'Massangeana' with predominantly yellow-streaked foliage. Right: *Dracaena fragrans* 'Lindenii' with central yellow streaks. Family *Agavaceae*

foliage. The strikingly beautiful *D. goldieana* originates from Guinea in West Africa; its slender stem bears stalked leaves with greenish white cross-banding, up to 25 cm (10 in) in length. It is not easy to propagate and is therefore quite rare. Suitable only for the heated greenhouse. *D. hookeriana*, from the Cape, is a trunk-forming plant, long-lasting and tough. Narrow green leaves, 60–80 cm (24–32 in) long; variegated strains are 'Latifolia', and 'Variegata'. *D. marginata*, a native of Madagascar, has distinctive narrow leaves, 30–40 cm (12–16 in) in length and edged with reddish brown; it grows to 2 m (6 ft) (illustration p. 10). Among the large decorative dracaena forms which like to form several trunks or branch freely, is *D. reflexa* from the island of Mauritius. Its variegated strain 'Song of India' (illustration p. 2, facing the title page) is quite famous. The smallest and most graceful of all dracaena species, a plant from the Congo to be grown in a heated greenhouse, is called *D. sanderiana*. It has fine white-variegated foliage and is also very suitable for a tropical plant window. Sections of trunk from an apparently dead dracaena are called 'Ti-plants', although this is not the original meaning of the word. Placed in water or in soil they will come to life once more, put out strong shoots and even occasionally flower. They are often for sale as a curiosity.

Care: when you buy a dracaena, try to establish whether the plant has been properly hardened off, that is to say, whether it has been acclimatised to the lack of moisture in the air with which it will probably have to put up in your living-room. If the plant has come straight from the very damp atmosphere of the greenhouse, it will begin by drooping a great deal of its foliage, which you will not like. In general the plant is best bought from a florist who has a reasonably large 'show case', where he can keep his stock of plants for a few months if necessary. Moreover a purchase in spring is much to be preferred to buying the plant in the autumn, because increasing daylight and the gradual reduction in indoor heating can only be of benefit. In autumn the situation would be the reverse. Particularly suitable for room culture are: *D. deremensis* with its cultivars, *D. draco*, *D. hookeriana*, *D. marginata* (which looks so fragile because of its narrow leaves, but is probably the toughest of them all!) and *D. reflexa*. All these plants in the first place require warmth and they should therefore live in a warm room summer and winter. *D. draco* and *D. hookeriana* can if necessary over-winter at 12–14°C (54–57°F), provided watering is restricted. The others need a minimum winter temperature of 15–18°C (58–65°F), which may well occur at night when it is freezing hard outside. If this should be the case, keep the plants away from the window and do not turn the central heating too low. It is better not to allow bright sunlight to strike the plant. The correct advice is: at most morning and evening sun; otherwise a place not more than 3 m (9 ft) from the window. However, as the plants are so strong, you will not kill them if you do not adhere strictly to this rule, but you will run the risk that the white or yellow of the striped forms will gradually turn green.

A high degree of atmospheric humidity is not essential for hardened plants. Of course it would be a good thing to moisten the foliage every day, or if you like several times a day, with a plant spray containing lime-free water, but the plant will survive even without this.

Where the plants remain in the room year in year out they do get rather dusty, particularly the narrow-leaved species which are less easy to clean. For this reason it is advisable to put them outside in summer in a mild shower. The pot may be protected from flooding by covering it with tinfoil. After their shower the plants must be taken indoors again; they are certainly not balcony plants. They really feel happy only if during the day the temperature rises to 22–28°C (72–83°F) (provided there is plenty of light of course) and this rarely happens in this country.

Watering: normally in summer, more sparingly during the dormant season. In general they do not tolerate surplus water left in the saucer; this may even cause the fleshy roots to rot. Drying out of the soil-ball is fatal for the variegated greenhouse species in particular.

Feeding: plants which are growing well must be given liquid feed at least once every two weeks from March to August – this should be poured only on to thoroughly damp soil. Older plants are better fed every week, but it is then particularly important to reduce feeding in time in preparation for the resting period.

Repotting: young plants every spring, preferably in May; older plants at longer intervals into only slightly larger pots or tubs. The soil mixture must be rich and humusy, but in view of the habit of the root system, porous rather than heavy. Normal pre-packed potting compost will fulfil these requirements.

Propagation: only *D. draco* from the Canary Islands can be grown from seed. Other dracaena seed on the market usually comes from *Cordyline*. A seedsman who carries unusual seed will stock *D. draco* seed. Bottom heat of 24–26°C (75–79°F) will encourage germination, which will take four to six weeks. The young plants are pricked out into small pots with J.I. potting compost No. 1, preferably in May–June. By autumn the little plants should be about 10 cm (4 in) high and should spend the winter in a warm place. Other species and strains are increased from cuttings or stem sections.

Diseases: incorrect treatment encourages pests (red spider mite, scale insect, mealy bug and thrips). The ends of the leaves will turn brown as a result of drying out, draught or overfeeding.

Dracaena godseffiana, fam. Agavaceae

Duchesnea fragarioides, Indian Strawberry, fam. *Rosaceae*

Duchesnea (Fragaria) – Indian strawberry

When people see the typical strawberry foliage and the red fruits of *D. fragarioides* (*Fragaria indica*) growing as a hanging or climbing plant in the living-room or on the balcony, or somewhere in the garden, they always want to taste the fruit, but they will be disappointed: the fruit is tasteless.

Origin: they are not true 'strawberries' in spite of the resemblance. The plant came from India to western Europe at the beginning of the last century. It occurs wild elsewhere in the world as well.
Appearance: it resembles our flowering, trailing garden strawberry, whose runners cover the soil and which might also serve as a hanging plant. The only difference is that the little flowers appearing from June to October are not white, but vivid yellow.
Care: very simple. Stand or hang the plant in a good light, in summer either in or out of doors; in winter at 10–12°C (50–54°F) rather than in a warm living-room, although if necessary it will grow there as well.
Watering: freely during the growing period; in winter water sparingly, depending on the temperature.
Feeding: feed normally in summer.
Repotting: see Propagation.
Propagation: older plants become unsightly and you would therefore do better to grow new plants every spring. There will be plenty of runners which will root. Raising from seed, if available, is also possible.
Diseases: none. Red discolouration of the foliage is the result of lack of water and nourishment.

Flowers of *Echeveria setosa,* fam. *Crassulaceae*

Echeveria

No doubt every plant and garden lover will know this attractive, rosette-forming plant. If you think that the rather grand-sounding name is derived from the Latin, you are mistaken; the plant was called after Atanasio Echeverria, a 19th century Mexican botanist.

Origin: more than 150 species occur in the proportionately small area from California in the north, via Mexico to Peru in the south.
Appearance: perennials, usually terrestrial, growing in rosettes. The wonderfully coloured fleshy leaves are often interestingly pilose.
Inflorescence: the flowers appear in umbels from the centre of the plant; as a rule they are orange-yellow, but may also be rose-red or scarlet. The flowering season of the indoor forms varies with the species, but generally occurs between winter and spring.
Species: for simple room cultivation the almost or completely stemless species are most suitable. These include: *E. agavoides*, whose pointed blue-grey leaves form beautiful rosettes; *E. carnicolor*, with a less compact habit; it develops so many runners that it may be used as a hanging plant as well; *E. setosa*, with long white hairs, and vivid yellow-tipped red flowers at the end of long stalks. Stem-forming, multi-branched shrubs or semi-shrubs, which grow to 70 cm (28 in) are more suitable to be grown in a plant window or at floor level. The following are three species of this kind: *E. coccinea*, white-

haired, summer-flowering; a native of Mexico; *E. gibbiflora*, large, fat rosettes on little trunks 30–50 cm (12–20 in) in height. In autumn it produces vivid red, white-frosted flowers on tall stems. There are a number of magnificent strains, for instance 'Carunculata'. *E. nodulosa* is the lowest of the rosette-forming shrubs, with trunks of only 20 cm (8 in). The leaves are edged in red and red-marked on the underside. From March onwards red flowers with yellow tips.
Care: plenty of light both in summer and in winter. In summer a sheltered place out of doors. In very good light a winter temperature of 6–10°C (43–50°F). Winter-flowering forms require more warmth.
Watering: sparingly in summer, even less in winter.
Feeding: a feed solution once a month from March to July.
Repotting: young plants every year, subsequently at longer intervals. Soil mixture sandy and rich in humus.
Propagation: species can be increased in three ways: 1. by removing side rosettes; 2. by leaf cuttings and even from sections of flower stems; 3. from seed; a good seed catalogue will certainly include it. The hybrids do not produce viable seed. These can therefore be propagated only from the rosettes or from leaf cuttings. However, the rosettes should not be removed from the parent plant too soon.
Diseases: too much water will cause root rot and mildew.

A number of well-known, strong echeveria species. In the front pan, left to right: *Echeveria elegans, E. agavoides, E. setosa*. Behind, left: *E. agavoides* 'Cristata'; centre: *E. nodulosa*; right: *E. secunda* 'Pumila'; at the back: *E. harmsii*

Echinocactus

Since the latest reclassification, only 7 species remain, of which only three are in cultivation. In Germany the most important species, *E. grusonii*, is sometimes called Mother-in-law's-Chair, an indication of how mothers-in-law are regarded in that country. In England it is called Golden Ball.

Origin: the species which still belong to the genus are natives of Mexico. They grow in stony ground, as well as on rock-strewn slopes at higher altitudes where the sun shines fiercely in the daytime while at night the temperature drops sharply. This explains why they are so hardy.
Appearance: the plants of the *Echinocactus* genus which grow rapidly into large globes or cylinders are found mainly in show collections, but *E. grusonii* may also be seen as eye-catching specimens in amateur collections. The yellow or pink flowers growing from the woolly crown usually keep us waiting for a long time. After flowering, scaly fruits containing black seed develop.
Species: *Echinocactus grusonii* forms a pale green globe with ridges covered in golden-yellow spines. The plant may reach a diameter of 80 cm (32 in). In the U.S.A. it is called Golden Barrel and is highly valued for its crested forms (comb or cristate formation). *E. horizonthalonius* is flatter in shape, pale green, with grey-white bloom and magnificent yellow spines. Can be grown only in a greenhouse; diameter 25–30 cm (10–12 in) *E. ingens* (formerly *E. grandis*) is an enormous plant with a dull green to blue-green, cylindrical body growing up to 2 m (6 ft) in height. Small young plants may be suitable for room cultivation.
Care: during the growing period in summer echinocactus requires plenty of warmth and sunshine, but in winter it must have a cool position, 8–10°C (46–50°F). *E. grusonii* should not be placed in full sunlight at the beginning of the growing season in spring.
Watering: during the growing period, up to mid-August, maintain an even degree of moisture, but beware of surplus water at the bottom of the pot. After mid-August gradually decrease the water supply; keep entirely dry in winter, when moisture may be fatal.
Feeding: generously during the growing period, using special cactus fertilisers. Decrease after mid-August. Gradually start feeding again in March.
Repotting: do not repot often, as the roots are very fragile. Use well-draining soil: special cactus compost or ordinary potting compost mixed with plenty of rubble. In summer frequently prick the soil surface with a stick in order to aerate it.
Propagation: very difficult for an amateur to grow from seed, as the roots easily break when the seedlings are pricked out.
Diseases: too little watering in summer is followed by red spider mite and mealy bug; too much moisture in winter will make the plant rot and die.

Echinocactus grusonii, fam. Cactaceae

Echinocereus

These plants have been split off from the collective genus *Cereus* and now form a separate group with about 60 species. Because of their small size and proportionately large flowers, *Echinocereus* species are very popular among cactus lovers. The plants are easy to care for.

Echinocereus pectinatus, fam. *Cactaceae*

Origin: their native habitat is in the south-western part of the U.S.A. and in central and northern Mexico. Most species are inhabitants of the desert, a few live among clumps of grass in the prairie, others among shrubs on mountain slopes.

Appearance: rarely taller than 20 cm (8 in). Small, column-shaped, branched plants, of which the green forms of the soft, densely spined or hairy species require different treatment from the others.

Species: among the species with few or no spines are: *Echinocereus knippelanius* and *Echinocereus pulchellus*. *Echinocereus delaetii* has fine spines or hair; the shoots, up to 30 cm (12 in) in length, partly recumbent, partly erect growing, are covered in long, bristly hairs, like those on an old man's head. The pink flowers rarely appear. *E. pectinatus* as a rule does not branch; it grows up to 15 cm (6 in) in height; the body is covered almost entirely in short, yellow-red to dark brown spines. Reddish purple flowers up to 8 cm (3 in) across.

Care: green species should be taken out into the garden in summer. Thorny and hairy species must be kept fairly moist in a warm, sunny, airy position. In winter cool and dry, but in a well-lit and sunny spot. Compost — see origin — rich in minerals but free of humus, loamy and sandy; both this and water should be slightly alkaline, pH 7–7.5.

Echinopsis

Don't let anyone tell you: a cactus is a cactus. The two genera represented on this page alone show enormous differences. If you want to take pleasure in your collection, you should pay attention also to the differences between individual plants — clear although not always striking.

Echinopsis eyriesii, fam. *Cactaceae*

Origin: the plants of the genus *Echinopsis* originate in South America, where they occur in the Argentine, Bolivia, Brazil, Paraguay and Uruguay.

Appearance: small, globe-shaped cacti; they have been cultivated by amateurs and collectors for over 100 years; today one finds practically only the hybrids. Magnificent flowers, long funnel-shaped, growing to one side of the crown, wide open and fragrant, white, salmon pink, yellow and scarlet. Some species flower at night and for that reason are sometimes called 'Queen of the Night'.

Species: *E. calachlora*, has grass green globes with short spines; flowers to 16 cm (6¼ in) in length and up to 10 cm (4 in) across, white, saucer-shaped; considered to be the finest. *Echinopsis eyriesii*, small globes with short, dark brown spines; flowers white, 20 cm (8 in) in length, to 12 cm (5 in) in diameter; night-flowering, opening late in the afternoon. Flowers of the hybrids are usually salmon pink. *E. kermesina*, to 15 cm (6 in) in diameter, multi-ridged, with variegated spines; carmine-red flowers 15–18 cm (6–7 in) long and 8–9 cm (3¼–3¾) across. Very fine.

Care: simple, provided the compost is right. It must be porous, rich and humusy. Feed therefore generously in summer; in winter keep cool and completely dry.

Epidendrum

The name is derived from the Greek and literally means 'one living on trees': in other words, an epiphytic orchid. The genus is considered to contain the largest number of species among orchids; it was classified by Linnaeus as early as 1737. Many species have been cultivated for a long time and new strains have enlarged the selection.

Origin: it is estimated that there are 500–600 species of epidendrum; in fact, according to some writers, more than 750. Their original habitat extends over all tropical regions of America, from Mexico and Florida to Bolivia and Paraguay, and even into Brazil.

Appearance: the species differ greatly in type, but they are all epiphytes. They have single- or multi-leaved pseudobulbs resembling those of *Cattleya*, *Coelogyne* and even *Dendrobium*. Some also have stem foliage. The inflorescence also differs in various species: in some the flowers grow singly, in others in short or long pendulous racemes or panicles; erect growing racemes also occur. The flowering season is from February to November, depending on the species. They are not exactly plants for beginners, but if selected with care and properly looked after, they can be grown in the living-room.

Species: *E. ciliare* occurs from Mexico to Central America; it flowers from autumn to winter, being greenish yellow and white; also suitable for smaller collections. *E. fragrans*, a native of the West Indies, is among the earliest-known epiphytes in Europe. Its cream and white flowers appear from late winter to May. *E. vitellinum* originates in Mexico; it is a particularly beautiful, readily

flowering orange-red orchid for the unheated greenhouse; flowering season from October to December. Various hybrids exist, for instance ×*Epidrobium*, a cross between *Dendrobium* and *Epidendrum* and also the 'Rainbow' hybrids.

Cultivation: very simple. The plants will grow well in moderate heat (orchid house). They will overwinter even where the air lacks moisture, provided they are kept cool and in plenty of light. The soil mixture must be porous. Water at pH 4–4.5. The new hybrids are preferable to many of the older, less strikingly flowered species. For the rest follow the general instructions for orchids.

Epidendrum 'Rainbow', fam. *Orchidaceae*

Epiphyllum

In the days when it was still called *Phyllocactus* everyone knew it as one of the best loved leaf cacti. Since it has been renamed *Epiphyllum* people have become confused as a result of frequent changes in names, and it is often mixed up with the Christmas cactus which now bears the botanical name of *Schlumbergera* (formerly *Zygocactus*).

Origin: the genus *Epiphyllum* embraces about 20 species, occurring in an area stretching from Mexico via Central America as far as the Argentine.

Appearance: like bromeliads and orchids, these species grow as epiphytic shrubs in the tropical rain forest. The sections, which at the base are round like stems, branch in elongated leaf-shapes, notched along the edges in various ways. The flowers appear only on sections at least two years old. Spindly sections, the result of lack of light in winter, do not bear flowers and should be removed in spring together with any older sections which have grown unsightly.

Species: the botanical species are no longer much in demand; they have been replaced by a large number of hybrids, easier to grow and with finer flowers. The main distinction is between night-flowering plants and fragrant day-flowering forms. The colours of the flowers range from salmon-pink and pink to scarlet and carmine-red. There are bi-coloured strains as well; in the United States these are called 'orchid cacti'. They are among the most beautiful and at the same time least demanding of houseplants.

Care: after overwintering at 8–10°C (46–50°F), epiphyllum requires increasing warmth and good light from April onwards. In late May it is best put out of doors in a warm and sheltered position out of direct sunlight. If the plant has to remain in the living-room plenty of fresh air is essential. When the nights grow cooler in the autumn, the plant must be brought indoors to start a resting period lasting from November to February in a cool position as mentioned. A short rest after each flowering period can also be recommended.

Humidity: only necessary during the growing and flowering seasons.

Watering: freely in summer, sparingly in the dormant season. In the transitional periods increase/decrease the water supply gradually. Tepid water, *p*H 4·5–5.

Feeding: when the plant is growing well feed generously with special cactus fertiliser.

Repotting: this is best done a few weeks after the spring flowering, but should not be too often, as the next flowering period will be delayed. Use a mixture consisting of loam, sphagnum moss and peat fibre.

Propagation: cuttings which have been left to dry out for a few days will root easily. Raising from seed is also possible, but flowering will not occur for several years.

Diseases: too dry air in summer leads to red spider mite and mealy bug. Too much moisture in winter may cause root rot.

Important: very suitable for growing in soilless cultivation, the plants will last for many years.

Epiphyllum, hybrid, fam. *Cactaceae*

Episcia

This plant belongs to the family of *Gesneriaceae* and, like many of its relatives, occurs mainly in regions of Central and South America where there are tropical rain forests. In this warm and humid atmosphere, shaded by trees, they feel most at home. They are therefore typical plants for a warm greenhouse.

The best known forms are *E. cupreata* from Columbia, with its many multicoloured strains, and the very similar, but slightly smaller species *E. reptans* from Brazil, sometimes marketed as *E. fulgida*. Flowering season from June to September. Both have red flowers. *E. dianthiflora* is a native of Mexico, it has oval green leaves and white flowers, lightly flecked purple round the centre and growing on long stems from the leaf axil.
They are all wonderful ground-covering plants, but may also be planted in bowls or hanging pots. Cultivation as for *Fittonia* (p. 215).

Episcia reptans, fam. *Gesneriaceae*

Erica – Heath

Considering the part played in our own landscape by the heather, it is difficult to believe that of the more than 500 species of the genus *Erica* only 16 occur in Europe. Almost without exception all others are natives of Africa; a few of these are known here as houseplants.

Species: it is not surprising that at one time dozens, indeed hundreds, of the African erica species belonging to this extensive genus were cultivated in botanical gardens and private collections. Amateurs now mainly confine themselves to *E. gracilis*, which produces rose-red flowers from September to December. The white-flowering *E. hyemalis* and its pink, salmon-pink and whitish pink hybrids follow in February and March. The *E. × willmorei* hybrids are magnificent spring-flowering forms. In summer there is *E. ventricosa*, in shades of lilac.

Care: plenty of light all the year round; in summer a sunny situation, preferably outside. Overwinter at 6–8°C (43–46°F). Water only with completely softened water (pH 4–4.5). Spray generously in summer; keep drier in winter. Feed from May to July.

Erica × willmorei, hybrid, Heath, fam. *Ericaceae*

Espostoa

In the United States this genus is known as Peruvian Old Man. This indicates that it is not a small cactus; in its native habitat on dry hills or in rocky soil it grows to a white-haired giant up to a few metres (feet) in height.

The chief species, E. lanata, slow-growing in cultivation, reaches a height of about 1 metre (3 ft). The long white hair gives this cactus a special kind of beauty. It is unlikely that the amateur will ever see the pale pink flowers which appear at night on the side of the column. They are only produced at an advanced age, as are the vigorous, twig-like offshoots which give the plant the appearance of a tree.
Care: a lot of heat and a sunny situation; it must be kept in a closed room in summer as well. Rich soil; water freely during the growing period. Otherwise treat as Cephalocereus.

Espostoa lanata, fam. Cactaceae

Euphorbia – Spurge

House plants of the Spurge family have undergone great changes. True, the Poinsettia is maintaining its place as a foliage plant, but increasing numbers of other interesting succulent Euphorbia species are now appearing on the market.

E. pulcherrima
Poinsettia

Origin: in 1834 the first specimens came to Europe from Mexico under their former name Poinsettia pulcherrima. At first they aroused little interest. The first great success was achieved in 1918 by the grower Paul Ecke with the variety 'Oakleaf' and in subsequent years world-renowned strains such as 'Eckespoint Pink', 'Eckespoint Red' and others were developed. The form available everywhere today is called after Ecke's Californian colleague Paul Mikkelsen. It is no longer a Christmas plant, for they now flower independently of the season.
Appearance: a long-lasting evergreen shrub which in the tropics grows naturally to 1·2 m (4 ft). As a rule not widely branched. Stemmed

Euphorbia pulcherrima, Poinsettia, fam. Euphorbiaceae

leaves large, pointed oval or lanceolate, or incised like an oak-leaf, with at the top magnificent bracts growing in rosettes, scarlet, salmon-pink or creamy white, depending on the strain.

Inflorescence: in the newest strains frequently double. The rosette of bracts, 20 cm (8 in) across, surrounds a circle of modest, greenish white spurge flowers. The colourful bracts serve to attract various insects for pollination.

Keeping: poinsettias are short-day plants (see p. 54). In order to flower and form the colourful bracts they require a so-called 'short-day treatment' with a ten-hour day for eight weeks. More light than this, even lamp-light, will prevent the formation of flower-buds and encourages leaf-growth; this will turn it into quite an attractive foliage plant. After the period of short-day treatment the bud should be sufficiently formed to enable the plant to flower and develop its colourful bracts within two to three weeks.

General care: keep the plant indoors throughout the year, although in summer it may be given a sheltered position on the balcony. Plenty of light, but no direct sunlight. After flowering, cut back to about 15 cm (6 in) leaving three to five of the best shoots. Resting period until May; when the normal flowering season is in winter the growing season occurs from May to November. Winter temperature about 18–20°C (65–70°F); a cooler position during the flowering season will make the 'star' last longer.

Watering: freely during the growing season; spray with tepid water. After flowering very sparingly until May.

Feeding: weekly from June to October.

Propagation: from shoot tips in summer.

Diseases: dry air will lead to attacks by mealy bug and scale insect, and sometimes by red spider mite.

Important: the white spurge juice which appears when the plant is damaged is very poisonous. Staunch the 'bleeding' with a burning match or candle.

E. milii
Crown of Thorns

Origin: its old botanical name was *E. splendens* – the splendid spurge. Only the Nomenclature Commission knows why it was changed into *E. milii*. Whatever the reason, *E. milii* was the first succulent spurge to be introduced into Europe from the granite mountains in the highlands of Madagascar: this happened in 1821. It soon established itself as a profusely flowering house plant with an almost unlimited life-span.

Species: for a long time only this species was known, but clever growers discovered the possibilities of cultivating variations and developed forms with even more brilliant red flowers and with denser growth. A yellow form was also introduced. The German grower Stirnadel achieved international success when, during a visit to Madagascar, he discovered a so-called natural *E. milii* hybrid with unusual properties as regards habit and flowering. He named it 'Somona' after the place where he had found it – the garden of a mission station. Since then this uncharacteristic Crown of Thorns and its many cultivars, including enchanting dwarf forms, have become almost world-famous. Novelties developed from the *E. milii* species were introduced on the market in the shape of the hybrids 'Petra' (red) and 'Victoria' (pink). They are cultivated chiefly in Tenerife, where the sun is stronger and the soil warmer.

Appearance: the original form of the old species consisted of pencil-thick, usually slightly angular, succulent and densely thorned shoots, long enough to enable today's cultivars to

Euphorbia milii, Crown of Thorns, fam. *Euphorbiaceae*

Euphorbia milii × *Somona* 'Gabriela'.
Somona. Fam. *Euphorbiaceae*

be trained into the shape of a 'crown of thorns'. The elongated to round bright green leaves develop between the thorns, becoming increasingly dense towards the tips of the branches; so do the usually multi-stalked little flowers. Like all spurges they owe their fine colouring to the bracts which surround them like petals.

Stirnadel's *E. milii* hybrid 'Somona', with the cultivar name 'Gabriela' is an erect growing plant; this also applies to the long-stalked inflorescence which grows from the tips of the branches. It reaches a height of about 60 cm (2 feet); each plant develops at most 3–4 branches. It does not matter if the lower leaves occasionally drop. The plant will continue to flower for many years.

The twin plants 'Petra' and 'Victoria', on the other hand, are bushy plants with graceful bright green foliage, from which in adequate light and warmth the innumerable little flowers appear in constant succession. In the course of time the almost spherical bushes may reach a height of about 60 cm (2 feet).

Inflorescence: As mentioned earlier, it is not the insignificant flowers themselves, but the surrounding bracts which create the colourful display. To this end the plants – in conformity with their tropical origin – require a great deal of light. Plant lovers should therefore accept the fact that their Madagascar *Euphorbias* will turn somewhat paler in winter.

Position. Their need of light makes them ideal plants for sunny situations and dry air; they dislike a humid atmosphere. Unlike many other succulents, including of course cacti, Crown of Thorns cultivars do not require a cool position in winter, but in addition to light and warmth particularly appreciate 'warm feet'. The summer can be spent out of doors only if the plants are given a very sheltered position, with the pots buried in the soil. Large, mature specimens need the support of a stick or a trellis.

Watering: because the plant flowers continuously, the soil must be kept somewhat moist; if the plant is kept in a sunny and warm position this applies even outside the main growing season, which occurs in summer; however, it must not be wet.

Feeding: during the main growing season – from spring until early autumn – give cactus fertiliser every 8–10 days.

Repotting: young plants every 1–2 years; older plants less frequently. Compost: sandy with humus.

Propagation: spring is the best time, but it can also be done later. Use the tips of mature, established branches rather than young tip cuttings. Leaving the cut surfaces to dry out for a day will encourage callus formation. Insert the cuttings in the smallest possible pots filled with special cuttings soil (cactus soil). It is also possible to raise plants from seed.

Important: leaf drop in winter generally indicates that the plant's environment is too cool, so that its feet are cold, or overwatering, or both. In spring it will usually put forth new shoots.

Caution: as in all *Euphorbia* species

Euphorbia pseudocactus, Cactus Spurge, fam. *Euphorbiaceae*

the white sap is poisonous. Wear gloves when handling a Crown of Thorns, for example when repotting it.
Note: excellent for hydroculture!

E. grandicornis
Cactus spurge

To give you an example of the often incredible resemblance between succulent spurge species and cacti we have chosen a species which even cactus experts can only identify by scratching the epidermis with a sharp needle in order to see whether the plant will exude the white spurge sap or not.

Actually this 'pseudocactus' with its enormous, always double, 'dorsal' thorns interspersed with tiny leaves is as attractive as it is easy to grow. In its native habitat in South Africa it grows into bushes several metres high, each shoot developing one of the three-jointed levels once a year. In older plants innumerable little yellow flowers will appear on the horny edges.

E. grandicornis is only one of several hundred succulent spurge species.

Species: many species resemble the cacti of Central and South America so closely that they were inevitably called the 'cacti of the Old World'. Some of the most attractive species, in indoor cultivation relatively limited in size, may be put out of doors in summer, but should have a minimum temperature of 12°C in winter. These include *E. caput-medusae*, a native of the Cape. Snake-like branches, surrounded by five-sided leaf-cushions, grow in all directions from a trunk thickening towards the tip. Plants do best in narrow palm pots.

E. globosa, also from the Cape, consists of clumps of spherical joints. Grown in shallow bowls in a well-lit and dry position they may flower. Self-propagating.

E. lactea, from India (Ceylon). Triple-jointed, candelabra-shaped habit. The trunks and branches are characterised by central white streaks. An imposing, but undemanding plant.

E. obesa, from the eastern part of the Cape. One of nature's puzzles, for this plant can be distinguished from the cactus *Astrophytum asterias* (sea urchin, p. 131) only by scratching it with a needle to see if it contains sap. The plant is, moreover, dioecious and barely reaches 8–10 cm (3–4 inches). Keep completely dry in the dormant season.

E. pseudocactus. Like *E. lactea* this plant is candelabra-shaped, but it has much thinner, 4–5-jointed, vividly green-marbled shoots and shorter thorns. Will grow well in indoor cultivation and is easily increased.

General care: generally simple, provided a dry resting season is observed in winter. In the growing season, from March to August, the plants should have a warm, sunny and well-ventilated position.

Exacum affine

It is difficult to imagine that this tropical herb, a counterpart of our good old Busy Lizzie, belongs to the family of *Gentianaceae*. Nevertheless it is a fact that gentians are found not only in the Alps or in European rock gardens, but also in Asia, in Madagascar, in Africa and elsewhere.

Origin: at the beginning of the present century *E. affine* was brought to Europe from Socotra, an island situated at the entrance of the Gulf of Aden and famous for its remarkable vegetation. It has ever since been cultivated over here.

Appearance: in this country the only available cultivar is *E. affine* 'Midget', an early summer flowering plant, more compact than the species. It is a herbaceous, juicy and freely branching plant, somewhat resembling the Busy Lizzie (p. 243), which is a balsam plant, but its flowering season is not so prolonged. The fragrant flowers are of purple with a yellow centre, underneath the anthers. They frequently appear as early as six months after the plant has been raised in a warm greenhouse. If they are sown in December, they may therefore be marketed from late June onwards as plants capable of flowering. They will

Exacum affine, fam. *Gentianaceae*

then flower until September–October. There is no point in trying to keep them longer, the plant is considered to be an annual.

Care: during its short sojourn in your home it is very easy to look after. Like the red Busy Lizzie, the purple, exacum requires plenty of light and not too warm a position, away from direct sunlight. A situation on the window-sill or on the balcony facing east or north-east, will hold the plant back. Water with reasonable soft water, pH 5·5–6, and take care not to let the soil dry out, as this will cause the flowers to shrivel up. Feeding will do no harm, but is unnecessary if the plant is grown in potting compost containing fertilising elements.

× *Fatshedera lizei,* fam. *Araliaceae*

× Fatshedera

It is merely due to the alphabetical order followed in this book that, contrary to the rules of biology, the cultivar × *Fatshedera lizei,* however famous in itself, is mentioned before the traditional species *Fatsia japonica,* but the order is justified, for the cultivar has certain advantages over the species.

Origin: × *Fatshedera lizei,* which resulted when *Fatsia japonica* was crossed with *Hedera helix* in 1912, is an achievement of the French tree-nursery of Lizé Frères in Nantes; their creation has since become world-famous, a star among foliage plants. The variegated strain 'Variegata', though slower-growing, is equally graceful.

Appearance: in the course of time × *Fatshedera lizei* may grow into a tub-plant of considerable size. It is erect-growing and has evergreen, usually 3- or 5-lobed leaves, up to 20 cm (8 in) wide and 25–30 cm (10–12 in) long. The leaf-stalks grow to 6–12 cm ($2\frac{1}{2}$–5 in). Older specimens occasionally develop greenish flower umbels in spring, but only if they have overwintered in a cool position.

Care: in general the same as for *Fatsia* (p. 208). The plant has great powers of resistance; provided the air is sufficiently damp it will even tolerate spending the winter in good light, away from sunlight, in a warm living-room. Cut back slightly every year, in order to encourage side shoots. A bushy growth may also be acquired by planting three cuttings in one pot. Easily propagated.

Fatsia japonica

The plant represents both genus and species and is one of those 'heroes of the living-room' with enormous powers of resistance against all possible errors in treatment. A plant to keep for many years. The popular name is False Castor-oil Plant. It has been cultivated as a pot or tub plant for more than 150 years.

Origin: as the name indicates, the plant is a native of Japan, where it grows as an evergreen shrub up to 5 m (16 ft) in height.

Appearance: it starts by forming a woody stem, but later reluctantly develops sideshoots. The leathery leaves grow on stems up to 40 cm (16 in) long; they are 7- to 9-lobed. Inconspicuous greenish white flower umbels only from older plants. The flowering season varies, but usually occurs in spring; the flowers are succeeded by dark berries, as in ivy.

Species: less frequently occurring strains are: the French 'Moseri', green foliage, even larger than that of the species a more compact habit and slower growth; 'Variegata', also slower-growing — the white and yellow variegated forms unfortunately tend to revert to green; 'Albomarginata', with white-edged green leaves; 'Reticulata', leaves marked with an obvious network of yellow veins.

Important: the variegated forms require more warmth, but are otherwise treated in the same way as others. Winter temperature 14–16°C (57–60°F).

Care: to grow *F. japonica* successfully, it should be kept throughout the year in a cool situation, preferably in good light near a north-facing window. In summer it likes a shady position out of doors; plenty of fresh air but not in a draught. To maintain the beauty of the foliage the plant should be kept at the temperature of an unheated greenhouse: 6–8°C (43–46°F) in winter. It is therefore particularly suitable for vestibules, shops and shop windows. Occasionally the plant can be cultivated out of doors.

Watering: freely during the growing period (April to August). A plant with such large leaves requires a lot of water, especially when the new shoots are appearing in spring and early summer. In the period of transition to dormacy, from August onwards, give less water and during the resting period adapt the amount of water to the temperature of the plant's situation. The soil ball must not dry out; avoid excess water in the bottom of the pot.

Feeding: once the plant is growing well, feed weekly from April to August.

Repotting: put into a larger pot only when the root system crowds the old pot. This is best done in March or April. Rich and somewhat loamy soil, with plenty of humus.

Propagation: the simplest method is to raise the plant in a heated indoor propagator from imported, guaranteed fresh, seed. Although it is not difficult to increase the plant by means of cuttings or air layering, this affects the growth of the parent plant; older plants in particular will suffer.

Diseases: too warm a situation or a dry atmosphere frequently cause foliage diseases resulting in shrivelling or curling of the leaves. Treat as for red spider mite; damaged leaves should be cut off and burned.

Fatsia japonica, fam. *Araliaceae*

Faucaria tigrina – Tiger's-chaps

This very decorative little succulent, a member of the family of *Aizoaceae*, belongs to an easily cultivated genus, of which a carefully chosen collection may be grown on the window-sill. Of the approximately 30 known species, about six are in cultivation. Their native habitat is South Africa.

Appearance: the thick, fleshy leaves, edged with erect teeth, grow opposite in pairs and are joined in the centre. The golden-yellow flowers appear from August onwards.

Faucaria tigrina, Tiger's-chaps, fam. *Aizoaceae*

Species: in addition to *F. tigrina* there are the following: *F. felina*, which has faintly stippled foliage; *F. lupina*, with fierce-looking teeth along the leaves, and *F. tuberculosa*, leaves covered in warts.

Care: the growing season occurs in summer; flowering follows. A position in full sunlight out of doors in summer will greatly benefit the plant. Pots should be small. As soon as shrivelling of the leaves indicates the beginning of the resting period, keep the plant dry.

Propagation: by division or from seed.

Ferocactus latispinus, fam. *Cactaceae*

Ferocactus

The genus *Ferocactus* consists of enormous, globular cacti originating in the south-western desert areas of the U.S.A. and Mexico, where in the course of dozens of years or even a century various species grow into immense columns. In room cultivations their dimensions will remain manageable.

Species: *F. latispinus* achieves a diameter of 40 cm (16 in); it is a squat globe, ribbed and with pale red to light brown spines. The flowers are small, reddish and fragrant but, unlike *F. hamatacanthus* and *F. stainesii*, it does not flower readily.

Flowering period: depending on the species from spring up to summer.

General care: a very light position in summer, in the warmest possible spot, for instance out of doors in direct sunlight. Water normally, spray when it is very hot. From October onwards place in a cool situation; resting period from November to February. The most desirable temperature is 8–10°C (46–50°F), do not water or feed at this time.

Propagation: mainly from seed, but also by removing the young offsets.

Diseases: drought in summer results in infestation by red spider mite and greenfly. Moisture in winter will lead to rotting of the crown.

Ficus — Rubber plant

The ficus has occupied a leading position in the world of houseplants for many years. There is no logical reason for this, for there are many other indoor plants, combining the same degree of modesty and vigour, as decorative as the ficus.

Origin: the enormous *Ficus* genus (the name simply means 'fig') includes more than a thousand species of tropical trees and innumerable epiphytic shrubs and climbers. The native habitats of species suitable for indoor or greenhouse cultivation are in southern and south-eastern Asia, from India via the Indonesian archipelago to Australia. They also occur in tropical regions of West Africa. *F. carica*, the domestic fig-tree, belongs in the area around the Mediterranean, both in southern Europe and in North Africa.

Appearance: the botanical link between the different species, which generally vary greatly in appearance, lies in the milky liquid, which at one time made certain species of rubber plant, such as *F. elastica* of great importance; this is hardly the case today. Another similarity is the unusual inflorescence, which in some species develops into edible or ornamental figs. The latter applies mainly to the Mediterranean *F. carica* already mentioned. Provided this plant is carefully acclimatised, it may thrive out of doors in mild northern regions as well, and even produce a considerable harvest. I can confirm this from personal experience. Of the rubber plant species suitable for room cultivation only *F. deltoidea* forms fruit. The other species of this group are really only found in botanical gardens and large showcases where, in a constant greenhouse temperature, they will develop their original fruit. They are: *F. aspera*, a native of the islands in the Pacific, and its hybrid 'Parcellii'; *F. cerasiformus*, from India; *F. edule*, which produces edible figs, also a native of the Pacific islands. The rubber plant species available for room cultivation never arrive at fruit formation, since even when they have grown large enough for tubs, they are still merely dwarfs compared to the giants of the jungle. The dimensions achieved by some of the species in their natural surroundings are mentioned in the section 'Species'.

The finest ficus species and their strains

Note: of the numerous species of the *Ficus* genus from different parts of the world, only relatively few are cultivated here. They are often among a plant lover's first selection. Although *F. elastica* undoubtedly heads the list, it is difficult to place them in their order of importance. We shall therefore mention them in alphabetical order, except that the small group of creeping and climbing forms are described together at the end of this section.

1. Erect growing evergreen rubber plants

F. benghalensis in its native habitat in Bengal grows into a 30-m (100-ft) tree with aerial roots. The leaves are 10–20 cm (4–8 in) long, stiff and leathery, an elongated oval, dark green. Young leaves and stems are covered in soft brown hairs. *F. benjamina* is a small-leaved rubber plant with gracefully arching branches and beautiful foliage (see illustration p. 42). It comes from India, where it grows into a large, broad-crowned tree. The leaves are of a glossy green, somewhat leathery, a pointed oval in shape. A valuable houseplant, also suitable for a large plant window. Can grow into a tub plant. *F. cyathi-*

Ficus deltoidea, fam. *Moraceae*

stipula is similar in appearance, but branches more freely and has larger leaves, 10–20 cm (4–8 in) long, a dull green in colour, broadest above the middle, abruptly pointed. For this species the best temperature is between 14 and 20°C (57 and 70°F). Like *F. deltoidea* described next, *F. cyathistipula* at an early age produces 3–4 pseudo-fruits. *F. deltoidea* (syn. *F. diversifolia*) comes from the Indian archipelago, where it grows as an epiphytic shrub in trees. In room cultivation it does not grow beyond 60–80 cm (24–32 in); it is freely branched. The leaves are grey-green, leathery, oval and pointed towards the stalk. The yellow, pea-sized berries are borne on erect growing short stalks, singly or in pairs. In adequate warmth and a moist atmosphere it will grow untiringly, if slowly, for dozens of years. The famous *F. elastica* in India grows to 20–25 metres (66–80 ft). It is an unwritten aesthetic law that as a houseplant its single trunk should grow absolutely straight and bear spotless leaves from top to bottom. Everyone knows what it looks like. It is also a fact that *F. elastica* is a reliable and relatively modest plant, although it may turn brown and lose its leaves at the end of the winter as a result of errors in cultivation. In addition it grows so rapidly that it may soon become too large for the living-room. It has been cultivated here since 1815. Among the advantages it offers in its long life is the fact that, if its top is cut off, it will form a dense crown of leaves, or vigorous young shoots can be trained round the window. In addition to the species, impressive, equally tolerant strains have appeared for sale in recent years. The first of these, soon after the war, was *F. elastica* 'Decora', which is the product not of crossing, but of specific seed selection. In the trade it is referred to simply as *F.* 'Decora'. It is a stately plant with large, broad-oval, deep green leaves. It does require some care to prevent the foliage of this stiffly erect plant from dropping. It is sensitive to sudden changes in temperature and to draughts, as well as to an excessive temperature in winter – it should not exceed 15–18°C (58–65°F) – and to cold feet, i.e. a soil temperature below 12–15°C (54–58°F), and to excess water in the bottom of the pot, especially during dormancy. The magnificent tri-coloured strain 'Decora Tricolor' (illustration on p. 212) was developed from the green form; approximately the same rules of treatment apply. On the other hand the greyish white strain variegated 'Doescheri', with pink mid-ribs to the leaves which was developed in Belgium more or less by accident, is much more accommodating. The 'Schrijvereana' strain, illustrated on p. 20 has very distinct and clear marbley white marking; it was developed in 1955, the latest of the ▶

Ficus elastica 'Variegata', fam. *Moraceae*

Ficus elastica 'Decora', one of the original 'rubber plants' cultivated as house plants. Fam. *Moraceae*

Ficus retusa, fam. *Moraceae*

modern *F. elastica* strains. The strain 'Variegata', developed nearly a century ago, has the original elongated narrow leaf-shape of the species, but is magnificently edged in yellow, and cream, and variegated dark and grey-green; this strain can easily be cultivated with more than one main stem.

F. lyrata, formerly called *F. pandurata*, a plant from West Africa, is the most monumental species among houseplants (illustrations on pp. 37 and 44). The thick, slightly rough and leathery, but succulent green leaves grow to 50–60 cm (20–24 in) long; they are violin- or lyre-shaped and have a wavy edge. It is a truly magnificent plant, but in order to develop fully it requires plenty of space, away from other plants, and likes to receive light from all directions, as shown in the photograph on p. 37. Moist air is particularly favourable to young plants. The plant does not branch but will continue to grow upwards. It has been known in this country since the beginning of the century. Propagation from shoot tops possible only in heated soil, 30–35°C (86–95°F); it is a job for the professional grower.

F. religiosa, known in India as the pagoda tree, the Peepul or Bo Tree, is rarely seen over here, since in the trade it is often regarded as a hot house plant. In actual fact it grows best at temperatures between 16 and 20°C (60 and 70°F), provided there is plenty of light; in this respect it differs from other species. The leaves are narrow, with very long leaf-stalks; the points almost resemble thin nails. Length 6–15 cm (2½–6 in).

F. retusa, which has blunt foliage, in India and the Indian archipelago grows as a large-crowned and well-loved tree. As shown in the photograph below left, it resembles *F. elastica*, but it tends to branch at an early age. In its natural habitat in Australia, *F. rubiginosa* is a freely branching shrub growing to 4 m (13 ft); the branches curve downwards and root in the soil, forming new shoots and thus covering large areas. The leaves of the species are short-stemmed, of average size, up to 10 cm (4 in) long, green, and in young plants covered in felty brown hair, as are the young stems. The undemanding white-variegated strain *F. r.* 'Variegata' has for several generations been a very popular plant. Both species and strain are typical greenhouse plants; however, they require a cool position in winter, 10–12°C (50–54°F), and in summer like a sheltered place out of doors. A great advantage is the fact that they grow very slowly and can thus become real heirlooms. Propagation only from cuttings; these root very readily during the growing season.

2. Creeping and climbing ficus

The geographical distribution of the climber *F. pumila* is spread between eastern Asia and Australia. It is self-clinging like ivy and can cover large areas; older plants produce pear-sized fruits. It is incredibly hardy and will even tolerate temperatures close to freezing point. There are several fine strains, for instance 'Minima', 'Serpylliolia' and the slightly more demanding 'Variegata'.

F. montana, formerly called *F. quercifolia*, has foliage resembling oak leaves; it is a native of Burma and develops short branches, later bending downwards. A hot house plant, very beautiful in hanging pots.

F. radicans, a creeping plant of unknown origin, develops roots from the leaf-buds of its runners and is therefore excellent as ground cover. The leaves are larger than those of *F. pumila*. Its strain 'Variegata' has white variegated leaves, some entirely lacking in chlorophyll.

Care: except where special instructions have been given for plants mentioned in the section 'Species', the following rules apply to all green-leaved species and strains. Variegated forms in general are somewhat more sensitive and require more warmth and humidity.

Position: usually indoors throughout the year. Ornamental plants, to be

Ficus rubiginosa 'Variegata', fam. Moraceae

placed individually. Older, branched specimens may also be used in a plant window. The rubber plant prefers to be left in the same position year after year.

Summer: a well ventilated situation, plenty of light, but no direct sunlight. If possible it should be lit from above. As soon as the topmost leaf reaches the unlit area above the window, normal growth will cease and the lower leaves will suffer as well. In general it is not advisable to place the plant out of doors, even on a balcony, since the conservative rubber plant will react unfavourably to being moved. If for some reason you have to put it outside, a warm, moderately sunny position, completely sheltered from the wind, is essential.

Winter: dormant period from October to February, preferably at a temperature of 12–15°C (54–58°F) although a healthy plant which is allowed to remain in the same position throughout, will tolerate a higher temperature. A temperature below 12°C (54°F) will in the long run cause damage, particularly if the plant is at the same time given more water than it needs.

Humidity: the simpler, all-green species like to be sprayed while the young shoots appear in spring. All variegated and climbing species and strains require regular spraying except in the dormant season.

Watering: freely from the beginning of the growing season until the end of August; the plant must never dry out. From late August onwards gradually reduce watering. In winter merely keep moist, even if the plant stands in a heated room. During this time the roots, which need a lot of oxygen especially where the large-leaved species are concerned, must be kept somewhat drier. Many rubber plants become diseased or die every year because they are given more water than the roots can deal with during this period of changed metabolism. The water should be about *p*H 6.

Care of the foliage: regularly clean the upper surface of the leaves with damp cottonwool. If you want glossy foliage, do not use household remedies such as milk, beer or salad oil, which will stop up the pores. There are special leaf-shine products on the market; these at the same time prevent the foliage being spotted by water, and even keep off dust.

Feeding: plants with good foliage development require a nutrient solution from the time they start into growth until mid-August. Less well-growing specimens will suffer if overfed. Stop feeding when no further leaves are formed.

Repotting: in February–March in only slightly larger pots, and only when the roots crowd the old pot. Mix leafmould and compost with the addition of sand, peat and pieces of charcoal. For larger, older specimens use a mixture containing a slow-acting fertiliser as for *Nerium oleander* (p. 270).

Propagation: an amateur who does not have the use of certain aids will find it difficult to raise ficus from cuttings. The cuttings will root fairly

▶

readily, but they require a rather high soil temperature, between 25 and 35°C (78 and 95°F). In an electrically heated indoor propagator such temperatures are easily achieved. As a rule eye-cuttings are used, that is, sections of the stem including a (dormant) eye in the axil of a leaf stalk. Cut the stem 1 cm (½ in) above and 1 cm (½ in) below the eye; do not remove the leaf and the leaf-stalk. In large-leaved species such as *F. elastica* it is customary to roll up the leaf, holding it in position with a rubber band. This considerably reduces evaporation. The soil used is the well known mixture of half sand and half peat fibre, thoroughly soaked. The propagator must be kept closed to maintain the humidity; watering will thus be unnecessary. Make sure that the cuttings have plenty of light, but keep them out of the sun. After a few weeks they will have rooted. Now the little plants must gradually be hardened off by leaving the propagator slightly open in the daytime. Once the dormant eye has developed into a small leaf, the old leaf can be carefully removed, but the young plant will now have to find its own nourishment and should therefore be repotted in more nutritious soil, for instance potting compost mixed with half its own volume of sand. For the first few weeks after repotting the tender plantlets should be kept under glass or polythene.

For air-layering, see the chapter on Propagation. Raising from seed is possible, but rarely done in room cultivation.

Diseases: a too warm, closed-in position leads to attacks by scale insects and mealy bug; too much sun or a draught will result in red spider mite or thrips. Many species are sensitive to pesticides; these should therefore be used with caution. Large brown spots spreading from the margin are an indication of gloeosporium, a fungus disease. Affected leaves should be removed and the rest of the foliage should be protected against spreading of the infection by spraying with systematic fungicide.

Important: drooping lower leaves indicate too warm a position in winter. If yellow or black blotches appear on the foliage, the roots have been damaged as a result of 'cold feet': soil too cold or too wet. Conclusion: most damage is caused by errors in cultivation.

Ficus pumila, fam. *Moraceae*

Ficus radicans, fam. *Moraceae*

Fittonia

Around 1850 Elizabeth and Sarah Mary Fitton published a book in London entitled *Conversations on Botany*. A genus discovered shortly afterwards was named after these two ladies; it belongs to the family of *Acanthaceae*. To owners of plant windows fittonia is known as a ground-covering plant.

Origin: the two species known originate in the tropical rain forests of Peru. They are typical hot house plants and will not thrive in a pot on the window-sill.

Appearance: herbaceous plants, partly creeping, partly erect growing. The leaves are round oval, 7–10 cm (2¾–4 in) long, placed opposite each other on short stalks. The insignificant yellow flowers appear in spring, as a rule only on the species, not on the hybrids.

Species: the form most frequently seen (in mixed bowls as well) is the silver-veined 'Argyroneura', a hybrid of the species *F. verschaffeltii*. The illustration on the left shows the conformity between the plant's name and its appearance. A rare strain is *F. v.* 'Pearcei', which has carmine-red veins. *F. gigantea*, a bushy, erect growing plant, may achieve a height of 60 cm (2 ft). The dark green leaves, up to 12 cm (3¾ in) long, have pale carmine veining. This species is not very suitable for amateurs and is mainly used by botanists in experiments concerning viable temperatures.

Care: to preserve the beautiful colouring it is necessary to put the plant in a reasonably good light, but not in direct sunlight. A high degree of atmospheric humidity in summer, slightly lower in winter. The best temperature is between 18 and 20°C (65 and 70°F).

Watering: from spring until autumn keep constantly moist (not wet!). In winter restrict the water supply somewhat; there is, however, no real resting period. Tepid water, pH 4–4.5. When spraying there is no danger of spotting the foliage.

Feeding: be careful with feeding. Give a weak, lime-free feed solution only, from March until early August; this might be a nutrient solution used in soilless cultivation.

Repotting: if necessary this is done when the plant starts into growth in spring. Use low plant bowls. Be careful, for the stems are fragile. Do not forget to provide good drainage in the bottom of the pot. Compost light and containing humus. Pre-packed potting compost should be mixed with

Left: *Fittonia verschaffeltii* 'Argyroneura'; right: *Fittonia gigantea*, fam. *Acanthaceae*

some sharp sand.

Propagation: very simple, from cuttings in spring and summer. The cuttings will root in an indoor propagator in a shady position at a soil temperature of 18–20°C (65–70°F).

Diseases: precautionary measures should be taken against the plant's two main enemies: woodlice and slugs. Sensitive to draughts.

Fuchsia

Linnaeus named this plant after the botanist Fuchs. Plant lovers often ask whether it is a garden plant or a balcony plant, and whether it is suitable for use as a houseplant. The answer to the last question is in the affirmative, especially for the species *F. procumbens*, which likes to grow indoors as a hanging plant.

Origin: around 1700 the French explorer Charles Plumier, to whom we also owe the begonia, discovered the fuchsia in the cooler, upper regions of the mountain forests of Central America, but the plant's geographical distribution reaches much further: all over South America, to the Falkland Islands, New Zealand and Tahiti.

Appearance: this shrubby plant with its pendulous flowers is found in all the regions mentioned. Growers have developed innumerable strains from the species. Two main groups are distinguished. The first includes erect growing and hanging fuchsias, with both single- and double-flowered strains. The second group consists of the *triphylla* hybrids cultivated especially by Californian growers. Worldwide, the genus includes over 2000 species and strains, of which only about 50 are grown over here.

Care: used as houseplants they should be given a very well ventilated, half-shady situation in summer and should overwinter at 12–15°C (54–58°F) in a good light, but out of the sun. In view of their origin in cool and damp mountain forests at altitudes of up to 2000 m (6000 ft) they require some atmospheric moisture indoors during their growing and flowering seasons. This should be borne in mind also for garden fuchsias.

Watering: since they are forest plants, they like to be kept constantly moist; a little less so in winter, but never dry. Water pH 5–6.

Feeding: from February to August feed every week.

Pruning: in late February: older strains and hanging plants only; newer strains will generally attain a bushy habit without pruning. The plants should, however, be cleaned.

Repotting: after pruning or cleaning, into only slightly larger pots. Pre-packed potting compost.

Propagation: this is best done in May and June from stem cuttings that are already slightly woody at their lower end; take 5–6 pairs of leaves and remove the bottom pair.

Diseases: dry summer heat will encourage greenfly, whitefly and red spider mite. A fungus disease occasionally occurs.

Fuchsia, hybrid 'Pink Galore', fam. *Onagraceae*

Gardenia — Cape jasmine

The plant owes its name to Dr. Garden, an 18th century American physician and naturalist. It enjoyed its greatest popularity in the last century, when it was frequently used as a buttonhole. It is now once more found in mail order catalogues.

Origin: its natural habitat, depending on the species, is either eastern Asia or tropical and sub-tropical regions of Africa. A number of species originating in China have been known since about 1750 and are still very popular; there are about 60 in all.

Appearance: evergreen, freely branching shrubs, with glossy, leathery leaves, elliptical or oval in shape, practically stem-less, usually growing opposite one another in pairs. Saucer-shaped flowers, 8–10 cm ($3\frac{1}{4}$–4 in) across, white, sometimes yellow; deliciously scented. The most important species, a native of China and South Africa, is *G. jasminoides*, from which in addition to other strains the smaller-flowered hybrid 'Veitchii', much used for cutting and for buttonholes, has been developed.

Use: today the gardenia is once more valued as a pot-plant. Only the winter-flowering *G. jasminoides*, or better still, the hybrid 'Veitchii', are suitable for indoor cultivation. Amateurs with some experience will find these plants not much more difficult to keep and bring into flower again than the camellia, which is cared for very similarly.

Summer: after careful hardening off it will keep from June onwards in a semi-shady, warm and sheltered position out of doors. As it flowers in winter, *G. jasminoides* must without fail be brought indoors before the first night frosts occur.

Winter: like the coffee plant and *Ixora*, the related gardenia is very sensitive to cold feet. A room temperature of 16–18°C (60–65°F), moist air and a constant soil temperature of 18–20°C (65–70°F) are vital conditions. Plenty of light and space and good ventilation; no direct sunlight.

Watering: during the growing and flowering season water normally, but only with soft, tepid water 20–22°C (70–72°F), pH 4–4.5.

Feeding: depending on the plant's growth rhythm, treat with a lime-free nutrient solution, if possible with organic plant fertiliser, for four months, from the time the first shoots appear.

Repotting: before the plant starts into growth. Older plants at most once every three to four years. Be very careful not to damage the fine roots; plant at exactly the same level as before; gently firm the compost, which should be rich and lime-free.

Propagation: from shoot tips, cut like those of the myrtle or the house lime; placed under a plastic bag and at a high enough soil temperature, 24–26°C (75–79°F) they will root fairly readily. Growing from seed is a professional's job.

Diseases: errors in cultivation encourage red spider mite, mealy bug and thrips. Too cold or wet soil leads to chlorosis, yellow blotches on the leaves.

Gardenia jasminoides 'Veitchii', fam. *Rubiaceae*

Gasteria

This is a large genus of succulent plants belonging to the lily family; there are several attractive species suitable for room cultivation; many of them are known by popular names. There are large and small species suitable for every taste and every collection of succulents. They are natives of South Africa.

Appearance: thick, fleshy leaves, often blotched with white; sometimes covered in pearly warts. They may form rosettes or be opposite growing. The bell shaped flowers are brownish-red and hang in loose clusters from a tall stem. Flowering season: spring to summer.

Species: small forms are *G. armstrongii*, 4–6 thick, dark green leaves, opposite growing, with crinkly indentation on both edges. *G. liliputana*, leaves in pairs, later forming rosettes, only 3–6 cm ($1\frac{1}{4}$–$2\frac{1}{2}$ in) long, white blotches; it has an inflated calyx. Larger species are: *G. candicans*, leaves arranged in rosettes, 25–30 cm (10–12 in) long, white-speckled; the inflorescence grows to a metre (3 ft) in height. *G. verrucosa*, one of the finest; opposite leaves with large pearly warts.

Care: tolerant of sun; suitable for a south-facing window. See also under *Haworthia* (p. 224).

Gasteria pulchra, fam. *Liliaceae*

Gloriosa — Glory lily

At one time they were offered as somewhat sensitive outdoor plants. Later they were cultivated as exhibition flowers and for cutting. This is still the case, but in addition their attraction as houseplants is now realised. Stress is always laid on the necessity for allowing the tubers a resting period.

Origin: six species grow in tropical regions of Asia and Africa.

Appearance: long-lasting plants with thick, tuber-shaped rootstock. It is a climbing lily with single stems, branching at a later stage; they grow to 1·5–2 m (5–6 ft). The 10–12 cm (4–5 in) long leaves embrace the stem; they are elongated oval, ending in thread-like points. Flowers up to 10 cm (4 in) long, growing singly or in groups on long stems emerging from the axils of the upper leaves. Petals reflexed, varying in colour. Flowering season from summer to autumn.

Species: the best known species are *G. rothschildiana*, flowers dark red with slightly wavy margins; *G. simplex*, flowers yellow at first, later turning red in the sun; smooth-edged; *G. superba*, inflorescence initially green, turning yellow and then red.

Care: during the growing and flowering season, from April to September, a warm situation in good light, but semi-shade. When the flowers have died, leave the tubers in the pot, keep them dry in winter at a temperature of 10–12°C (50–54°F). In February–March remove the old roots and plant the tubers in groups of two to four 2 cm (¾ in) below the surface in fresh soil in larger well-drained pots. Place in a warm position and, until the flowers appear, ensure adequate humidity. Provide canes to which the climbing plants can be tied.

Watering: in summer normally, using tepid water, pH 5–6. From early September onwards gradually reduce watering, allowing the plant to die down in October.

Feeding: from the beginning of new growth until August feed normally.

Propagation: by division of the tubers (risky), from offsets and from seed (both lengthy processes).

Diseases: draught and lack of warmth in spring lead to greenfly infestation.

Gloriosa rothschildiana, Glory Lily, fam. *Liliaceae*

Glottiphyllum

Every expert or interested amateur knows that *Glottiphyllum*, *Fenestraria* and *Ophthalmophyllum* are among the most 'difficult' *Aizoaceae* genera. However, this belief may be questioned, for it is quite possible to keep them in good condition.

Problems: if one studies the manner of life of these succulent plants, natives of the arid regions of South and South-west Africa, one will realise that correct treatment essentially means correct watering and feeding, or rather abstention from both. Many people are blind to the fact that the glottiphyllum with its thick, fleshy, tongue-shaped leaves can never be treated too meanly.

Basic rules: these plants should not be planted out, but left in small pots; use very porous, poor soil (2/3 sand-1/3 loam); keep practically dry even during the growing period in summer and do not feed. Plenty of sun is favourable, but an excess of sunlight may stunt the foliage and cause it to turn mauve, whereas it should be green in colour. Provided these rules are followed, the healthy plants and their usually yellow flowers will give you a great deal of pleasure. Compare also the *Aizoaceae* on p. 128 (*Argyroderma*), 174 (*Conophytum*), 209 (*Faucaria*) and 252 (*Lithops*). Together they form a truly interesting group.

Glottiphyllum oligocarpum, fam. *Aizoaceae*

219

Graptopetalum

This succulent, whose Greek name, freely translated, means 'painted petal', is related both to the *Echeveria* and to the *Sedum* genus. In fact it has been suggested that it should be robbed of its independent status and be regarded as a bastard of *Echeveria* and *Sedum*.

Origin: the species described originate in Mexico.
Appearance: the botanist calls the species *Graptopetalum paraguayensis* (syn. *Sedum weinbergii, Graptopetalum weinbergii*), a small succulent shrub, widely branched; the branches are up to $\frac{1}{2}$ cm ($\frac{1}{4}$ in) thick. The grey-white succulent leaves appear in rosettes at their extremities. The Heidelberg Professor Werner Rauh, one of the greatest experts in the field of succulents, at one time described *G. filiferum* as follows: 'Sessile; the rosettes, 5–6 cm (2–2$\frac{1}{4}$ in) across, grow close together. The green or brownish leaves end in a long, bristly-haired point. The 5–8 cm (2–3$\frac{1}{4}$ in) long flower stalks are freely branching; the flowers are proportionately large with white petals shading to red towards the extremities'. But his most important statement is: 'a very decorative, easily cultivated species'.

G. amethystinum has greyish white leaves, growing in loose rosettes, mauve-bloomed. The flowers appear in summer, but are insignificant.

Care: this plant requires no special treatment. You might follow the rules given for *Echeveria* (p. 196) and *Sedum* (p. 329). Water sparingly.

Graptopetalum paraguayense, fam. *Crassulaceae*

Grevillea robusta, fam. *Proteaceae*

Grevillea robusta – Silk bark oak

For use indoors, grevillea is known only as a foliage plant, although it has the name of being one of the most beautiful flowering shrubs in the world. The reason is that, like so many 'foliage plants', it does not grow large enough as a pot plant to produce flowers. It may occasionally be seen in flower in a botanical garden.

Origin: *G. robusta* is a member of the family of *Proteaceae*. In South Africa it is cultivated for its magnificent flowers, which are flown to all parts of the world. In its native habitat in Western Australia, grevillea is a tree growing to a height of 50 m (165 ft).
Appearance: the genus includes about 170 evergreen shrubs and trees. *G. robusta* is the only one suitable for room cultivation. It is an undemanding plant with fern-like feathered foliage. Unfortunately it grows so fast that it usually reaches the ceiling within three to four years. A number of smaller-growing species, which at the same time flower magnificently, can be kept only in botanical collections.

Care: in summer a half-shady situation; in winter cool and plenty of light; best temperature 6–10°C (43–50°F). Once the plants are acclimatised they will tolerate 12–15°C (54–58°F). When the new shoots appear in spring provide some atmospheric humidity.
Watering: too much and too little are equally harmful.
Feeding: only in summer.
Repotting: in spring, when it starts to put forth shoots, in not too large pots.
Diseases: too high a temperature in winter frequently encourages greenfly.

Guzmania

The scant information we have of Mr. Guzmann relates that he was of Spanish origin, a pharmacist, botanist and collector of 'historical objects'. He must have been an important man, since one of the most beautiful and widespread bromeliad genera has been named after him.

Origin: the native habitat of this genus is spread from Central America to the farthest regions of South America and to the Antilles.

Appearance: all *Guzmania* species come from tropical rain forests, where they are both epiphytic and terrestrially growing plants. Their funnel-shaped rosettes may grow to a height and diameter of 50 cm (20 in) depending on the species. The smooth-edged leaves, deep green in some species, variegated in others, can be clearly distinguished from the bracts developed at the base of the rosette, which surround the inflorescence. In some species the flower grows at the end of a tall stem. Flowering season as a rule from November to January.

Species: the forms now most in demand over here are undoubtedly the hybrids developed about twenty years ago, including world-famous strains such as 'Intermedia' and 'Magnifica'. *G. lingulata* forms large, dense rosettes; the modest little flowers in the centre are surrounded by pale red bracts. Magnificent strains of this species are 'Broadview' and 'Splendens'. *G. minor* is an attractive, small bromeliad; it grows to only 20 cm (8 in); small white flowers surrounded by red bracts. *G. monostachya* (formerly *G. tricolor*) grows to 40 cm (16 in) and has a three-coloured flower at the end of a tall stem. *G. zahnii* has tongue-shaped, pointed leaves of an unusual red/green colour with bright red longitudinal veins. The flower stem grows to 40 cm (16 in) and is covered along its entire length in red leaves; the flowers and the bracts are golden-yellow. All the plants mentioned are epiphytes.

Care: apart from a few of the newer strains, most guzmanias are happiest in a tropical plant window with constant humidity and temperature, in a shady position. In winter the temperature should not be below 16–18°C (60–65°F).

Watering and spraying: only with soft, tepid water. During the growing period, usually from March until August, pour the water into the funnel; spray frequently while the leaves and flowers are appearing. After flowering, when the resting period begins, pour the water from the funnel; restrict watering and do not spray.

Feeding: during the growing season pour very diluted nutrient solution into the rosettes at long intervals, or use a foliage fertiliser, as for aechmea (p. 117).

Repotting: unnecessary, since the parent plant will gradually die.

Propagation: from young offshoots, as for *Aechmea* (p. 117). Although *Guzmania* is known to germinate readily, it is not easy to grow from seed.

Diseases: there is less risk of diseases and pests than of damage caused by errors in cultivation: scorch marks from sunlight and rotting inside the funnel in winter.

Guzmania lingulata, fam. Bromeliaceae

Gymnocalycium

This globular species of cactus is by now known to most plant-lovers. It used to be very expensive, but is now more easily obtainable. However, the red or yellow globe-shaped form is only one of the many easily-grown *Gymnocalycium* species which flower magnificently.

Gymnocalycium denudatum, fam. *Cactaceae* (left)

Gymnocalycium mihanovichii 'Friedrichii', fam. *Cactaceae* (right)

Origin: the genus *Gymnocalycium* includes about 50 species and belongs in South America. In the Argentine in particular they grow in widely varying conditions: from rich soil in full sun to crevices in bare rock-walls just under the snow-line.
Appearance: they are all small globular cacti, 5–15 cm (2–6 in) in diameter. As a rule the ribs are not very wide; the spines often have beautiful colours. The flowers are frequently as large as, or larger than the plant itself, disc-shaped and growing on long stems.
Species: among the most beautiful and interesting species are: *G. denudatum*, with spidery yellow spines; white flowers up to 5 cm (2 in) in length. An indestructible plant is *G. gibbosum*, which is also white-flowering. A favourite among the smaller cacti is *G. mihanovichii* with its pink-flowered variety *friedrichii*. From this strain a mutation has developed, which contains no chlorophyll and is, therefore, clear red. It lives entirely on its green sub-stem. Since 1970 the same phenomenon is available in yellow. These cultivars do not flower, but they do develop offsets at the base.
Care: depending on the species, a very light to half-shady situation. Keep dry in winter at 6–10°C (43–50°F).
Propagation: from seed.

Gynura

For more than a hundred years there has been among the houseplants a little known, erect growing, tropical shrublet. A few years ago the climbing *Gynura scandens* (syn. *G. sarmentosa*) appeared on the market and was well received, but nevertheless many incorrect statements are still being made about it.

Origin: it is one of about 25 species from the warmer regions of Asia and Africa. Its native habitat is in the sparsely wooded mountain forests of tropical East Africa.
Appearance: the angular stems grow to 80–100 cm (32–40 in) in length. Like the dentate leaves, green on their upper surface, they are covered in soft hair on the underside, and the upper surface when young, deep violet-purple in colour. The plant's chief value lies in this wonderful colouring, which can only be preserved in plenty of light. This is also essential for the stem which will otherwise grow unnaturally long, making the distance between the pairs of leaves too great. If cut back regularly they will branch freely. They are most attractive as hanging plants or trained along bent wire.
Even professionals sometimes maintain that gynura is purely a foliage plant, but in late summer it will produce elongated dark lilac buds which develop into orange-red flowers. Un-

fortunately these smell rather unpleasant for the first few days, but after this you will be rewarded for months on end by the silver-haired, bell-shaped capsules inside which the seeds ripen. Towards the end of the flowering period flowers and capsules are often seen simultaneously.

Care: the plant should live throughout the year in the best possible light. In less sunny situations they will grow out of shape and barely flower; although this means that you will not be bothered by the unpleasant smell of the flowers, it also brings about attacks by greenfly. Water and feed freely in summer; in winter water sparingly, depending on the temperature. Repotting is unnecessary; older plants will become unsightly. It is therefore advisable to take cuttings in the autumn.

Propagation: the cuttings will root rapidly and in spring you will once more have fine plants. Topping the young plants once or twice will improve their shape.

Gynura scandens, fam. *Compositae*

Haemanthus

We have a choice of two species. One of them, *Haemanthus albiflos*, is a regular customer on the window-sill, where it flourishes in the simplest of conditions. The other, *H. katharinae* is an important greenhouse plant.

Haemanthus albiflos, fam. *Amaryllidaceae*

Origin: both belong to the genus *Haemanthus*, which includes about 50 species and is found everywhere from Ethiopia to the Cape.

Species: *H. albiflos*, which in some countries has the popular name of Elephant's-ear, is one of the evergreen species with thick, fleshy leaves that really look like drooping elephant's ears. They emerge in pairs from the slightly flattened bulb, grow to at least 20 cm (8 in) in length and 10–12 cm (4–4¾ in) across, dark green, hairy along the margins. The spherical white flower umbels appear in groups on a short, thick stalk; the protruding white stamens are covered in yellow anthers. Flowering season from summer into autumn. *H. katharinae* and its now frequently cultivated hybrid 'King Albert' belong to the thin-leaved type; the leaves appear at the same time as the inflorescence or a little later and drop in autumn. The flower-stem is flecked at the base; it grows to 30 cm (1 ft) and bears a dense umbel of long-stalked, bright red flowers. In older plants the flowers may be up to 25 cm (10 in) across. The flowering season occurs in summer. A number of species secrete an innocuous blood-red juice when damaged.

Care: *H. albiflos* is a robust and very accommodating plant. It is happiest if

▶

left undisturbed throughout the year in a sunny spot. During the dormant season in winter a moderate temperature, 12–15°C (54–58°F) is better than normal room temperature, but this will nevertheless be tolerated. The plant has no special requirement as regards humidity. *H. katharinae* and its hybrid 'King Albert' are really hot house plants, but will also grow in a warm, but slightly shaded position in the living-room. From mid-August onwards gradually withhold water, to give the plant a rest. *H. albiflos* may also be kept almost completely dry in winter, depending on the room temperature.

Feeding: sparingly and only in summer.

Repotting: *H. albiflos* only when obviously necessary. *H. katharinae* in early spring; the bulb should be put only half-way into the soil.

Propagation: new plants can be grown from offsets, but you would do better to allow *H. albiflos* to grow into a richly flowering community in one pot. *H. katharinae* can be raised from seed, but the plants will flower only after several years.

Haemanthus katherinae, fam. *Amaryllidaceae*

Haworthia

The immense lily family includes numerous succulent plants of all sizes. Among the most attractive and undemanding small succulents are the *Haworthia* species, named after the English botanist A. H. Haworth, ±1800.

Origin: there are about 100 species in South Africa, most of them in the Cape Province.

Appearance: all haworthia species are small plants or shrublets with thick, fleshy leaves forming rosettes. The small bell-shaped flowers which appear in spring and summer at the end of long and slender stalks, are insignificant and have no decorative value. As compensation there is an enormous choice of different species, from which attractive strains have been developed. We distinguish several groups.

Species: the most beautiful and at the same time the most durable species are those which bear white, pearly nodules on the dark green lower surface of the foliage. The plants do not need an actual resting period. We should mention in the first place *H. fasciata*, which has a spherical, wide open rosette; the leaves are triangular and pointed and have fairly large nodules on their lower surface. *H. margaritifera* is the species most valued as a houseplant; the rosette is 15 cm (6 in) across and slightly closed; large round nodules appear on both sides of the leaves. This is a species with numerous forms and varieties. It closely resembles *H. papillosa*, which, however, has larger nodules. The photograph shows *H. reinwardtii*, narrow and erect growing, with 10–15 cm (4–6 in) small trunks and many small nodules on the underside of the leaves. Species without nodules, sometimes with a transparent upper or lower leaf surface, are somewhat more tender. Among these are *H. cymbiformis*, with short, erect, grey-green succulent leaves and *H. planifolia*, a superb plant with spreading larger and flatter leaves, transparent mainly along the edges. Another worthwhile plant is *H. tessellata*, in which the upper surface of the leaves has reticulate marking, while the lower surface is covered in nodules, as are the serrated edges.

Care: all these attractive small succulents require a great deal of light throughout the year, but must be kept out of strong sunlight. They must be kept in a closed room. Haworthia species with pearly leaves have a resting period from October to February, if possible at 10–12°C (50–54°F). Species with transparent leaves, on the other hand, are dormant from April to September and during this time must be kept drier; do not feed. In winter they require more warmth, 14–16°C (57–60°F), since this is their growing season. Keep the soil moderately moist and every two weeks give a greatly diluted liquid cactus fertiliser. They do not have special requirements as regards atmospheric moisture.

Repotting: when they outgrow their pots. Use flat plant-bowls.

Propagation: by removing new offshoots or from seed.

Haworthia reinwardtii, fam. *Liliaceae*

Hebe

The veronica is known chiefly as a herbaceous garden perennial, but there are also a number of winter-hardy, evergreen shrubs and a group of tender shrubby species suitable for growing in pots or tubs. The latter used also to be called veronica, but they have now become an independent genus under the name *Hebe*.

Hebe andersonii, hybrid, fam. *Scrophulariaceae*

Origin: the complicated family relationship has led to confusion in the botanical nomenclature. However, it is certain that the approximately 140 hebe species are natives of New Zealand, while some occur in Australia as well.

Appearance: the small woody plants of some species, with their needle-like foliage, resemble conifers to a large extent. There are also species with evergreen, leathery leaves. The erect flower racemes appear in summer on stems growing from the axils. Depending on the strain the flowers may be white, carmine-red or blue. Through careful selection many hybrids have been developed from *H. andersonii*. The best known is 'Variegata', which has white variegated foliage. Seen from a distance the carmine-flowering forms such as 'Evelyn' and 'La Seduisante' might almost be taken for *Callistemon*, the bottlebrush. Unfortunately the flowers do not last very long.

Care: in summer the plant is best placed in a sheltered position on the balcony or in the garden, but it should be brought indoors in good time for it is sensitive to frost. Overwintering at 8–10°C (46–58°F). Otherwise it should be treated in the same way as similar shrubby cold greenhouse plants (bottlebrush, myrtle, etc.).

225

Hedera — Ivy

Now that climbing and trailing plants have once more conquered our homes in large numbers, the ivy is again among the most popular of indoor plants. Actually it has never ceased to be man's companion ever since the classical era.

Origin: the original species, which in Egypt were dedicated to the sun god Osiris and in Greece to the merry god of wine Dionysius, were common wild plants, probably *Hedera helix* ssp. *canariensis* (formerly *Hedera canariensis*), which occurs widely in North Africa and the Canary Islands, and *H. colchica*, which grows in eastern Europe and in Asia Minor. The geographical distribution and historical importance of these plants is indicated by the fact that there are also Japanese and Irish ivies.

Appearance: the ivy is an evergreen climber, whose adhesive roots enable it to cling to vertical walls. Any attempt to detach the plant causes loss of vigour and of foliage. It is still not clear in how far the forms now on the market are botanical varieties or cultivated strains of the original species *H. helix*. The names also indicate that family relationships have to some extent become confused. However, like all aralia plants, the ivy is very variable where its inherited attributes are concerned. In a textbook for professional growers we read: 'Numerous strains are being cultivated by horticulturalists to be sold in pots for use as houseplants as well as in plant communities. The enormous demand for ivy plants both in Europe and in America, as well as the plant's inborn variability, leads to the constant occurrence of new forms with deviating habits.' The number of strains available is therefore constantly changing.

Ivy: a theme with many variations

Professional ivy growers all over the world of course never lack fine sounding names for their new strains. Among the all-green varieties there is a form with cristate foliage called 'Crispa'; an American strain with a maple-shaped leaf has been named 'Maple Queen'; a dwarf form with small foliage is 'Minima'; while 'Glacier', 'Goldheart' and 'Luzii' are among representatives of the variegated strains. But as an exception to the rule that such names are generally soon lost, we should mention the standard strain 'Gloire de Marengo' ('Variegata'), a climbing or trailing, very decorative variegated plant with practically unlobed foliage derived from *H. h. canariensis*; it is occasionally cultivated in tree-form. Nobody has been able to explain why this peaceful plant was given a name commemorating one of Napoleon's bloodiest battles in the war against Austria (June 14, 1800). Whatever the reason, 'Gloire de Marengo' still flourishes. Perhaps it was developed in France?

Care: although it is considered as a proverbially simple plant and has the somewhat denigrating title of 'domestic hero', it nevertheless presents problems in care in the long run. Is it really such an undemanding and hardy plant, able to tolerate any conditions? It is a fact that it puts up with a great deal. Its all-green varieties are modest in their light requirements, but dislike surplus water in the pot and in winter prefer a cooler situation. The latter often presents difficulties where, for instance, the ivy has been trained in an attractive pattern round the pictures on the walls of a heated living-room. In such a situation it must certainly not lack light, since otherwise it will develop long and spindly offshoots with widely spaced foliage.

In general hedera should be cared for in the same manner as *Fatsia japonica*. The ivy is self-clinging and attaches itself by means of its aerial roots. Detaching branches leads to loss of vigour. It therefore requires a position where it can grow undisturbed. There have been some remarkable experiences with these adhesive roots. It appears that many forms of floor covering, hardboard and other modern materials used to cover window-sills ▶

Hedera helix ssp. *canariensis* 'Gloire de Marengo', fam. *Araliaceae*

Hedera helix ssp. *canariensis*, fam. ▷ *Araliaceae*

contain elements which the ivy dislikes and that it prefers woodwork and masonry. It would be useful to compare experiences of this nature with those of other plant lovers. If a wall proves to be unsuitable, one might use a wooden or bamboo climbing frame attached to the wall.

H. h. spp. *canariensis* 'Gloire de Marengo' does not develop aerial roots. Nevertheless the young shoots like to grow upwards and where a plant is intended to grow against a wall it is necessary to provide some support in the form of a frame or nails.

Summer: put in shade or half-shade; the variegated forms especially should be kept out of the sun. Cool shade provides the best environment.

Winter: according to the rules, the plant should be kept in an unheated but frost-free room during its winter resting period. This lasts from October to January.

Watering: during the growing period keep constantly moist. Be careful not to let the plant dry out – it is a forest plant! Keep drier in winter, especially if the plant stands in a cool room. Water pH 5·5–6·5.

Feeding: from April to August give a nutrient solution every week.

Repotting: once the ivy is growing well it should be repotted at least once a year. It is not a bad idea to use plastic pots for the purpose, since the roots enjoy constant moisture. A clay pot may dry out quite rapidly. Ordinary prepacked potting soil is adequate, but a mixture containing leafmould and well-rotted cow manure is even better. The vigour with which the ivy grows up old walls proves that it is not a lime-hater, so the potting soil should not be too acid.

Propagation: the ivy can easily be increased from cuttings throughout the year; nevertheless this is best done in the autumn. Shoot tips or 10-cm (4-in) sections of branches are inserted straight into small pots at a soil temperature of about 20°C (70°F). The cuttings should be kept under glass or plastic until they have rooted. They are then hardened off and transferred to larger pots containing rich soil.

Diseases: excess warmth and dry air will encourage attacks by scale insects and red spider mite. Suddenly occurring foliage diseases may turn healthy plants into barren skeletons in a few days. Prevention is by spraying from autumn to spring.

Hedera helix 'Pittsburgh', fam. *Araliaceae*

Hemigraphis

Only two of the numerous species of this tropical plant are in cultivation; both are natives of Malaysia, which means that they belong in a hothouse. They are long-lasting plants with decumbent stems and beautifully coloured foliage. They are sometimes used as trailing plants as well.

H. colorata; leaf metallic green with a violet sheen on its upper surface; the lower surface is reddish purple. In *H. repanda* the upper surface of the leaves is dark olive-green, the reverse side is red. Both have white flowers growing in terminal spikes, but above all they are ground-covering plants for use in a tropical plant window.
Care: warm, damp air throughout the year. It is a typical plant for a shady position in the hot house. Water only with completely demineralised water; too much is equally as harmful as too little. See *Fittonia*, p. 215.

Hemigraphis alternata, fam. Acanthaceae

Hibiscus

Thanks to its modest requirements, its readiness to flower and the variety of its blooms, *H. rosa-sinensis*, a shrub well-known under the name Chinese rose, has become one of the best loved plants on the Continent for the balcony and the window-sill. It shares some of its characteristics with *Abutilon*, described on p. 114.

Origin: the reason for this similarity is the fact that both plants belong to the family of *Malvaceae*, whose 85 genera and more than 1500 species are represented all over the world. The most beautiful and interesting occur in tropical areas. But whereas abutilon, which in recent times has rapidly made itself at home in this country, came from the west and occurs between Mexico and the northern part of South America, *H. rosa-sinensis*, as the name indicates, is probably a native of southern China. It is said that it was discovered there in the early 18th century. In all tropica countries it is now regarded as one of the most valued plants.
Multilateral relationship: the genus *Hibiscus* includes not only tropical, but also winter-hardy ornamental shrubs, of which *H. syriacus* is regarded as one of the finest in central Europe. This species also originates in eastern Asia and is therefore incorrectly named. Among the domestic tropical plants is *H. cannabinus*, a hemp-like plant whose fibres are still used in industry. The species formerly known as *H. esculentus* has now, together with *H. abelmoschus*, been classifed under the genus *Abelmoschus*.
Appearance: the Chinese rose is a somewhat bushy evergreen shrub with receding side-branches and shiny green dentate leaves, oval at the stem end and finishing in a point. Growing wild in tropical regions the plant may grow to as much as 5 m (15 ft) in height. Grown initially as a potplant and subsequently in a tub it may, if not pruned too often, in the course of time exceed 2 m (6 ft). Apart from the well-known shrubby habit, it is also simple to grow the plant as a standard. This form may be seen every year in the Italian rose garden on the island of Mainau, Lake Constance. Since the hibiscus has a natural tendency to some extent to form a main stem, it is only necessary to remove the side-branches up to the required height to achieve this tree-form.

▶

Hibiscus rosa-sinensis, Chinese rose, fam. *Malvaceae*

Twigs growing at the top are trained and pruned to achieve a good shape. For your information: the oldest of these little trees on Mainau, including the various *Abutilon* species, were grown from cuttings about 1942.

Inflorescence: even young plants will produce the large funnel-shaped flowers with their wide open calyces and protruding stamens. As a rule the flowering season is said to be from early summer until autumn. However, in practice this is not always so. Apart from a resting period from November to January, the small shrubs, if well cared for, will untiringly produce one flower after another from every leaf axil. No mean achievement, this, for each flower keeps for only a day or two, but new buds open every day.

Species: the original colour of the hibiscus was a bright rose-pink; single flowers, 8–10 cm (3$\frac{1}{4}$–4 in) across. Meanwhile there have been some changes: apart from the single-flowered species there are now semi-double and double strains with flowers up to 12–15 cm (3$\frac{3}{4}$–6 in) across. To give some examples: *Hibiscus rosa-sinensis* 'Flore Pleno', rose-pink, double; *H. r.* 'California Gold', golden yellow, single; *H. r.* 'Miss Betty', clear yellow, single; *H. r.* 'Laterita Variegata', a Danish cultivar, orange-yellow with a scarlet centre, single flowers and deeply incised foliage; *H. r.* 'Flamingo', enormous flowers, yellowish pink in colour, single; very large, dark almost black leaves. In comparison to the foregoing, *H. r.* 'Cooperi' has inconspicuous small red flowers, but on the other hand its striking, multi-coloured asymmetrical foliage makes it a valuable possession. A species apart is *H. schizopetalus*, discovered in a tropical region of East Africa in 1879. Because of its remarkable lantern-shaped small flowers it has some resemblance to *Abutilon megapotamicum*. The elongated long-stalked red or orange-yellow flowers hang from somewhat arching branches. The reflexed petals are deeply incised; the pistil and stamens hang downwards. This species requires more warmth than *Hibiscus rosa-sinensis;* it must never be placed out of doors in summer, but is confined to the living-room throughout the year. In winter the temperature should not fall below 16–18°C (60–65°F).

Situation: *H. rosa-sinensis* is happiest in a warm place in a good light but without direct sunlight. As a rule a hibiscus grown indoors will flower less profusely, but for a longer period than those grown permanently in the garden or on the balcony; when the plants are brought indoors in the autumn, the transition is too sudden, which affects flowering. The most favourable temperature in winter is 12–15°C (54–58°F). If kept cooler, the plant will frequently be attacked by pests; at higher temperatures the foliage will turn yellow and the plant will grow spindly. In both cases flowering will be affected.

Watering: because of its extensive root system, its vigour and its untiring flowering, the hibiscus naturally requires generous and regular watering in summer. Make sure that the soil

Hibiscus rosa-sinensis, fam. *Malvaceae*

does not dry out, for in that case the buds will drop. From September onwards decrease the water supply. Occasional spraying is desirable.
Feeding: the plant needs a great deal of nourishment. From the beginning of the growing period until mid-August give a nutrient solution every week; then at longer intervals until the dormant period.
Repotting: in February, when new shoots are put forth. Young plants every year, older plants at longer intervals. Use prepacked potting compost with extra humus added.
Pruning: when buying young plants observe whether there is more than one stem. If not, induce bushy growth by shortening the upper branches. Older plants should be cut back drastically in February. Shorten branches which have grown too long, in order to improve the shape. For plants grown as standards see under Appearance.

Propagation: soft tip cuttings are rooted at a soil temperature of 24–26°C (75–79°F) and a high degree of atmospheric humidity; use an indoor propagator or the propagating bed in the greenhouse. May–June is the best time.
Diseases: diseases and pests are easily avoided. A draught may encourage the appearance of mealy bug and red spider mite. Dropping of the buds is not necessarily the result of over-dry soil, but may also be caused by insufficient nourishment or a change in temperature or situation. Leaf-curl indicates that the air is too dry.

Good advice: in cool and rainy summers the hibiscus should not be sacrificed to the vagaries of our climate; at such times it is happier indoors.

Hibiscus rosa-sinensis 'Cooperi', fam. *Malvaceae*

Hippeastrum – Amaryllis – Vallota

We usually speak of *Amaryllis*, but actually we are concerned with three genera, entirely different in origin and living requirements: *Hippeastrum*, *Amaryllis* and *Vallota*. All three belong to the family of *Amaryllidaceae*.

Origin: the genus *Hippeastrum* with its 60–70 species occurs in the area between Mexico, Peru and Brazil. Their natural habitat is in steppes or wooded regions with a distinct dry season. The genus *Amaryllis* from the Cape Province has only one species: *Amaryllis belladonna*; so has the genus *Vallota*, of which only the species *Vallota speciosa*, originating in the Cape, is known.
Appearance and species: 1. In its specific form hippeastrum has long since been lost, but it survives in new generations of hippeastrum hybrids, sometimes called 'growers' Amaryllis'. These descendants of the South American species *Hippeastrum aulicum* and *H. vittatum* have almost spherical pale brown bulbs, rounded at the bottom, in sizes varying with the strain and the quality. The leaves are 50–70 cm (20–28 in) long, narrow strap-shaped and pale green in colour; they appear at the same time as, or after the flowers and fade in the autumn. Depending on vigour, 1 to 3, occasionally 4, 50–70 cm (20–28 in) tall, hollow round stems, are produced, each ending in 2 to 4 large, funnel-shaped flowers, 18–24 cm (7–9½ in) across. Colours vary from white via salmon-pink to pale and dark red; bi-coloured flowers, streaked or with a differently coloured margin or throat, also occur. Flowering season from December to May.

Most garden centres and mail order firms offer amaryllis bulbs by colour, for instance red, pink, white, etc. These have been propagated from seed and their attributes are therefore not constant. However, there are also a large number of named strains among hippeastrum hybrids in the most wonderful colours, such as 'Apple Blossom', white with pink marking, 'Happy Memory', white with red marking, 'Fire Dance', red; not to forget the 'Picotee' strains in various colours, all edged in a contrasting

▶

colour. All these strains have been increased from offsets.

2. *Amaryllis belladonna* has firm, narrow, channelled dark green leaves. The flower-stem is solid, in contrast to that of the hippeastrum and bears its flowers mainly in summer. At one time the species was relatively unknown and difficult to obtain, but the situation changed when in the early sixties the South African amaryllis hybrids came into the shops. Thanks to their flowering season, which occurs as early as the beginning of November, these plants with their large, pear-shaped dark brown bulbs and their firm, compact, slightly broader foliage, fill a real gap in the season. The first flower-stem usually appears before, or at the same time as the leaves; the second follows soon afterwards. Colours vary from shades of soft pink to a velvety bright red, the original colour. After flowering the foliage dies down.

3. *Vallota speciosa* grows to only 30–40 cm (12–16 in) and shows to best advantage if planted in groups in a flat bowl. The bulbs are small and brown, an elongated oval; the leaves are up to 40 cm (16 in) long, 3 cm ($1\frac{1}{4}$ in) broad, dark green. The flowers grow in umbels of 3 to 10 on a stem about 30 cm (12 in) tall.

Hippeastrum, hybrid 'Apple Blossom', fam. *Amaryllidaceae*

Flowering occurs in late summer. Vigorous bulbs will produce several flower-stems, one after another. They develop a large number of offsets which, planted separately, will take three years to flower. To achieve rich flowering every year, older plants should rarely be repotted. The foliage does not fade; the plant is kept in cultivation throughout the year.

Cultivation-hippeastrum: newly bought bulbs should be started into growth from the end of December onwards. If they are planted in soil, use fairly large pots with a good drainage layer of crocks and pebbles in the bottom. Fill with loam and humusy soil; prepacked potting compost is satisfactory. To prevent fungal infection treat the bulbs with a product sold for this purpose. Plant the bulbs firmly up to half their depth, and water. Place in a warm position, at about 25°C (78°F). Water sparingly until the first shoot appears; do not increase the water supply until the plant is growing well and the stout flower-buds or the tops of the leaves (or both simultaneously) are visible. Use tepid water, pH 5–6·5. Now give the plant a position in good light and feed every fortnight until the end of the flowering period with a nutrient solution. It will take about eight weeks from the time the bulb starts into growth. The flowering season may be prolonged by placing a plant in full bloom in a cooler spot. Prevent seed formation. Do not cut the flower-stem until it has completely dried out. The bulbs can be planted at intervals until the end of February, to ensure flowering plants until May.

Summer care: after flowering place in a sunny, well ventilated position indoors or, from late May onwards, outside. Water and feed regularly. The foliage should now grow vigorously, in order to enlarge and feed the bulb, which has shrunk as a result of flowering. To be able to rely on a second flowering, the bulb must at the end of summer be visibly larger than when planted in winter. Properly cared for bulbs increase 1–2 cm ($\frac{1}{2}$–$\frac{3}{4}$ in) in size from year to year. My oldest hippeastrum bulb now has a circumference of 37 cm (15 in); by way of comparison a skittle ball has a circumference of 50 cm (20 in).

Resting period: there will be no subsequent flowering unless the bulb has a strict resting period from October to December. To prepare it for dormancy, start by gradually decreasing water and feed from early August onwards, finally ceasing altogether, so that the leaves will die off within four to five weeks. Now place the pots in a not too cool position 10–12°C (50–54°F). Towards the end of December take the first bulbs out of the pot, remove old compost and remnants of roots and plant in slightly larger pots with fresh compost as described previously, in order to start the bulb into growth once more.

Propagation: from the offsets mentioned above, or from specific seed (obtainable from a good seed merchant). Both methods take a long time. It is not advisable to grow plants from seed won from old specimens, as this will never be true. In any case it is better to prevent seed formation, which detracts from the bulb's vigour.

Diseases: apart from the fungus disease mentioned under Cultivation, these plants rarely suffer from other diseases or pests. This disease results in red spots and streaks on the leaves, and breaking of the flower-stems. The only method to combat the disease is by prevention – treating the bulbs before planting with one of the fungicides currently available.

Cultivation-amaryllis: bulbs of the South African amaryllis hybrids are treated in practically the same manner, but their growth rhythm starts about three months earlier. This means that the first bulbs are planted in mid-September, at the latest from early to mid-November. The bulbs have clusters of long roots which must not be damaged or removed, since these enable the plant to develop rapidly to the point of flowering within three to four weeks. After the foliage has died, a strict resting period should be maintained from July onwards, in order to be sure of subsequent flowering.

Soilless cultivation: these plants are extremely suitable for soilless treatment; magnificent results have been achieved. The difference in treatment only applies to the preparation for dormancy and during the resting period. The foliage will die if the nutrient solution is gradually withheld, finally leaving only a little water in the bottom of the hydro-pot to provide some moisture. The roots will thus be preserved and, after the resting period, will once more ensure vigorous growth, when fresh nutrient solution is provided and the pot is placed in a warm position, 25–30°C (78–86°F). Vigorous growth and four flower-stalks every year are no exception. Propagation from offsets may often provide new flowers after three years. In this method the fungus disease hardly ever occurs.

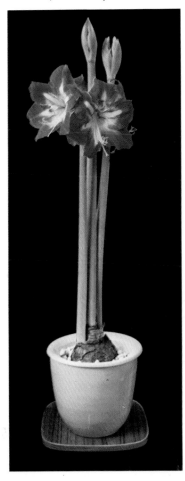

Hippeastrum hybrid in soilless cultivation, fam. *Amaryllidaceae*

Howea – Kentia palm

The family relationships of this beautiful palm, very suitable for room cultivation, are somewhat confusing. The genus *Howea* contains only two species. The botanical name *Howea* is derived from Lord Howe Island in the Pacific, of which Kentia is the capital.

Origin: the above island is the native habitat of the plant. Botanists have concluded that the two species of the genus *Howea* occur only there. So far they have not been found elsewhere.

Appearance: the two species are very similar; the difference lies chiefly in their respective size. *H. belmoreana* grows to 8–10 m (26–33 ft); *H. forsteriana* may achieve more than 15 m (48 ft). However, when grown in a palmhouse, where they belong, the difference between the two is hardly noticeable. An amateur growing them for many years as houseplants will not be aware of it at all.

In the course of time these palms can grow from pot plants into tub-plants. The leaves, 2 m (6 ft) long in the case of *H. belmoreana*, 3 m (9 ft) in *H. forsteriana*, are of equal length in room culture. Older specimens of howea in particular differ from other palms by the strikingly thickened base of the stem. The graceful dark green leaves on slender stems are spread fanwise; in *H. belmoreana* they are more erect and have a clearly visible central nerve; in *H. forsteriana* they are slightly curved; the central nerve is visible on both sides and is scaly on the lower surface. For room cultivation both species are usually cultivated on multiple stems. In older, thick-stemmed specimens the spadices, which may grow to more than a metre (a yard) in length, hang among the lower leaves like flaccid sausages.

Care: at one time the kentia palm was one of the most popular indoor palms, not least because of its modest requirements. The plant needs little light, which might be explained by the fact that in the jungle young plants lead the life of undergrowth and have to survive in deepest shade. They are kept indoors throughout the year and should overwinter in a moderately warm temperature, 14–18°C (57–65°F). Night temperature 12–14°C (54–57°F).

Watering: palms are not very thirsty and as they do not actually become dormant in winter, they must be watered sparingly throughout the year, depending on the room temperature; a little more frugally in winter. The water must not be too hard, as this would lead to calcification of the soil: the plants are rarely repotted. The first signs of calcification are the well-known brown leaf points. It is advisable to use water with a pH of 5–6.5. In summer the palms may occasionally be submerged in lukewarm water, to which a little nutrient has been added. In a mild shower, place the plants out of doors to rinse the dust off the foliage.

Feeding: weekly from March to July. Apply the nutrient solution only on to damp soil.

Repotting: if absolutely necessary this should be done in March or April. Check the pH of the soil. Use a humusy, moderately heavy mixture. Pre-packed potting compost is also suitable.

Like most other palms, howea grows best in a deep palm pot. This need not necessarily be of clay; possibly one of those modern plastic plant cylinders is even better (see photograph on p. 16), provided one is very careful when watering, making sure that no water can collect in the bottom.

Propagation: only from seed that is guaranteed fresh. Depending on the time it has been imported, it will take 2–9 months to germinate. The most suitable time to sow is March. The seed will germinate at a temperature of 25–30°C (78–86°F).

Diseases: the plant is subject to various foliage diseases which may result from too sunny a situation or too low a temperature. Too much sun in summer may also cause scale insect, mealy bug, red spider mite and thrips.

Howea belmoreana, Kentia Palm, fam. ▷ *Palmae*

Hoya — Wax flower

To start once more with the name: although hoya sounds as if it might be Latin, the name is of English origin. Thomas Hoy, after whom the plant has been named, lived in the second half of the 18th century and was gardener to the Duke of Northumberland as well as a plant grower.

The correct name for this plant should have been *Asclepias*, but this has long since belonged to another plant: *Asclepias curassavica*.
Origin: the genus *Hoya* includes about 200 species, all occurring from southern China, via the Indonesian Archipelago to deep into Australia. In this country only about six species are found as ornamental greenhouse plants or in show collections, while only two are generally cultivated as houseplants. This in contrast to the United States, where a whole series of species and their magnificent strains form part of the current assortment of houseplants.
Appearance: in this country *H. carnosa*, the Wax flower, is best known. It is an evergreen climber with supple, greenish brown stems which subsequently become woody; they grow to several metres (feet) in length. The leaves are broad, a pointed oval, 5–8 cm (2–3½ in) long; the main nerves, as it were, divide them into two. They are slightly suc-culent (fleshy), and a glossy dark green in colour. The white spots which occasionally occur on the leaves are inherent and do not indicate disease.
Inflorescence: star-shaped waxy flowers, white or pale pink, with a bright red centre; they are fragrant, particularly at night. They grow in groups of 12–15, forming fairly large, short-stemmed umbels, which last for a long time. In warm weather they exude honey-sweet drops of nectar. The flowering season occurs between the end of May and the autumn and lasts for weeks. Never remove flowering or faded umbels, for the hoya flowers not only on new shoots, but also on short offshoots on the old wood. Hence the rule: the older the plant, the more profusely it will flower. Unfortunately there are plenty of wax flowers which, in spite of excellent care for no apparent reason never flower and spend their lives as foliage plants. A white-variegated form, 'Variegata', which has yellowish white, red-edged leaves, is rarely-flowering by nature; it is slightly less vigorous, but very beautiful. The Australian *H. australis* rarely occurs here. A typical feature of these plants is that they have no red spot in the centre, but fade from white into red from the centre outwards. Flowers in the autumn.
H. bella is a dwarf form with horizontal stems bending forwards, and densely covered in small pointed leaves, only 2·5 cm (1 in) long. Flowers in pendulous umbels from the top of the stems; they are waxy white in colour, with a star-shaped red spot in the centre. Summer-flowering. This is a magnificent plant, happiest in hanging pots or in orchid baskets. It requires a great deal of warmth.
Care: keep indoors throughout the year. Experiments in placing the plant out of doors are usually unsuccessful. In summer a warm place in good light, without direct sunlight; well-ventilated, but definitely out of a draught. Spray when the plant begins to put forth new shoots, but not again before and during flowering. *H. bella* is best placed in a warm plant window with a high degree of humidity. After flowering, from autumn onwards, the resting season starts (October to February). Winter temperature 10–12°C (50–54°F); good light, no draught. A higher temperature will cause the plant to develop lanky winter stems which do not flower and also have a detrimental effect on the vigour and flowering of the rest of the plant. Warmth in winter will further lead to badly developed young foliage and attacks by pests.
When, after overwintering in a cool spot, the first tips of the young leaves appear, the plants should not immediately be transferred to a heated living-room, but should have a transitional period of about two weeks at a temperature of 15–18°C (58–65°F). In an ordinary home this requirement cannot always be met. As a rule the problem can be solved by placing the plants behind a curtain out of the sun on a window-sill where there is no heating radiator underneath.
Watering: too much or too little water are the chief causes of bad growth. The correct dose is: freely from the time the plant is growing well until after flowering; then gradually decreasing in preparation for the resting period, and from October on-

Hoya bella, fam. *Asclepiadaceae*

wards, when the plant is kept in a cool position, it should be kept only just moist, so that the soil-ball does not dry out. When it is moved to a warmer position at the end of the winter, slowly increase the water supply. The water need not be soft, pH 6·5–7.

Feeding: vigorous plants which flower regularly should be generously fed every fortnight from the time they are growing well until the maximum flowering season. This applies in particular to older plants, which will not have to be repotted, since otherwise they will fail to flower in the following year. Slowly growing specimens should be fed carefully only once a month. Excess feeding will interfere with flowering and is harmful rather than beneficial.

Repotting: older plants should be repotted very rarely, since, as mentioned above, this will interfere with the bud formation. If repotting becomes unavoidable, it should be done very carefully in an only slightly larger pot. To encourage drainage supply a good layer of large pebbles in the bottom of the pot. The best time to repot is before the new shoots appear, in February–March.

Soil mixture: these plants require a sandy-loam, very porous soil with neutral acid-reaction, pH 6·5–7. The addition of peat is therefore incorrect; normal, undecalcified sand is desirable and the addition of pieces of charcoal is recommended. Larger specimens planted in tubs can tolerate a soil mixture containing slow-acting nutrients, like oleander (p. 270).

Propagation: in adequate heat and humidity, for instance in a plastic indoor propagator, plants may easily be raised from cuttings. The best time is May–July. Use young shoots, already slightly woody at the base. Cutting 1–2 cm ($\frac{1}{2}$–$\frac{3}{4}$ in) below a leaf-bud, taking one pair of leaves and a terminal bud, is adequate. Use small pots with special cuttings.

Diseases: too warm a situation in winter leads to severe attacks by greenfly on the underside of the leaves. The fungus disease downy mildew often follows. Mealy bugs may occur in the axils of the leaves. It is better to dip the plant in a solution of pesticide than to spray it.

Important: the larger wax flowers need to be trained along a cane or bent wire. The shoots should always be tied on the same side, since the plant is very sensitive to the direction of the light. Unfortunately this means that the flowering side always faces the window, away from the room. From the time the buds appear the plant must not be turned or moved, otherwise the buds will drop. To prevent this indicate the pot's position. Finally it should be mentioned that wax flowers can grow to a ripe old age.

Hoya carnosa, the Wax Flower, fam. *Asclepiadaceae*. This species has larger and more upright flower clusters than *Hoya bella*

Hyacinthus — Hyacinth

In climatically suitable countries all over the world the demand for hyacinth bulbs for forcing in the appropriate season every year far exceeds that for other pot plants, even though the hyacinth is not a true houseplant at all. No doubt the delicious scent plays an important role in this connection.

World famous: before we talk about the practical instructions, we should first say something about the history of this plant. *Hyacinthus orientalis*, a bulbous plant from Asia Minor, has been known in western Europe since the middle of the 16th century. It appears that the bulb was gradually introduced to these regions after the capture of Constantinople by the Turks in 1453.

The Netherlands: by a roundabout route via Italy and France, the hyacinth arrived in the Netherlands where, after the tulip mania had abated, it was greatly valued among growers and became one of the best-loved flowers of the baroque era, even though its career was less spectacular and speculative than that enjoyed earlier by the tulip. Whereas around 1600 only four types were known, by about 1700 there were already 2000 strains. Calculated at our present rate of exchange there were bulbs which sold for over £600 each! Large areas in the neighbourhood of Haarlem were reserved for the cultivation of hyacinth bulbs. After 1700 the enormous interest for the plant gradually faded. After 1800 the best quality bulbs cost 'only' a few (golden) guineas.

The present: the cultivation of hyacinths is once more concentrated round Haarlem and Lisse, but it is now aimed at the international market. If you look in a bulb catalogue anywhere in the world, the famous old strain 'Delft Blue' will undoubtedly be included.

Species: some catalogues offer indoor hyacinths simply by colour, but this way you will not know whether you will receive the strains most suitable for forcing. The following are some of the strains which will certainly not cause problems: 'L'innocence', pure white; 'Jan Bos', red; 'Ostera', blue and 'Pink Pearl', pink. Other bulbs sometimes prepared for early flowering: 'Amsterdam', dark salmon-pink; 'Anna Maria', pink, and 'Tubergen's Scarlet', bright carmine-red. The so-called multiflora hyacinths which, instead of having one large flower spike, produce a large number of graceful little racemes, are also used for forcing. These strains are available by colour only.

Hyacinthus orientalis 'Blue Jacket' (forced), fam. *Liliaceae*

At one time the so-called Roman hyacinths, little bulbs producing small white or blue flowers, were much used for forcing; they are still available though usually only through specialist bulb nurserymen.

Forcing hyacinths for pots

Prepared early, half-late and late bulbs are used for forcing; these are planted at intervals in order to have flowering plants from Christmas until Easter. The bulbs must not be damaged, bruised or blotched; they must feel firm and, above all, have an undamaged base.

Planting: the earliest time for planting is in early September, the latest mid- to late-October. If the bulbs are potted separately, use proportionately tall pots; if planted in groups place them at distances of 6–8 cm (2½–3½ in). Provide a good drainage layer in the bottom of the pot. Use sandy, porous compost, not too rich. Before the bulbs are planted they should be dusted with powdered charcoal; they should then be pressed in the soil to such a depth that 1 cm (½ in) only of the nose is exposed. Cover the surface with a 1 cm (½ in) layer of sand and soak once.

Bringing into growth: circumstances frequently make it impossible to bury the bulbs in the garden to begin with. In that case the pots may be kept in a closed wooden container filled with compost and placed in a cool, but frost-free position (north-facing balcony, unheated cellar). The compost should be kept constantly moist. If plastic containers are used, be careful when watering. Rooting takes eight to ten weeks.

Forcing: the pots are taken out of the container and surplus compost is removed; keep the plants in a cool and dark position for a transitional period of eight to ten days. Place little paper cones over the now visible growing points and wait until you can feel the flowerbud. Now the plant may be placed in good light at a temperature of about 15°C (58°F). From then onwards the flower will develop.

Important: the quality of the flower is determined in the period between removal from the container and transfer to a warmer position. During this time changes in temperature, transition from dry to humid air and from dark to excessive light (in sunlight) should be avoided. After flowering, continue to water and at the same time provide feed for the bulb until the leaves have withered completely, when the bulbs should be cleaned and kept in a dry place until the time for potting again.

Forcing hyacinths in water

The same general rules apply as given for cultivation in compost. The prepared bulbs should be placed on their glasses in their order of flowering; later flowering bulbs are kept in a cool, but frost-free place until forcing can start.

Bringing into growth: in mid-September at the earliest, place the bulbs in semi-darkness in a warm, well ventilated spot, for example on top of a bookcase. A circle of rootlets must be formed round the base of the bulb; it will look like a string of small white beads.

Then the bulbs are placed on hyacinth glasses in such a way that their base is a few millimetres above the surface of the water and does not get wet. Cover the nose with a paper cone.

Water: must not contain chlorine, nor be too hard. It is advisable to use rapidly boiled water. Pure rainwater is even better. The water must not be changed throughout the period of cultivation.

Rooting: place the glasses with the bulbs in a cool and dark position. If no dark room is available, surround the glasses with paper cuffs or tinfoil. As soon as the roots have reached the water, increase the distance between bulb and water to 2 cm (¾ in); after this fill up to this level. Now wait until the flower-bud can be felt. Gradually accustom the plant to light and warmth (see also under pot cultivation). Do not remove the cone until it is slightly pushed up by the bud. The bulbs should now flower. After flowering treat the bulbs as described under pot cultivation, but plant in the garden in autumn, as hyacinths grown in water cannot be forced again.

Note: if the flower-bud stops growing, this is due to too rapid forcing or too dry air. It can be counteracted by returning the plant to a cool position and shortening the longest leaves to 8–10 cm (3¼–4 in).

Hyacinthus orientalis 'Delft Blue' in water, fam. *Liliaceae*

Hydrangea

Among the few saxifrages proved to be suitable for room cultivation, the hydrangea is undoubtedly the most important in every respect. Without exaggeration it may be said that it has by now had a well-deserved place in house and garden for nearly two hundred years.

Origin: hydrangeas belong to the oldest flower nobility of Asia Minor. Of the approximately 90 species, in addition to which a number were subsequently found in North and South America, *H. macrophylla* in particular became a valuable plant.

Appearance: in their original habitat they are deciduous shrubs, compactly branched and growing up to 4 m (12 ft) in height. As a pot plant it is now usually cultivated on a single stem. It may later grow into a robust tub plant; as a garden shrub it grows to over 1 m (3 ft) tall. The leaves are 7–15 cm (3–6 in) long, fairly thick, large-toothed along the edges; they are beautifully veined and opposite-growing. The leaf-stems are short. The flowers grow in spherical groups, 15–20 cm (6–8 in) across, in the colours white, pale and deep pink to carmine red and blue; the latter is achieved by artificial means. The natural flowering season is in summer, but as a result of special methods of cultivation this has been brought forward to the time just before and just after Easter. The flowers keep for a long time.

Care in summer: a hydrangea bought in flower should be placed in shade in a cool and ventilated position. Some degree of air moisture is desirable. After mid-May put the pot into the ground in a shady and sheltered spot in the garden or place it on a balcony out of the sun. If neither is possible, the only solution is a place by an open window, with plenty of fresh air. Faded flowers must immediately be removed. In the compact, single-stemmed new strains, a pair of leaves must be removed together with the inflorescence. In long-stemmed, multi-flowered older strains, the flower-stem is shortened by 2–3 pairs of leaves.

After flowering: if after flowering you wish to keep the hydrangea as a garden plant, it should be planted in a shady and sheltered position after the faded blooms have been cut. The best method is to dig a fairly large hole, filling it with a mixture of damp peat and some sharp sand. If, on the other hand you want to keep the plant as a houseplant, it should not be moved from its position until the foliage dies.

Care in winter: after the leaves have dropped, the plant must be brought indoors at the first sign of night frost, cleaned, and placed in its winter position. It likes to spend the dormant season in a cool room at 4–8°C (40–46°F). To bring forward its flowering, the hydrangea should gradually be moved to a warmer position in good light. Otherwise await the natural start of growth; as a rule the terminal buds start to swell towards the end of February.

Watering: the botanical name indicates that the hydrangea, with its enormous inflorescence and large, smooth foliage requires a great deal of water during its growing and flowering season. In normal circumstances it will empty its water-filled saucer twice a day and as soon as the soil-ball dries out even slightly, the plant will droop. When this happens the only remedy is to place the pot temporarily in a large outer pot filled with soft, tepid water up to the rim of the plant-pot; leave it until you see no more air bubbles. From August until the foliage has dropped gradually decrease the water supply. During the resting period you should regularly check whether the soil-ball has dried out. Gradually increase watering only when the new leaves appear and the plant is put in a warmer position.

Hydrangeas tolerate lime to a certain degree only. For plants with flower colours between white and carmine red, fairly soft water with a pH of 5·5–6 is adequate. Blue-flowering hydrangeas, whose colour has been brought about in the nurseries by the use of special soil mixtures and the addition of sulphate of aluminium, must definitely be watered with demineralised water with a pH of 4–4·5. The water must always be lukewarm, even during the resting period in winter.

Feeding: from the time the plant starts into growth until mid-August feed every week, following the instructions. Use a lime-free solution.

Repotting: initially every year, later every two to three years, in large pots; use tubs for older plants. The compost should be light, humusy and peaty. Pre-packed potting compost mixed with some extra peat fibre is also suitable.

Pruning: the flowers appear from the terminal buds of the new wood and pruning this would of course prevent flowering. Remove only unsightly branches. The best time to do this is April–May.

Propagation: from shoot tips in a warm propagating frame. It is a job for the professional.

Diseases: greenfly are encouraged by draughts. Thrips and mildew should be treated in accordance with instructions. Yellowing of the leaves (chlorosis) is the result of too limy soil. Flowers turning green or deformed indicate a hydrangea virus, to which some species are particularly sensitive. Affected plants should immediately be destroyed.

Hydrangea macrophylla, fam. *Saxifragaceae*

Hypocyrta

With one exception the nine known species of this graceful gesneria genus, of which three are in cultivation, belong to the large group of Brazilian jungle plants. From this fact we may conclude that they are happiest in a half-shady position in a heated plant window with a high degree of humidity.

H. strigillosa is an unbranched semi-shrub which may grow to 60 cm (2 ft). *H. glabra*, which from time to time is sold in supermarkets as a 'fashionable' plant, can be recognised by its glossy dark green leaves, resembling those of *Buxus* (box). The relatively unknown *H. nummularia*, which has trailing, red-haired main stems with short, thick side-stems, is an attractive hanging plant.

Inflorescence: in Germany the plant has the nickname 'kiss me': the somewhat inflated corona, usually scarlet with a paler margin, resembles a mouth puckered for kissing.

Care: as for *Columnea* (p. 173).

◁ *Hypocyrta strigillosa*, fam. Gesneriaceae

Hypoestes, fam. Acanthaceae ▽

Hypoestes

This plant was formerly called *Eranthemum* and even now is often sold under this name. The two species introduced to cultivation, *H. sanguinolenta* and *H. aristata* (syn. *H. taeniata*), are so similar in appearance that even experts have difficulty in distinguishing these two unusual acanthads.

Both are shrub-like plants originating in Madagascar, with beautifully marked foliage and insignificant lilac or pale purple flowers. As the little shrubs tend to lose their lower leaves, they should be pruned in order not to let them grow beyond 40–50 cm (16–20 in). The cuttings can easily be rooted in a heated propagating frame.

Position: an airy plant window without direct sunlight; warmth and humid air during the growing period. Water liberally and spray with soft water. Care as for *Fittonia* (p. 215).

Impatiens wallerana — Busy Lizzie

Like many other plants in cultivation, the Busy Lizzie or Balsam has undergone a long period of development. In grandmother's day the balsam plant was confined to the window-sill, but today the many hybrids may be used in several ways in the home and in the garden.

Origin: the ancestors of our Busy Lizzie, *I. sultanii* and *I. holstii*, can probably still be found in the mountain forests of tropical East Africa and the adjoining islands, especially Zanzibar. Our cultivated Busy Lizzie was given the specific name *I. wallerana*.

Appearance: it is an herbaceous semi-shrub, growing to heights of 30–60 cm (1–2 ft). The thick fleshy stems are freely branched and very succulent which, together with the plant's tropical origin, indicates that it dislikes cold. The pale green leaves, in many hybrids white-variegated, grow on stems; they are pointed-elliptical in shape, with slightly dentate edges. The smooth surface of the foliage and of the stems indicate a great need for water.

Inflorescence: with the exception of the darkest period of the year the flowers, in various shades of red, appear in abundance throughout the year; they grow from the axils of the leaves.

Species: some compact, low-growing strains suitable for pot culture are: *I. wallerana nana* 'Firelight', bright red and orange; Baby Hybrids ('Rosa Baby', 'Orange Baby', etc.); the American 'Blaze', orange-scarlet; all these are 20–25 cm (8–10 in) tall. Among forms suitable both for pot cultivation and for use on balconies or as bedding plants we should mention the F_1 hybrids belonging to the Imp series, with white, salmon pink or violet flowers. Among the older, taller-stemmed strains are the carmine red 'Ammerland' and the orange-red 'Liegnitzia'.

General care: in summer a position in light shade is best, that is, out of direct sunlight. In autumn and spring the situation should be half-shady. If kept in good light in winter the plant requires a fairly high temperature 20°C (70°F); in a darker spot 15°C (58°F) is adequate. If it is exposed to temperatures of 12–14–C (54–57°F) for long periods, the foliage will drop and the plant will fail to flower. If the soil is too wet all the stems will rot.

Watering: the warmer and lighter its position, the more the Busy Lizzie should be watered. Always use soft water. Opinions vary on whether or not the plant requires moist air. In any case you should not spray directly on to the flowers, as this will cause staining.

Feeding: feed weekly only during the growing period from March to September.

Repotting: if really necessary this should be done in spring or summer, using light, humusy soil (pre-packed potting compost).

Propagation: from cuttings or from seed, which will take ten days to germinate.

Diseases: sun in summer leads to attacks by red spider mite; if the air is too dry greenfly and whitefly will also occur.

Impatiens wallerana, fam. *Balsaminaceae*

Iresine herbstii

As already mentioned in the description of the *Amaranthaceae* family, *I. herbstii* and its numerous strains are as yet little known as pot plants, but in Holland it is among the most valued houseplants and in the United States it is available everywhere, together with the species *I. lindenii*.

The beautiful wine-red, clearly heart-shaped leaves with paler veining and the red stems make this native of Mexico an ornamental foliage plant of the first rank. More than this it cannot be, for the flowers are very small and insignificant, unlike those of other members of the family, such as *Amaranthus* and *Celosia*.

Care. Unjustifiably the *Iresine* is still regarded only as an annual bedding plant, kept in shape in summer by pruning. I have now tried it out for myself: potted in hydroculture, kept in a warm, somewhat humid environment in the growing season, cooler in winter, the plant will keep for at least two years. In any case pruning will provide cuttings for constant new supplies. It is important to avoid cold water and drying out of the root-ball.

Iresine herbstii, fam. *Amaranthaceae*

Ixora coccinea

This magnificent greenhouse plant, a native of India, is now frequently found among other tropical beauties at the florists'. The ixora has the advantage over most of the others in that it keeps better. It grows best and flowers most readily in a densely planted flower window.

Appearance: of the more than 200 species of the genus *Ixora* which occur in the tropics of every continent, only *I. coccinea* is cultivated here. It is an evergreen shrub; as a pot plant it grows to about 1 m (3 ft) in height. It has glossy, leathery leaves, 5–10 cm (2–4 in) long; the scarlet flowers appear in broad umbels. Some of the hybrids are internationally famous, for instance the profusely flowering dark orange 'Biers Glory', 'Morsai', with large, pale orange flowers, often streaked with red, and the salmon-pink 'Shawii'.

Flowering season: ixora is summer-flowering; a well cared for plant will flower without interruption from May to September. Professionals frequently complain that the plant occasionally drops its buds without apparent reason. It is also annoying that in pruning, the terminal flower-buds are always lost.

General care: it belongs in a tropical plant window or in a greenhouse. During the growing period, from March to August it requires a slightly shaded position out of direct sunlight; if possible it should not be moved during its winter dormancy. Soil temperature 18–20°C (65–70°F) throughout the year; air temperature according to the season: not below 16–18°C (60–65°F) in winter, otherwise the foliage will drop and the next season's flowering is checked.
Watering and spraying: use only softened water (pH 4·5–5·5); a high degree of atmospheric humidity in spring and summer is essential. When the resting season begins gradually reduce the water supply and the atmospheric humidity.
Feeding: feed generously in the growing season.
Propagation: this is a job for the professional.
Diseases: in too sunny a situation the leaves will curl. Too dry air or too cool a position may lead to an attack by scale insects.

Ixora coccinea, fam. *Rubiaceae*

Jacaranda

In a flowerpot the jacaranda is of course not so large as the purple-flowered tree in its native Brazil. We only know it as a young plant with feathered foliage resembling filigree. It reminds us of *Mimosa pudica* and also a little of *Grevillea robusta*, the Australian silk bark oak.

Jacaranda mimosifolia, fam. *Bignoniaceae*

Because of its resemblance to the former, it has been given the name *Jacaranda mimosifolia*. In the description of *Grevillea* on p. 220 you can read how it should be cared for, but it must be remembered that, as it originates in the tropics, this attractive plant is much more sensitive. It is popular as a decorative plant in a plant community, but if you want it to give you pleasure for a long time, it must be put in a tropical plant window, in good light but semi-shade, warm and humid. The winter temperature should not fall below 16–18°C (60–65°F).
Watering: naturally only with soft water; sparingly in summer, even less in winter.
Feeding: from spring until autumn give a lime-free nutrient solution every two weeks.
Propagation: at a soil temperature of about 25°C (78°F), the jacaranda can relatively easily be raised from imported seed.

Jacobinia

This is one of those plants which in the last 150 years have occasionally been forgotten and have then once more returned to favour. We do not know whether or not this is due to a change in name, but it is certain that when its name was changed from *Justicia* to *Jacobinia*, it was introduced as a new plant.

Origin: the distribution of about 50 species of the genus *Jacobinia* is restricted solely to tropical areas of the New World. Only two of them, both natives of Brazil, proved with some extra care to be suitable as long-lasting houseplants.

Selection: it should first be mentioned that *J. carnea*, a vigorous, freely-branched shrub which, bought as a flowering plant 40–60 cm (6–24 in) tall, can with good care grow to 1½ m (4½ ft). The short-stemmed leaves are smooth-edged and downy. The flowers are flesh-coloured, pink or orange and are slightly sticky to the touch; they appear in large groups at the top of the stems. Flowering season from spring until autumn. Secondly there is *J. pauciflora*, occasionally still sold under its old name *Libonia floribunda*. This is a small, shrubby plant with small, leathery leaves; it grows to 30–60 cm (1–2 ft) in height and from February to April has many small red and yellow flowers hanging from short stalks placed in the axils of the leaves.

Care: *J. carnea* is best kept throughout the year in a half-shady plant window with humid air; it should not be closed in by other plants. Temperature in winter 18–20°C (65–70°F). For *J. pauciflora*, which has a different growth rhythm, the best temperature is 8–10°C (46–50°F) until the beginning of flowering and then 16–18°C (60–65°F). Only *J. pauciflora* tolerates, and in fact requires, a sunny position in the garden in summer.

Watering: water very liberally during the growing period, for if the soil dries out the foliage will drop. Tepid water, pH 5·5–6. A higher degree of atmospheric humidity for *J. carnea* in summer and for *J. pauciflora* during the flowering season. During the dormant season in winter a not too dry living-room atmosphere will be tolerated. Spray the plants from time to time.

Feeding: weekly during the growing season.

Repotting: *J. carnea* should be repotted in spring in roomy pots. Older plants, which have become bare at the base, should be cut back vigorously. *J. pauciflora* is repotted after flowering into small pots; prune as required. Pre-packed potting compost.

Propagation: in spring from cuttings in an indoor propagator, temperature 20–22°C (70–72°F).

Diseases: in *J. carnea*: leaf curl, greenfly or red spider mite may be the result of excessively dry air.

Jacobinia carnea, fam. *Acanthaceae*

Jasminum — Jasmine

The garden jasmine is widely known, but there are also species on the borderline between garden and house plants. These include somewhat sensitive plants for the unheated greenhouse, which overwinter in a cool, frost-free position, as well as species belonging in the hot house. They are all worthy of consideration.

Species: the most hardy is *J. officinale*. It is a native of China, but also occurs in southern Europe. It is a deciduous plant, with fragrant white flowers from early summer to autumn. *J. sambac*, an evergreen from India, flowers in the same season and also has fragrant white flowers. Its hybrid *J. s.* 'Plena' is occasionally available. This climber belongs in a hot house. *J. polyanthum*, a climber from eastern Asia, flowers from January onwards; it has pink flowers, is deliciously scented and has feathered foliage. It requires an unheated greenhouse.

Jasminum officinale, fam. *Oleaceae*

Kalanchoë

The family relationships of kalanchoë have in recent years become more and more confused. Formerly only the red-flowering kalanchoë was known as a house plant, but since it has been joined by the *Bryophyllum* genus, the situation is not always clear to the amateur.

Kalanchoë blossfeldiana, fam. *Crassulaceae*

Origin: since the most recent botanical classification more than 200 species belong to the genus *Kalanchoë*. All the *Bryophyllum* species are now included in this group. The most beautiful and varied forms come from Madagascar, but many other interesting species occur in other regions as well: all over Africa into southern Arabia and also on the island of Socotra, which lies at the entrance to the Gulf of Aden and is famous for its varied flora.

Appearance: the genus includes small perennials, not more than 10 cm (4 in) high, small shrubs, climbers and even trees which may grow to 3 m (9 ft). For the amateur it is the succulent species, making undemanding houseplants, that are of greatest interest. Not only do they have attractive evergreen foliage, but the inflorescence is also of value.

Inflorescence: all these plants have a special characteristic in common: all *Kalanchoë* species are short-day plants. Their natural flowering season is in the winter months, but it can be arbitrarily extended by the short-day treatment now frequently applied by growers; as mentioned on pages 54 and 204 this is also used for the poinsettia.

Species: *K. blossfeldiana* is the best known species, and since it is fairly inexpensive it is much in demand. This plant, which grows to only 30 cm (1 ft) originates in the highlands of Madagascar. The foliage is dark green with a red margin and may be smooth-edged or indented. The compound pale red flowers appear from January onwards. In addition to the species there are now various strains on the market, larger or smaller forms, in colours ranging from orange-yellow to deep dark red. Thanks to the short-day treatment, flowering kalanchoë plants are now available throughout the year. For instance, there is the smallest, *Kalanchoë blossfeldiana* 'Compacta Liliput', only 10–12 cm (4–5 in) tall, with large leaves and red flowers. The 'Tom Thumb' series is of medium height and spherical in shape; the flowers are larger, some fairly early flowering, others late-flowering. 'Orange Triumph' is another attractive cultivar. 'Goldrand', up to 45 cm (1½ ft) in height, is one of the finest strains; exceptionally large flower umbels, each individual flower bright red with a yellow margin. Very suitable for room cultivation. An interesting novelty is *K. miniata*, with larger, bell-shaped red and yellow flowers. Height 20–25 cm (9–10 in). It pays to bring this plant through the winter — it is sure to flower again in the following year. A particularly beautiful species is *K. tomentosa*. a native of Madagascar, where it grows to 80 cm (32 in). It is a semi-shrub with a woody main stem, herbaceously branching. The leaves vary in shape, thick and fleshy with a silvery sheen; the crenate edges are covered in reddish brown hairs. A fine houseplant, although it rarely flowers. Among the hanging plants *K. jongmansii* is one of the best; separate, short-stemmed flowers appear from the thin, somewhat woody, hanging branches.

Care: to prevent the colour of long-lasting flowers fading, all kalanchoë species and hybrids should be kept in the best possible light during the winter; in summer they should be given a half-shady position; a sheltered spot on a balcony is also acceptable. If they are not regarded as 'disposable' plants, they should have a resting period from October until the new flowering season. Temperature not below 15°C (58°F) but at the same time not living-room temperature or dry air.

Watering: normally in summer, more sparingly in winter, depending on the temperature.

Feeding: from May till August feed approximately every two weeks.

Repotting: if necessary put in slightly larger pots.

Kalanchoë tomentosa, fam. *Crassulaceae*

Pruning: larger species, in particular, should be cut back after flowering. Smaller forms usually branch spontaneously. Overlarge plants become unsightly in the long run.
Propagation: from seed or cuttings, but hardly worth the trouble.

Diseases: diseases and pests rarely occur, but excess moisture may cause root rot.

Kalanchoë (Bryophyllum)

Although *Bryophyllum* now officially belongs to the genus *Kalanchoë*, the amateur will find it difficult to discover the relationship between the two groups of houseplants. Notwithstanding botanical regulations we are therefore treating them separately.

Origin: there are more than 20 species of former bryophyllums which now bear the name *Kalanchoë*. Most of them originate in Madagascar. Three of them are very easy to grow and even easier to propagate — two reasons for their popularity.
Species: *K. laxiflora* (formerly *Bryophyllum crenatum*), to 50 cm (20 in) tall, has pointed oval leaves, up to 12 cm (5 in) long, grey-green and fleshy; the edges are notched. The little buds along the edges grow into independent, rooted plantlets. Reddish brown flowers in umbels. *Kalanchoë daigremontiana* (formerly *Bryophyllum daigremontianum*) resembles the latter in many respects, but the leaves are narrower, more pointed and flecked with brown. The flowers are pink. *Kalanchoë tubiflora* (formerly *Bryophyllum tubiflorum*), to 70 cm (28 in) tall, leaves are a faded grey-green colour, brown-flecked, grouped irregularly round the main stem. The plantlets appear in small groups at the extremities of the leaves. Flowers red, bell-shaped. *Kalanchoë pinnata* (formerly *Bryophyllum pinnatum*) was a curiosity even in Goethe's time. It is no longer seen so frequently, probably because it grows so tall. The new strains developed since the beginning of this century all remain much smaller.

Care: in summer warm and airy; in the afternoon near a screened, south-facing window. To induce the plant to flower freely, it should be kept at 10–12°C (50–54°F) in winter. Bryophyllum plants are among the hardiest and least demanding of all house plants. Dry air is tolerated better than humidity. As with many other succulent plants, damage is caused by excessively damp soil; this may cause rotting of the main stem. To avoid this risk, grow the plants under soilless cultivation; they will flourish.
Watering: normally in summer, in winter so little that the soil-ball appears to have almost dried out.
Feeding: every two to three weeks in summer.
Repotting: not applicable, since older plants become unsightly.
Propagation: from rooted plantlets grouped together in small pots or bowls. Prick out after the middle of August to ensure a first flowering in late winter. New shoots can be rooted in spring.

Kalanchoë daigremontiana, fam. *Crassulaceae*

Kalanchoë tubiflora, fam. *Crassulaceae*

Lilium — Lily

At one time plant lovers regarded the pot plant lily with some suspicion, but this changed when towards the middle of the present century Jan de Graaff, an American of Dutch descent, started to cultivate it. The lily as a pot plant entered upon a great career.

Origin: the lily belongs to the oldest and most distinguished plant aristocracy. For Mediterranean peoples, to begin with the most ancient Egyptians, followed by the inhabitants of Greece, Byzantium and Rome, this noble bulbous plant became a symbol. In later ages the lily was planted in gardens and its many species conquered the world.

Modern lilies: the hybrid lilies we now know were developed as the result of many years of intensive work. In addition to the important cultivars grown by de Graaff and other American specialists, Dutch, Canadian and German hybrids were developed as well. The low and medium types are the most suitable for pot cultivation; as a rule they grow to 70–90 cm (28–35 in), do not need too much time to develop, and in addition to abundant flowering have attractive foliage. They are stem rooters: roots form on the stems above the bulbs; for this reason the bulbs must be covered with a 5–10 cm (2–4 in) layer of soil; they are best planted in groups. For planting bulbs separately it is best to use tall palm pots.

Species: *Lilium longiflorum* hybrids, with elongated, greenish white, funnel-shaped flowers are very suitable for room cultivation, but *L. auratum* hybrids are rather more exotic. Jan de Graaff picks out 'Golden Chalice', 'Mid-century' and 'Rainbow' hybrids and in particular his famous orange-red strain 'Enchantment' with which the lily's career started around 1950. It grows to 75 cm (2½ ft). The illustration on this page gives only the merest indication of the beauty of this lily. Other suitable forms to be grown indoors are: 'Destiny', 90 cm (3 ft); the golden-yellow 'Golden Chalice' itself, 50–75 cm (20–30 in); the red and yellow 'Harmony', 50–75 cm (20–30 in) and the bright red 'Paprika', 60–90 cm (2–3 ft). And this is only the beginning, for the development of lilies is by no means at an end.

Potting: lily bulbs can be planted at various times, even to achieve normal summer flowering, but as a rule it is preferable to force them for late winter or spring flowering and in that case the bulbs are potted in the

Lilium, hybrid 'Enchantment', fam. *Liliaceae*

autumn. The main thing is that the pot should be large enough and that good drainage be provided by means of an adequate layer of pebbles. In large plant bowls make a number of extra drainage holes. Recommended compost: 1 part leafmould, 1 part sand and 1 part peat with the addition of a tablespoonful of bonemeal per pot. Normal potting compost can also be used. First half fill the pot with the mixture, add a thin layer of sand, place the bulb on top and cover it entirely.

Forcing: to make them root, place the bulbs in a cool but frost-free place in winter. In February–March move to a warmer position, at 10–12 °C (50–54 °F). As soon as the first shoots appear give the bulbs a very airy position in good light, at 12–16 °C (54–60 °F). The first flowers will appear towards the end of April.

Watering: very sparingly from the time the bulbs are planted until they are moved to the warmer position; then gradually increase the water supply in accordance with the growth, but do not keep the plant too moist.

Feeding: every 10–12 days from the time the first shoots appear until flowering; then at longer intervals.

Keeping: the simplest method is to plant the bulbs in the garden after they have flowered. To bring them into growth for a second time, continue watering and feeding until mid-August. The leaves will wither in early autumn. After the dormant period it is possible to force the bulbs again, but this is not always successful with prepared bulbs.

Liriope

This plant is hardly known in this country, but elsewhere it is greatly valued as it is so undemanding. In the United States it is grown everywhere as the big Blue Lily-turf. It is high time that this most attractive member of the lily family, growing no taller than 30 cm (1 ft), should be introduced into the houseplant selection here.

Like *Ophiopogon*, which it greatly resembles, it is a native of eastern Asia. *Liriope* belongs to a small genus of only five species, of which *L. muscari*, a plant from China and Japan, is occasionally available. This species with its broad, grass-like foliage and up to 40-cm (16-in) tall flower-stems covered in small violet-coloured flowers, is often sold under its old name *Ophiopogon muscari*. It is the progenitor of many fine strains, one of which, *L. m.* 'Variegata', is illustrated here.

Flowering season: in late summer; the flowers, resembling those of the grape hyacinth, are succeeded by attractive, long-lasting berries. Position in summer either in good light or in semi-shade; in winter the plant is best kept at a temperature of 10–12 °C (50–54 °F) although a higher winter temperature, up to 18 °C (65 °F), is tolerated.

Liriope muscari 'Variegata', fam. *Liliaceae*

Lithops — Living stone

The multiple forms of this interesting species with its numerous strains look just like pebbles. Of all plants the lithops species provide the most striking examples of mimicry.

Mimicry: this means that lithops adapts itself to its surroundings better that any other succulent plant. Even in show collections it is quite difficult to distinguish the plants from the pebbles among which they grow. The protective colouring adopted as camouflage against a hostile world is only relinquished for a short time in the autumn, when the spherical leaves divide and the glossy, silky, white or yellow flowers appear.

Origin: like all the other remarkable species of the family of *Aizoaceae*, the living stones originate in South and South-west Africa, where the soil in the sandy deserts, consisting of loamy sand and gravel, provides the right living conditions. During the dry season and in the resting period, which here occurs in winter, the plant is able to withdraw completely into the soil.

Care: as with all succulents, the correct amount of water is the most important aspect of care. In this case watering is restricted to the growing season in summer, but even then they should be watered very sparingly as otherwise the leaves will swell up to such an extent that they burst open. From September until the first shoots appear they must be kept entirely dry.

Propagation: easiest from seed. The first flowering will occur two or three years later.

◁ *Lithops fulleri*, Living Stones, fam. *Aizoaceae*

Lobivia

Cactus experts do not entirely agree on the question of whether or not the lobivia can be regarded as a genus in its own right. However, they do confirm that the globular cacti now classified among lobivia are some of the most suitable plants for beginners, especially as these small cacti flower so readily.

Origin: this extensive genus, including very diverse species, originates in South America, especially in Bolivia, the northern part of the Argentine and Peru.

Appearance: medium-sized, absolutely globular to cylindrical plants. The ribs are often obliquely grooved or may be divided by swellings. The spines vary considerably in different species and classes: they may be very dense, fine and short, but also strong, hooked and intertwined; there are even forms with snow-white hairs. As

Lobivia densispina, fam. *Cactaceae*

Lobivia, fam. *Cactaceae*

a rule the flowers are 5 cm (2 in) in diameter, but larger and smaller forms, greatly varying in shape and colour, also occur.

Species: *L. densispina* has a short, cylindrical body, 10–15 cm (4–6 in) tall and tending to branch. The yellow to brown spines grow densely. Silky yellow flowers, 5 cm (2 in) in diameter. *Lobivia densispina* f. *leucomalla* is similar in shape, but has snow-white hair and dense, bristly spines. The flowers are yellow. *L. haageana* is also cylindrical, but less branched; it grows to 30 cm (1 ft) in height. Strongly spined; funnel-shaped flowers up to 7 cm (3 in) across, yellow to red with a dark throat. *L. jajoiana*, the queen among lobivia species, is closely covered in single flowers whose dark throat is enclosed by a dense ring of colourful stamens. The petals are yellow, wine or tomato-red. *L. pentlandii* develops runners and is a very popular species since it is very versatile. It flowers at night in wonderful colours from creamy white to carmine red. *L. tegeleriana*, a vivid green globular cactus, grows to 5–6 cm (2–2½ in) in height. Long, golden-yellow spines. The flowers are of a silky rose-red.

Care: in summer airy and sunny, in winter cool and light, 4–6°C (40–43°F). Slightly more humusy soil and a little more moisture than for most other cacti, but keep dry in the dormant period (from October until the buds are formed). pH of the water 5–6. Feeding well in summer encourages flowering; flowering period from May to July. Easily propagated from seed.

Important: lack of sunlight harms the beauty of the spines. Too warm a situation in winter leads to an attack by red spider mite.

Lophophora

This interesting individual is not without danger as you will realise when we mention the word 'mescalin'. The flattened globular plant with its long taproot is in fact the intoxicating poisonous plant which the Mexican Indians call 'peyotl' (English: mescal).

The desiccated 'buttons' on the plant contain the drug mescalin. The largest supplier of this drug is *L. williamsii*.

Care: a very sunny and warm situation with adequate moisture in summer; in winter it should be kept cool and completely dry. The soil must be mineral-retaining and light. The plant flowers every year and is easily grown from seed won from the long pink fruits.

Lophophora williamsii, fam. *Cactaceae*

Lycaste

This exquisite plant from Guatemala is one of the most easily cultivated orchids. The leaves are about 50 cm (20 in) long, the flowerstalk grows to 30 cm (1 ft); the soft rose-red flower, darker at the base of the lips, is 15 cm (6 in) across. Flowering season from November to March. Epiphytic orchid for the cold greenhouse.

L. virginalis (trade name usually L. skinneri) has large egg-shaped pseudo bulbs and forms 2–3 leaves, veined lengthways and dropping off every year. Cultivation in winter, cool and in good light; in summer in halfshade. After the leaves have dropped it is strictly dormant until it begins to put forth shoots once more. Keep moist in the growing period, do not spray; give liquid feed. The plant should never be kept so dry that the pseudo bulbs will shrivel. Also recommended for cutting.

Lycaste virginalis, fam. Orchidaceae

Mammillaria

On the subject of the genus *Mammillaria* all professional literature will mention the almost innumerable species, and at the same time the value of these plants for amateur growers and collectors. Various 'subgenera' have been created to bring some order into the classification.

Mammillaria rhodantha cristata, fam. Cactaceae

Origin: mammillaria species are natives of an area extending from California via Texas, Arizona and Mexico to the West Indies, from Cuba to Trinidad. However, most of the species occur in Mexico.
Appearance: they are mainly globular in shape, growing separately or in groups; some are short and cylindrical. A typical feature of all species is the warts which occur in a large variety of shapes and arrangement, and their equally varied spines. The spines may be short, or long and barbed, but may also be replaced by white hairs, as in the case of the Old Man cactus. Compared to the form and colour of the interesting thorns, the flowers, which in summer encircle the crown, although beautiful in colour, are small and insignificant. Many species bear scarlet fruits until the following spring; these are often more attractive than the flowers. When they are ripe, they will drop spontaneously, or else they can easily be detached. See also under Propagation. Some species tend to be cristate, forming a crest or cockscomb, in which the normally pointed embryo appears in a plane along a line or a widened band; botanists call this development 'fasciation'.
Species: all mammillaria species are worth cultivating. The following is a selection of the most beautiful: *M bocasana* forms groups of globes, each only 5 cm (2 in) in diameter,

Mammillaria wildii, fam. *Cactaceae*

covered partly in fine white, partly in brownish yellow spines. The flowers are creamy white with pink stripes. Fruits up to 4 cm (1¾ in) in length, an unusually pale carmine. *M. geminispina* develops white-thorned globes, up to 8 cm (3½ in) in diameter, which flower magnificently but also tend to form crests. *M. gigantea* is one of the largest: up to 18 cm (7 in) in diameter. *M. hahniana* is among the finest white-haired species; it has carmine flowers. *M. humboldtii* is also covered in snow-white spines. The small flowers are of a clear carmine red. Unfortunately it is relatively rare. *M. pennispinosa* develops a single globe from a thick taproot; it has unusual feathery spines and pale yellow-pink flowers with a pink throat and 2 cm (¾ in) long carmine-red fruits. *M. plumosa* forms large groups of little globes covered entirely in even longer and softer spines. Insignificant flowers and reddish green fruits. *M. prolifera* is an indestructible species, very suitable for beginners. It will grow and flower in the most unfavourable conditions. *M. rhodantha* is a very variable species, occurring in yellow, brown and red. The flowers are carmine to purple. *M. schiedeana,* with its silky spines, white flowers and pale carmine fruits, up to 2 cm (¾ in) in length, is one of the forms most valued by cactus experts. *M. wildii* likes to form groups of individually placed, pale green, short cylinders. It has beautiful spines and flowers profusely; reddish brown fruits. In addition to the species illustrated there is a variety bearing pink flowers. A hardy and undemanding plant. *M. zeilmanniana* is also among the species most suitable for a beginner. It is a globular cactus, 5–6 cm (2–2½ in) in diameter; white, often hairy spines; the deep pink to purple-carmine flowers grow in dense circles round the crown; green, insignificant fruits.

Care: apart from the green, soft-tissued species, such as *M. wildii,* which do not tolerate direct sunlight, all species require a very sunny situation in summer, since the colour and beauty of their thorns depends on it. This great dependence on light manifests itself in the way they grow towards the sun; this happens even with older plants. To prevent a lopsided shape, these plants must be turned occasionally. This will not harm either the flowers or the fruits. In winter the position should be light and cool; temperature 6–8°C (43–46°F). The soil mixture must be rich and contain mineral elements; regular feeding is recommended.

Watering: freely in summer; in winter keep practically dry.

Feeding: feed generously in summer, using a special cactus fertiliser.

Mammillaria, fam. *Cactaceae*

Manettia

The Florentine botanist Xavier Manetti (1723–1784) gave this beautiful South American plant its name. Although the climber was introduced into Europe more than 150 years ago, it is practically unknown in this country. It deserves more attention.

At most two of the approximately 40 species of trailing plants and small shrubs are occasionally found in cultivation, namely *M. bicolor*, with two-coloured flowers, and the very similar *M. inflata*, in which the red part of the petals is inflated, as can be clearly seen in the photograph.

Care: manettia species are plants for a temperate greenhouse. After flowering profusely from spring until autumn they require a moderate temperature during the dormant season in winter — not below 12–15°C (54–58°F). This can also be provided behind the curtain in a north-east facing window in a normally heated room. If you can provide these conditions, you will derive a great deal of pleasure from this graceful trailing plant. It is advisable to spray as a prevention against pests, using a harmless pesticide based on pyrethrum. Treatment otherwise as for *Bougainvillea*.

Manettia inflata, fam. *Rubiaceae*

Maranta

Although at one time the marantas could be grown only in a hot house or tropical plant window, they have now, as a result of modern methods of cultivation, become suitable for room cultivation. They are not easy to grow and require extra care, but the unequalled beauty of their foliage will compensate you for your trouble.

Origin: they are plants that can be cultivated for a variety of purposes. For instance, there is a green-leaved species with tuber-shaped roots from which arrowroot is produced in several tropical countries. Because of this the useful plant became known in Europe as early as the beginning of the 18th century, whereas the various ornamental species did not come to notice until nearly a century later. Most of them come from the tropical rain

Maranta leuconeura 'Kerchoveana', Prayer Plant, fam. *Marantaceae*

forests of Brazil, where they occur in open spaces rather than in the shade.

Appearance: the approximately 30 species of the genus *Maranta* as a rule do not grow beyond 20–25 cm (8–10 in). There are larger plants — 60–80 cm (24–32 in) and more — available under the name maranta, but these belong to other genera which are very similar in appearance. See *Calathea*, p. 145, and *Ctenanthe*, p. 181. The latter are somewhat more tender and sensitive than the true maranta, whose vigour is truly impressive, especially as ground cover in a plant window. They are all perennials, with short-stemmed leaves attached to horizontal runners. Depending on the species, the plants may be spreading or erect in habit. There are also species whose leaves flatten out during the daytime, but stand up at night. Only older plants produce the inconspicuous white flowers, growing in terminal spikes, from spring to early summer.

Species: only two variegated species with a number of hybrids are in cultivation. One of these is *M. bicolor*, whose leaves, including the stem, grow to 30 cm (1 ft) while the leaf alone is 15 cm (6 in) long and about 11 cm (4½ in) wide. On the upper surface the marking is dark green, with 6–8 large pale brown blotches on either side of the principal vein and the pale central streaks; the lower surface is reddish purple. It was introduced in 1823. The species *M. leuconeura*, a white-veined maranta, is now rarely cultivated, as its cultivars are more beautiful and less demanding. The most important strain is still *M. leuconeura* 'Kerchoveana', the Prayer Plant. It resembles *M. bicolor*, but the emerald-green upper surface lacks the pale central streaks flanking the principal vein and between the brown blotches. The lower surface is moreover blue-green, with red blotches. This is the strain considered most suitable for room cultivation. Even more showy is *M. leuconeura* 'Massangeana', with its beautifully marked, slightly smaller olive-green leaves. The white and dark brown marking is grouped along the central vein; in addition the veins have a silvery herringbone pattern. The lower surface of the leaves is dark red. A very unusual development has occurred in *M. l.* 'Fascinator'; here the silvery white lateral veins have turned orange-red on a velvety dark green background.

Care: naturally the chances of success are greatest if the plant is placed in a tropical window or a plant window with constant soil heat. In any case its situation should be in good light, but out of the sun; in the daytime the temperature should be constant at 22°C (72°F); at night it may drop to 16–18°C (60–65°F) in accordance with natural conditions. In winter a daytime temperature of 18–20°C (65–70°F) is sufficient; at night 16°C (60°F) but definitely not lower.

Watering and spraying: always use tepid, softened water, with a *p*H of 4–4.5. Water liberally during the growing season, but avoid excess water in the bottom of the pot as well as drying out of the soil-ball. From September to February water more sparingly.

Feeding: from April to August once a fortnight at most; even so, halve the concentration indicated in the instructions.

Repotting: if they cannot be planted out, repot into shallow pots or bowls before the plants start into growth. Well-draining, crumbly loam or ordinary pre-packed potting compost is suitable.

Propagation: only by division when repotting at the end of winter.

Diseases: beware of slugs in the peat mixture in the plant window. Treat preventively.

Maranta leuconeura 'Fascinator', fam. *Marantaceae*

Medinilla

The species discussed here is justly called *M. magnifica* — the magnificent. If you are not yet convinced that the name is fully justified, just look at the photograph on p. 21, which shows an adult plant in full flower. The plant owes its name to a Spanish official who worked in the Philippines around 1820.

Origin: needless to say a plant like this can only be native in regions where, in addition to a tropical climate, all other conditions are equally favourable. Other species, of which more than a hundred are known, occur in the extensive archipelago stretching from Malacca to far into the Pacific, but *M. magnifica* grows only on the island of Luzon in the Philippines.

Appearance: in the jungle medinilla grows to a freely branched shrub, up to 1½ m (4½ ft) tall, with angular branches divided by nodules and with a rough and crusty surface. The opposite growing leaves are thick, leathery and dark green, up to 30 cm (1 ft) long.

Inflorescence: the flower trusses grow on foot-long stems appearing from the leaf axils; they consist of numerous florets surrounded by large very pale pink bracts. The natural flowering period occurs from February to August; the photograph on p. 21 was taken in April. As a result of modern methods of cultivation there are now of course other flowering seasons as well; these are possibly better for young plants.

Position: be careful when buying this fairly expensive plant, often available now. Medinilla is not a houseplant, although a flowering specimen is occasionally very carefully brought into the living-room as a show piece. Young plants which are intended to be kept also definitely belong in a very roomy tropical window. It is even better — and more in accordance with its vigour — to grow it as a hanging plant in a small heated greenhouse where it can develop unrestricted; it should have plenty of light but be out of the sun. At a later stage a tall pot will be needed to keep the heavy pendulous trusses away from the soil. Provide support in the form of canes to prevent the stems breaking.

Care in summer: from March to August a temperature of 20–22°C (70–72°F) or even higher. Water liberally with demineralised water (pH 4–4.5) and provide a high degree of atmospheric humidity. Feed generously (lime-free fertiliser!) in view of its vigour.

Care in winter: after a transitional period, the plant must have a dormant season from November to early February at a temperature of 15–17°C (58–63°F). During this period the flowerbuds are formed. Water more sparingly and do not feed until the flowerbuds are visible. If the dormant period ends too soon, the plant will develop a great deal of foliage, but will fail to flower.

Repotting: if it is necessary at all, early spring is the best time. Soil mixture as for bromeliad species. From time to time remove old wood.

Propagation: a professional operation.

Diseases: inhibited growth and pests are usually the result of lack of moisture in the air.

Medinilla magnifica, fam. *Melastomataceae*

Melocactus

To be honest, this is a difficult group of cacti, however beautiful and attractive they may appear. Only very experienced amateurs will be able to grow them.

Origin: the genus includes about 30 species, occurring mainly in the coastal areas of Central America and in the West Indies.

Appearance: they are globular cacti, 10–12 cm (4–5 in) in diameter. Older plants have an erect melon-shape, with pronounced ribs and decorative spines, usually reddish brown. Woolly and bristly tufts are developed on the crown (cephalium). The flowers appear only on older plants. Once a plant flowers, it ceases growing. The cephalium continues to grow and eventually exceeds the body of the plant in height. Fine, usually red, small flowers, followed by club-shaped red fruits, half hidden in the cephalium.

Species: the photograph shows an older specimen of *M. maxonii*, which has already outgrown its globular shape and has begun the formation of the cephalium; two red fruits on the crown. Other species from the coastal regions are *M. obtusipetalus* and *M. violaceus*. *M. peruvianus*, which may occasionally be seen at an exhibition, is a native of Jamaica.

Care: experience has taught us that imported plants find it difficult to adapt themselves and, in contrast to plants raised from seed, are difficult to grow. Since seedsmen stocking exotic seeds provide a choice of seeds of at least 10 different species, an experienced cactus collector would do best to try growing the plants by this method. The usual rules obtain for growing the seedlings. Well-rooted plants will do well and will even eventually flower.

Compost: two parts beech-leaf-mould, some garden compost and some garden soil, to 1 part coarse, sharp (i.e. lime-free) sand. Compost reaction pH 4·5.

Watering and spraying: only with softened water, pH 4·5. Freely in summer; keep dry in winter.

Feeding: if the plant is growing well, feed normally in summer, using a cactus fertiliser.

Repotting: these cacti have an unusually large root system, which in their native habitat spreads widely and grows close to the surface. As pot plants they are unnaturally restricted and are endangered every time they are repotted, especially after the formation of the cephalium, since afterwards there will be hardly any further root growth. Whenever possible fertilising should therefore replace potting.

Situation: in summer best under glass, warm, but not too sunny. Air should be kept moist by overhead syringeing. From mid-August onwards, gradually prepare the plant for the dormant period in winter. Keep light and dry at a temperature of not less than 10°C (50°F).

Important: these instructions apply only to melocactus species from coastal areas.

Melocactus maxonii, fam. *Cactaceae*

259

Microcoelum (Syagrus) – Coconut palm

This attractive little coconut palm from Brazil has long been popular among amateurs. Its new botanical name is not yet widely known. Long ago, even before it became fashionable to include myrtle branches in bridal bouquets, it was the custom to give this plant to young girls.

Its unusual desire to grow in a tall and narrow pot and, in contrast to other plants, constantly have water in the saucer underneath, has always been accepted.
Origin: as already mentioned, *Microcoelum weddellianum* (syn. *Syagrus weddelliana*) originates in tropical Brazil, quite unlike the areas of barren deserts and oases which provide the usual environment for palms. This plant is accustomed to the climate of the tropical rain forests and even in the most favourable conditions it can be kept under room cultivation for only a few years. The stately specimen shown in the photograph is an exception to the rule.
Appearance: in its natural habitat this small palm can have a trunk of up to $1\frac{1}{2}$ m (5 ft) tall with a diameter of 3 cm ($1\frac{1}{2}$ in). The pinnate leaves growing on stems also reach to $1\frac{1}{2}$ m (5 ft). In room culture these dimensions are rarely achieved. As with an ordinary coconut palm, the trunk is covered in brown fibres.
Care: experts are of the opinion that the regrettably short life-span of the coconut palm is chiefly the result of incorrect treatment in winter. Although the plant likes to be kept in the living-room in summer, it definitely does not tolerate the usual domestic temperature in winter. Overwintering at a slightly lower temperature, 16–18°C (60–65°F) and reasonably moist air would prevent the tips of the foliage turning brown as well as the loss of the lower leaves. After having studied the subject thoroughly, I can confirm these conclusions from my own experience. The fact that my latest little coconut palm perished after eight years was due to my clumsy handling of the roots which are more fragile than they look. The following rules for correct treatment therefore apply: do not place the plant out of doors or on the balcony in summer, but keep it indoors throughout the year. Good light, but out of the sun. Moist air from the onset of growth until about mid-August is essential. In winter the development of new leaves should be avoided by placing the plant in a cooler position, although unlike most other palms, the coconut palm does not require an actual resting period.
Watering: freely from the start of new growth until July. Do not use hard water, the most favourable acid reaction is at pH 5–5.5. During the growing period always leave a little water in the saucer, or plunge the pot into damp peat fibre. At any rate provide sufficiently damp compost even during the winter, in accordance with the room temperature. Drying out of the soil-ball is one of the main causes of browning of the leaf-tips.
Feeding: during the growing period (spring to mid-August) pour a nutrient solution at half strength on to thoroughly moist compost once every eight to ten days.
Repotting: most of the difficulties experienced in the cultivation of these small palms occur when they have to be repotted, for it is often difficult to remove the soil-ball from the necessarily tall and narrow pot. If in the old pot the roots have grown so long that they emerge from the hole and must be broken off, this may be the death of your plant. It is often possible to break the old pot with a hammer and tongs in such a way that the roots remain undamaged. The next problem is planting the soil-ball firmly into the new pot. The best compost consists of a mixture of equal parts of rotted turves and leafmould, with some sharp sand added. Pre-packed potting compost is also suitable.
Propagation: difficult, but nevertheless possible from freshly imported seed.
Diseases: much damage may be caused by stubborn attacks of scale insect and thrips. Control should start at the source of the infection: the base of the leaf-stems and the heart of the plant. If stumps of old leaves, which have been cut off, remain on the plant, these should be removed immediately. Thoroughly sponge the entire trunk, or dip the whole plant in a pesticide solution, having first covered the compost with polythene.
Important: 1. If you buy a coconut palm at the florists', always check first whether the pot has a drainage hole. 2. When repotting, use a pot only 2–3 cm ($\frac{3}{4}$–$1\frac{1}{4}$ in) larger in diameter. 3. Frequently spray and sponge the foliage even in winter.

Microcoelum weddellianum, Coconut Palm, fam. *Palmae*

Miltonia

This lovely South American plant owes its name to the English orchid expert Lord Fitzwilliam, Viscount of Milton, and has nothing to do with the poet Milton (1608–1674). Miltonia was first described in 1837. It is not exactly a plant for beginners.

Origin: about 20 epiphytic species are divided into two groups: the Brazilian miltonia and the Columbian miltonia. A number of species were discovered further west in Costa Rica, and further south in Paraguay.

Appearance: in general there is a great resemblance to the genera *Brassia*, *Odontoglossum* and *Oncidium*. Most species develop a large number of pseudobulbs, oval or oblong in shape and growing close together. Their length varies between 6 and 10 cm (2½ and 4 in). The leaves are long, narrow and strap-shaped and so limp that they curve downwards when only 50 cm (20 in) in length.

Flowers: much valued for their beauty, size and variety of shape; they last for weeks on the plant, but when cut soon fade. Flowering season varies with the species, but is distributed over the entire year.

Species: one of the finest is *Miltonia candida*, which has pure white flowers with two reddish purple blotches on the lip. Flowering season from August to November. Cultivation moderate to warm. *M. clowesii* in September develops flower-stems bearing 8–10 splendid flowers, each up to 8 cm (3¼ in) across. The marking is chestnut-brown with yellow tips and dark reddish purple cross-banding; the lip is white above, violet below. Cultivation moderate to warm. *M. roezlii*, from Colombia, is an orchid for the cold greenhouse. Flower-spike with 2–6 fragrant flowers, 8–10 cm (3¼–4 in) across, pure white or colourfully marked. Flowering season October to November. *M. vexillaria* is another native of Colombia for the cold greenhouse. In its native habitat it grows at altitudes of 1500–2000 m (4500–6000 ft) on the edge of the forest; it is considered one of the finest and most popular species. The leaves are reddish to blue-green in colour. Often develops several flower-stems, each bearing 3–9 flowers 8–12 cm (3¼–4¾ in) across. Their colour varies greatly; usually it is rose-red, darker in parts and with a spotted throat. Flowers in spring. Because of its small, violet-coloured flowers, *M. vexillaria rubella* is sometimes called the pansy orchid. Apart from the species there are many beautiful hybrids, easy to cultivate.

Care: see under other comparable orchids.

Miltonia vexillaria, fam. *Orchidaceae*

Mimosa — Sensitive plant

It is a pity that these beautiful, abundantly flowering plants play such a minor role in the selection of houseplants. The mimosa is so easy to care for and to propagate that it would even be a good plant for children to look after, in order to encourage their love for flowers.

Origin: of the more than a hundred species only *M. pudica* (the sensitive plant) has achieved the position of houseplant. It is a native of tropical regions of America, but was introduced a long time ago in the tropical areas of the Old World, where it spread almost like an unwanted weed.
Appearance: the plant grows to about 30–50 cm (12–20 in); its stems become woody. The feathery leaves are so sensitive that they fold at the slightest touch, either singly or in pairs, while the stem bends down. This phenomenon occurs only in daylight and at temperatures of more than 18–20°C (65–70°F). When the stimulus has been withdrawn, after a pause, the leaves gradually flatten out once more. At night it sleeps with folded leaves.
Inflorescence: small pink to lilac-pink flowers, growing in dense, ball-like clusters at the end of the branches. Properly cared for the plant will flower without interruption from May to September.
Annual cultivation: as older plants become unsightly, it is better to sow every year in February–March; 6–8 seeds are sown in small pots; kept at a constant temperature of 20°C (70°F) they will germinate after 12–14 days. Leave the three or four strongest seedlings and discard the others. Keep them moist and warm, perhaps under a plastic bag or in an indoor propagator; prick out twice. The advice frequently given to prune the plant in order to encourage the formation of new branches is definitely incorrect. Only if left unpruned will the plant develop so successfully that the first flowers will have started to appear by the end of April; they will form a great deal of seed. Good light and a half-shady position, at a constant temperature of 20–22°C (70–72°F) and adequate humidity is essential. Old plants should be discarded in the autumn.
Watering: until August keep constantly moist; do not let the soil-ball dry out. If the air is dry spray frequently.
Feeding: from spring until late summer give a normal nutrient solution every fortnight.
Diseases: these are rare. If mimosa stands in a draught, greenfly or red spider mite may appear.

Mimosa pudica, Sensitive Plant, fam. *Leguminosae*

Monstera — Swiss cheese plant

To prove the point that this magnificent plant is one of the finest of all we should like to refer you to the illustration on p. 13. This gives an obvious example of the powers of expression and the harmony that may be achieved in a prolonged friendship with one's plants.

Origin: more than 25 species of the genus *Monstera* occur in the morasses of tropical Central and South America. Apart from innumerable cultivars, four of the species are available here. In some countries, for instance in the United States, the choice is much wider, possibly up to 25 species.

Appearance: evergreen, semi-epiphytic climbing shrubs with frequently very deeply incised leaves which often have holes as well; some have aerial roots. The shape of the leaf varies with age.

The inflorescence, with its large white bract and fragrant cone-shaped fruits, which later become edible and taste of pineapple, rarely appears under room cultivation, but is regularly produced by older plants grown in a greenhouse. Even our finest indoor specimens are small compared to the enormous, robust jungle plants, whose irrepressible vigour makes them easily grow to over 10 m (30 ft) in botanical greenhouses.

Species: *M. acuminata*, a plant with pointed, perforated leaves, originating in Guatemala, is rarely grown as a houseplant. Young plants have smooth-edged foliage; the leaves become deeply slashed only when they are 25–30 cm (4–6 in) long.

M. deliciosa, of which a magnificent specimen is shown on p. 13, is a native of Mexico and in this country has been a much loved foliage plant for more than a hundred years. The enormous shield-shaped leaves growing on 30–50 cm (12–20 in) long stems are leathery, dark green and deeply slashed. They become perforated as well. The leaves may grow to 40–70 cm (16–28 in) width and 60–80 cm (24–32 in) in length. The shape of the foliage of young plants is entirely different from that of older plants. Hexagonal, aromatic berries develop on the spadix; they have a delicate pineapple taste but leave a disagreeable burning sensation on the tongue.

The cultivar 'Borsigiana' is slightly smaller in all its parts. *M. obliqua* from Brazil, also somewhat smaller than *M. deliciosa*, is a striking species with large holes leaving the edge of the leaf intact. This phenomenon occurs particularly in young plants of the strain *M. obliqua* 'Leichtlinii'. *M. pertusa*, also from Brazil, is a rapid climber with dense foliage. As in *M. acuminata* the holes in the leaves do not appear until the plant is older.

Important: because of the considerable outward resemblance, a number of monstera species are often taken for *Philodendron*. It is true that certain biological differences between the two genera occur only in the inflorescence. Treatment is the same for both: see *Philodendron*, p. 294.

Monstera obliqua 'Leichtlinii', fam. *Araceae*

Myrtus — Myrtle

The compass and versatility of myrtle plants has already been mentioned in the chapter on botanical families. The most beautiful representative of the genus is the species *M. communis*, the common myrtle, which in Germany is called 'bride's myrtle'. Forms with variegated foliage are no longer in demand.

Origin: the plant's native habitat is in the areas round the Mediterranean, where it grows to a height of 3–5 m (9–15 ft). The plant was brought across the Alps by monks. The medicinal properties of the ethereal oil produced from myrtle leaves were already known here at the beginning of the Middle Ages. The use of myrtle branches in bridal bouquets is an ancient custom.

Appearance: every plant lover is familiar with the myrtle and knows that there are both standard forms and shrubs. Both of these can be kept in shape by means of pruning and tying. Myrtle plants can live to a ripe old age.

Foliage: evergreen, leathery, small and oval; the leaves give off a pleasant aroma when rubbed between the fingers.

Inflorescence: the very numerous flowers appear singly on short stems growing in the axils of the leaves; they are white, sometimes with a pink bloom. The flowering season is from May to June. Blue or white pea-sized berries are often produced in autumn; these are edible.

Summer: a roomy situation near an open window, good light but not too sunny. From May onwards it may also be put in a sheltered position on the balcony or planted, pot and all, in the garden. It must be brought indoors before the first night frosts.

Winter: if the plant is kept in a heated or in a dark, rarely ventilated room, the foliage will turn yellow and attacks by greenfly and other troubles will occur. The myrtle must over-winter at a temperature of 4–6°C (40–43°F) in good light and a well ventilated position. Moderate daytime temperature, 10–12°C (50–54°F), will be tolerated, provided the temperature is very low at night. It is also essential that the plant should have plenty of space and not be crowded by other plants. Standard forms may be turned round from time to time; this can only improve their shape.

Watering: use tepid, lime-free water, approximately pH 5. Water freely in summer, especially when the buds are about to open. Beware of water in the base of the pot, as well as of drying out of the soil-ball. In winter the plant's water requirement depends on its situation: the cooler its position, the less water. At a temperature of 4–6°C (40–43°F) the soil should be kept constantly moist.

Feeding: from the time the plant starts into growth until the end of the flowering period feed liberally, using a lime-free fertiliser.

Repotting: if necessary, repot in spring; at longer intervals as the plant grows older. The compost should be lime-free, rich and humusy. Be sure to plant at the correct depth; if the myrtle is planted too deep it will wilt.

Propagation: in early summer from cuttings of profusely flowering parent plants. Seeds may take a whole year to germinate.

Myrtus communis, Myrtle, fam. *Myrtaceae*

Narcissus — Daffodil

The daffodil, a bulbous plant belonging to the family of *Amaryllidaceae* is undoubtedly first and foremost a garden plant. However, many plant lovers are ignorant of the fact that, apart from the well-known 'Paperwhite', flowering at Christmas time, nearly all other species and cultivars can be forced in pots or bowls. Bulbs grown indoors will not flower again.

Origin: no doubt you know the Greek myth: a young man called Narcissus saw his face mirrored in the water and, falling in love with his own beauty, could not be induced to move, so pined away and died. A white flower grew on the spot and was named after him. This story tells us that the native habitat of the more than 30 species is in the southern regions of Europe, the countries of the eastern Mediterranean, and Asia Minor.

Appearance: bulbous plants with strap-shaped leaves, usually appearing at the same time as the angular flower-stem and the flowers. Eleven groups are distinguished, of which the main ones are: trumpet daffodils, small-cupped narcissi, poet narcissi, tazetta narcissi, jonquils and botanical narcissi. For indoor cultivation prepared bulbs with double 'noses' are used: having more growing points they will flower more profusely. They are brought into growth in the same way as hyacinths, but without using paper cones. The most interesting form is *N. tazetta*, the 'Paperwhite' which flowers at Christmas time. To achieve this the bulbs are placed, in mid-October, in groups of three, five or six in low, pebble-filled dishes. The water should not touch the base of the bulbs. Place the dishes in a good light in a well-ventilated, cool position and replenish the water as required. Too warm a situation will result in lanky stems. Bulbs set up after Christmas until the end of January flower if possible even more profusely than the first group. After flowering the bulbs should be discarded, for they are not winter-hardy.

Narcissus tazetta 'Totus albus', fam. *Amaryllidaceae*

Neoporteria

The most beautiful feature of these small spherical cacti, which become cylindrical at a more advanced age, is provided by the magnificent spines, whose colours vary in different species from white to yellow and brown to a deep black, a colour rarely seen in the plant world. Opinions about treatment differ.

Origin: about 25 species are known, all originating in Chile and northern Argentina. They were collected together in one genus only a few years ago. The botanical name refers to the entomologist Porter.

Opinions: cactus specialists differ in their opinions on the characteristics of neoporteria. The famous specialist Krainz describes the genus as 'very slowly growing plants with great power of resistance, suitable for extremely warm situations', and for room cultivation he recommends only grafted forms. Professor Buxbaum in his no less famous book *Kakteenpflege biologisch richtig* considers the species *N. nidus* and *N. napina* with their long tap roots — these two are generally regarded as the most beautiful — as 'distinctly difficult to grow'. On the other hand the German expert Dr. Cullmann is of the opinion that 'in mineral-containing soil and a light position they will grow and flower satisfactorily'. He considers *N. senilis* (formerly called *N. gerocephala*) to be the finest; because of its long, entangled white spines this cactus resembles the Old Man cactus.

General care: approximately the same as for *Coryphantha* (p. 176), *Mammillaria* (p. 254) or *Rebutia* (p. 312).

Neoporteria, fam. *Cactaceae*

Neoregelia and Nidularium

There has been many a family split in the circle of *Bromeliaceae* or pineapple plants. Genera were created and abolished, only to arise once more like the phoenix from the ashes. Neoregelia and nidularium both belong to this group and are described here together.

Origin: about 30 species of these genera occur in Brazil. In spite of their striking similarity, they are nevertheless different in character. Neoregelia simply means 'New Rule', the second word being the translation of the name of a one-time curator of the Botanical Gardens in St. Petersburg, E. A. von Regel (1815–1892). The name nidularium on the other hand, is derived from the Latin *nidus*, or nest; the suffix makes it into 'little nest'. To the neoregelia the 7 species of the genus *Aregelia* were added, but this happened so long ago that few reference books mention the fact.

Appearance: both genera belong to the epiphytic plants growing in tropical rain forests. They form closed rosettes, consisting of thorny dentate or saw-edged leaves. The differences in outward appearance and in treatment which must be taken into consideration are described under Species.

Inflorescence: in all the species of the two genera the inflorescence is developed from the centre of the rosette. The flowers grow from the axils of the vividly coloured bracts, which indeed creates the impression of a small nest.

Species: the list of the most beautiful species habitually starts with *N. binotii*, a plant whose rosette consists of 30-cm (12-in) long, leathery leaves. Their edges are covered in strong thorns; they are green above, with red

Neoregelia, hybrid 'Volckhartii', fam. *Bromeliaceae*

▶

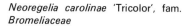
Neoregelia carolinae 'Tricolor', fam. *Bromeliaceae*

tips, but, as with *Aechmea chantinii*, the lower surface in addition has white cross-banding. *N. carolinae* forms flat rosettes of 40 cm (16 in) long leaves, glossy green on both sides. They are edged with fine spines; the inner leaves are vivid red at the tip. The shiny red flowers are surrounded by deep violet bracts. The magnificent colouring of the best known cultivar, *N. carolinae* 'Tricolor' is evident from the illustration above. In addition there are beautiful cultivars such as 'Marechalii' and 'Meyendorffii'. *N. concentrica* with its 30 cm (12 in) long and 10 cm (4 in) broad leaves, dark red and green streaked on the upper surface, and ivory or violet bracts, may be said to have the most remarkable 'nest' of them all: the small, white-edged blue flowers appear on what looks like a pincushion. The smallest rosette is found in *N. tristis*, whose leaves, brownish purple, spotted towards the centre, are only 15–20 cm (6–8 in) in length; its bracts are coloured in various shades of brown. Of the genus *Nidularium* one of the best is *N. innocentii*, with leaves dark red above, very dark green below; after and during its inconspicuous flowering the inner leaves surrounding the 'nest' retain a beautiful coppery colour for many weeks. Other strains, particularly popular in the United States, for instance 'Paxianum', 'Purpureum' and the dwarf form *N. innocentii* 'Nana' confirm the value of these beautiful plants. The cultivar 'Maureanum' has been developed in Belgium. In *Nidularium striatum*, a species which has been known since 1888, the bright green, yellow-striped leaves form a less dense rosette than do other nidularias. Bracts with brown tips; small white flowers.

General care: here the considerable differences between the botanically so similar plants become apparent. Although both genera are essentially hot house plants, neoregelia, in particular, includes species which with some extra care will grow in the living-room as well. Apart from *N. carolinae* and its cultivars, *N. princeps* is also very suitable for this purpose. To develop their beauty fully all neoregelia species require plenty of light, but no direct sunlight. In winter the most favourable temperature is 15–18°C (58–65°F), while in the growing season and during bud formation, usually from March to August, the plants require temperatures of 22–25°C (72–78°F). Thanks to their tough leaves, neoregelia species will tolerate dry air during their dormant season, which starts fairly early in winter. On the other hand nidularium species with their more tender and even more colourful foliage, are dependent upon a half-shady situation in a climatically correct tropical plant window. During their growing period they require a constant high temperature and a high degree of humidity.
For the remaining rules of treatment see under *Aechmea* (p. 117), *Guzmania* (p. 221), *Tillandsia* (p. 342) and *Vriesea* (p. 348). If you have to give the plants a less favourable position, this does not matter too much, for old rosettes will die in any case.

Propagation: growing plants from new, young rosettes will only be successful in adequate soil heat. Propagation from seed – if, indeed it is obtainable from a dealer in exotic seed – frequently leads to disappointment: the foliage will be all-green.

Diseases: errors in cultivation may lead to scale insect and thrips infection. The vase may rot.

Nephrolepis

Literally translated, the botanical name means kidney-shaped scaly fern. The popular name is Sword Fern or Ladder Fern. It is one of the most hardy and most popular house ferns. It can decorate your living-room for a long time: it easily lasts for twelve years or more.

Origin: for a long time the nephrolepis belonged to the family of *Polypodiaceae*, but together with a practically unknown tropical fern it has now been classified in the family of *Oleandraceae*. About 30 nephrolepis species occur in tropical and sub-tropical regions.

Appearance: there are terrestrial as well as epiphytic forms. Some species, including those most suitable for use as houseplants, develop long runners, which not only serve as a simple method of propagation, but also enable the plant to be used as a hanging plant.

Species: *N. exaltata*, a tall growing form, has practically disappeared as a species, but it survives in strains cultivated for house and garden use. There are pennate forms as well as doubly-feathered forms in which the individual leaf section is pennate in turn; there are also types in which the feathered leaves are wavy-edged or spiralled. In addition there are very robust and vigorous cultivars such as 'Teddy junior', and small types like 'Dwarf Boston'. The frequently encountered strain 'Whitmannii', with densely crisped foliage, is considered to be somewhat less hardy than the 'Bornstedt' cultivar it resembles; the latter is the only *Nephrolepis exaltata* hybrid which can be propagated not only from runners, but also from spores. Among other species one of the most interesting is *N. cordifolia*. It is terrestrial as well as epiphytic growing and like some marantas develops long runners in its stems, with scaly tubers underground.

Care: general care is like that for *Pteris* (p. 310), although most nephrolepis species tolerate more light; some even require some sunlight. Temperature in winter at least 14–16°C (57–60°F).

Feeding: like most ferns, nephrolepis reacts to good nourishment. In the growing period it should be given a feed weekly. After a resting period, cut the plant back, repot and bring into growth once more. Feed again after six weeks.

Nephrolepis exaltata 'Teddy junior', fam. *Oleandraceae*

Nerium oleander – Oleander

Until recently the oleander was known only as a tub plant which kept on growing bigger and bigger. In the Netherlands people got tired of the large size and of the problems of overwintering and decorative plants of more acceptable dimensions were cultivated for indoor use. The photograph shows an example which, together with its container, barely reached table-height. Unusual, but perhaps not to everybody's taste...

Origin: its native habitat is in areas round the Mediterranean, where the shrubs, to 5 m (15 ft) in height, grow in sunny places, where the soil is damp, for instance along brooks or rivers, like alders and willow in this country. Knowledge of the conditions in which a plant grows in its original state, contributes greatly to its correct treatment. Oleanders are often found drooping on the window-sill.

Appearance: spreading, evergreen shrubs with pliable shoots, which at a later stage grow woody; they bear leathery, narrow lanceolate grey-green leaves growing opposite in pairs or in threes. Among the many strains now available are the 'Variegatum' forms, with white or yellow variegated foliage; however, these flower less profusely.

Inflorescence: the flowers appear in terminal clusters. There are many species and strains with single or double flowers in the colours creamy white, pink, red, and, more rarely, tea-rose yellow. The flowering season is from June to September.

Summer: since the oleander will tolerate a short frost period at minus 4–5°C (22–24°F) without being harmed it can, in temperate climates, often be moved from its winter position as early as April, without fear of night frost. Choose the sunniest and warmest position available; it will tolerate very dry air. It feels happiest against a white-painted, south-facing wall in burning heat, provided the soil is damp. Leave it outside as long as possible; there may often be fine autumn days.

Winter: a resting period at a temperature of 4–6°C (40–43°F). If absolutely necessary it may be put in a not too dark cellar, but of course a position in good light with plenty of fresh air is much better. Only in cases of extreme necessity should the plant be put in even moderately heated spaces, such as vestibules, corridors or landings. As a rule the low temperature required cannot be provided in such a place. More warmth will cause the plant to grow lanky and outgrow its strength, which in turn affects its flowering.

Watering: from the time it starts into growth until autumn water freely to very liberally, depending on the weather conditions. If possible use lukewarm water, 25–30°C (78–86°F)! Watering with warm water even in winter encourages the plant's readiness to flower. The pot or tub should be placed in roomy saucers so that during the growing period it can be watered from below every day in warm weather. The plant likes to have a 'footbath' for days on end; it certainly is no desert plant. From mid-August onwards gradually decrease the water supply, allowing the stems to ripen thoroughly. In winter, when the plant is indoors, merely see to it that the soil-ball does not dry out; from time to time clean the leaves with a dry duster. The oleander does not like rain, damp air or spraying.

Feeding: from the beginning of the growing period until mid-August give a strong nutrient solution every week; then cease feeding, to allow the wood to ripen. In former times oleanders were given cow dung solution; dried manure can serve the same purpose.

Compost: good, loamy soil, mixed with compost, peat and sand, or a good quality pre-packed compost.

Repotting: only when the root system is crowding the pot. The best time is May–June. Loosen the roots (see also p. 68) and shorten them by 2–3 cm ($\frac{3}{4}$–1$\frac{1}{4}$ in). Naturally the new pot or tub should be larger than the old. Since the plant likes to be watered from the bottom, a good layer of crocks or pebbles must be provided; this should be covered with a slow-acting nutrient mixture, for instance 1 kg (2·2 lb) of hoof and horn meal mixed with 250 g (9 oz) bonemeal and 250 g (9 oz) sulphate of potash; the correct dosage is 50 g (1$\frac{3}{4}$ oz) of this mixture to 10 l (2·2 gal) of compost.

Pruning: an oleander which has grown bare at the base, too old, or too large, may be rejuvenated by pruning the old wood. This is best done in the autumn or in late February. Loss of early flowers will have to be accepted.

Propagation: cuttings may be taken from May until the end of July. They will root in water in dark bottles and should then be planted in small pots containing cuttings compost. Be careful, for the roots are very fragile.

Diseases: mealy bug and scale insect infestation in winter may be encouraged by too warm a position or a lack of fresh air. Stubborn attacks of greenfly on diseased plants, where the infection starts near the neck of the root, should be treated with special remedies available for this purpose.

Nerium oleander 'Oleander', fam. *Apocynaceae*

Nertera — Bead plant

Among plants whose ornamental value lies in the fruits rather than in the flowers, the bead plant, sole representative of its genus, occupies the leading position. It is usually, quite unjustly, regarded as a short-lived disposable plant. Nertera can, however, be kept permanently and very successfully.

Origin: the native habitat of this little plant is in the Andes mountains between Peru and Cape Horn. It has been known in Europe for more than 200 years.

Appearance: *N. granadensis* is a perennial, creeping herbaceous plant with 15–25 cm (6–10 in) long stems bearing tiny leaves, 0·7 cm ($\frac{1}{4}$ in) long on 0·4 cm ($\frac{1}{8}$ in) long stems. The small, star-shaped white flowers appear in April–May, developing until August into the pea-sized berries, in such large numbers that the pale green foliage is hardly visible. The berries may remain on the plant until the winter.

Care: in summer a cool position in good light, without strong sunlight. Plenty of fresh air. Overwinter at 8–10°C (46–50°F) in good light and an airy situation. Dormant season from October to February.

Watering: in summer maintain constant moisture; in winter the plant should be kept a little drier. Before and after flowering spray regularly.

Feeding: very sparingly, with a nitrogen-free fertiliser, in order not to stimulate leaf formation.

Repotting: only if the plant needs to be divided; in August (this is a better time than when the berries have dropped). A sandy-humusy, porous compost is best.

Propagation: by division; also from seed in February–March, in high air temperature. A pleasant job, but it takes time.

Diseases: draughts will lead to greenfly at the tips of the stems; slugs will often occur in the foliage.

Nertera granadensis, Bead Plant, fam. Rubiaceae

Notocactus

This large group of globular cacti formerly belonged to the genera *Echinocactus, Malacocarpus* and *Brasilicactus*. They have now been more clearly arranged into one genus; they are very popular, beautifully spinate and freely flowering cacti for beginners. Suitable for simple room cultivation.

Origin: the genus *Notocactus* embraces about 15 species, occurring in the central part of South America. They grow on boulder-covered mountain slopes and in sandy prairies in southern Brazil, Paraguay, northern Argentina and Uruguay.

Appearance: they are small cacti, spherical in shape, subsequently also becoming cylindrical. The numerous ribs covered in a variety of spines, give the plants a most distinctive appearance even when not in flower. No

Notocactus leninghausii, fam. Cactaceae

Notocactus ottonis, fam. *Cactaceae*

other genus flowers as freely; even young plants are easily brought into flower. The flowers appear summer after summer, singly or encircling the crown in various shades of silky yellow; the typical red style emerges from the yellow stamens.
Species: one of the smallest is *N. apricus*; it has 15–20 knobbly ribs with red spines. The flowers are up to 8 cm ($3\frac{1}{4}$ in) in length, with a diameter of up to 6 cm ($2\frac{1}{2}$ in), yellow inside, red-streaked on the outside. A very popular standard form is *N. concinnus*, a glossy green, flattened sphere with a wide indentation on the crown, which is interspersed with warts; the spines are yellow or red, the large flowers silky and yellow. A great favourite among notocactus species is *N. haselbergii* (formerly *Brasilicactus haselbergii*), a globular cactus with a diameter of up to 10 cm (4 in), with fine, very dense white spines. The crown turns towards the light and is dotted with yellow/red or blood-red flowers. *N. leninghausii* grows from a sphere 10 cm (4 in) in diameter into a column of up to a metre (3 ft). It has a slanting yellow crown even when young; beautiful golden-yellow spines and pure yellow flowers up to 5 cm (2 in) in diameter. Much in demand, but very sensitive to dampness, is *N. ottonis*; its diameter rarely exceeds 6–7 cm ($2\frac{1}{2}$–$2\frac{3}{4}$ in). Yellow and brown thorns; deep yellow flowers, about 4 cm ($1\frac{3}{4}$ in) in diameter. A most grateful cactus is *N. submammulosus*, which has individual globes, 8–9 cm ($3\frac{1}{4}$–$3\frac{3}{4}$ in) in diameter; the ribs are subdivided by notches. Long or short creamy-white spines; pale yellow flowers 4–5 cm ($1\frac{3}{4}$–2 in) in diameter. A valuable recently imported species is *N. uebelmannianus*, whose glossy dark green globes may achieve a diameter of 17 cm (7 in). It is distinguished by thick, flattened ribs flanked by large spines. The flowers, up to 5 cm (2 in) across, occur in various shades of pale and dark red, a delight to the eye.
Care: normal summer treatment in a light and sunny position out of doors or simply on the window-sill. If the plants have been kept out of the light during winter they must, like other cacti overwintered in this way, first gradually get hardened off and acclimatised to the change in their environment. To avoid scorching of the as yet too delicate tissue, the plants should be screened against strong sunlight during the first two weeks. Where circumstances allow, you could simply paint the outside of the window with whitewash which can easily be removed afterwards. For notocactus species the winter position should not be quite so cool as for other cacti. 10°C (50°F) is considered to be the minimum temperature.
Watering: in summer, growing and freely flowering plants must be kept distinctly moist, but excess water at the bottom of the pot should be avoided. Slightly acid-reacting water, pH 5–5.5 is desirable. Light syringeing is also necessary, especially on hot summer days (it has been recommended to spray at night with softened hot water!). In winter, on the other hand, the plant must be kept entirely dry, regardless of room temperature.
Feeding: as usual only in summer, with nitrogen-free cactus fertiliser. From August onwards, feed at longer intervals and gradually cease.
Repotting: where necessary, in a rich compost made light by the addition of coarse sand and given a slightly acid reaction by the addition of leafmould.
Propagation: very simple, from seed. Exotic seed catalogues contain more than 50 species and strains.
Diseases: the usual warning concerning the effect of draught in summer and excess moisture in winter applies.

Odontoglossum

Everyone interested in orchids is familiar with the genus *Odontoglossum*; understandably so, since some of the numerous species and especially of the hybrids — particularly the hybrids resulting from bigeneric crossing with the closely related *Oncidium* — are suitable plants for amateurs and beginners, even for room cultivation.

Origin: about 250 species are known, all, like *Oncidium*, growing as epiphytes usually at altitudes above 1500 m (4500 ft). They are typical plants of the rain forests of the slopes and plateaux from tropical regions of Mexico to the northern part of South America. Some species grow at altitudes of up to 3000 m (9000 ft). They are therefore plants for the unheated greenhouse, with few requirements.

Appearance: like the *Miltonia*, the genus *Odontoglossum* has egg-shaped, flattened pseudo bulbs growing close together and forming a kind of nest; each puts forth two narrow leaves. The long, sometimes very long, flower stem is developed from the young leaves of these pseudobulbs. The stem may be erect growing or pendulous and often bears dense sprays of magnificent large flowers of many shapes, which will keep for weeks even without water.

Species: innumerable forms are available. Some of the so-called Mexican species and their hybrids are particularly suitable for room culture. *O. bictoniense*, pseudobulbs 10 cm (4 in) in height, flower stem a metre (3 ft) long, bearing a spray consisting of 10 flowers, each up to 5 cm (2 in) across. Flowers in autumn. Place in a sunny spot out of doors in summer. *O. cervantesii*, pseudobulbs no larger than a pigeon's egg. Pendulous spike consisting of 6 flowers, each 6 cm (2½ in) across, fragrant. Winter-flowering, easy to cultivate. *O. citrosmum*, pseudobulbs up to 12 cm (3¾ in) in height. Flower stalk up to 90 cm (3 ft), pendent, bearing 30 flowers. Flowering season May–June. *O. grande*, the favourite among this group; flowers up to 16 cm (6½ in) across. Flowering season autumn to March. It requires a strict resting period to be able to form new flowers. The most beautiful species and the easiest to grow. Other excellent species are: *O. pulchellum*, flowering in late winter; *O. rossii*, an attractive winter-flowering dwarf species, and *O. uro-skinneri*, flowering in early summer.

Care: since the flowering seasons of the various species are spread over the entire year, it is not possible to give general rules. Moreover, some species behave in an 'unnatural' manner; they want to be kept cool and humid in summer, and dry and warm in winter. You should therefore ask for rules of cultivation for each new plant you purchase. As regards watering and feeding follow the general instructions for orchids. Water *p*H 4–4·5; use a weak solution of nutrients at a strength of 0·5%, not alkaline.

Odontoglossum grande, fam. *Orchidaceae*

Oncidium

Around 1800 the Stockholm professor Olaf Peter Schwarz in the course of an expedition to the islands of the West Indies discovered the first oncidium. Since then the genus has provided us with a large number of fine species, which also proved to be suitable for room cultivation.

Origin: in the course of the years so many other species were found in the huge distribution area between Mexico, Bolivia and Paraguay, that 350 are now known, forming one of the largest genera of the *Orchidaceae* family.

Appearance: the above indicates that there are great differences between the various species, as regards both appearance and requirements. We have already pointed out the close relationship existing between this genus and the genus *Odontoglossum*. They are excellent partners for crossing in order to produce beautiful, usually easily grown hybrids. Nearly all plants belonging to the genus *Oncidium* are epiphytic; only a few of these live, like *Odontoglossum*, in damp, cool and misty forests at altitudes of up to 3000 m (9000 ft); other species prefer the warm and dry coastal areas with more sunshine; a third group occurs in tropical rain forests and should be treated accordingly in room cultivation.

Species: among the species suitable for room cultivation are: *O. bicallosum*, an orchid with a small double protruberance on the lip. Thick, erect flower stems, up to 40 cm (16 in) in length, branched at the tip. The flowering season is in winter. To grow these orchids one definitely needs green fingers.

O. forbesii, a native of Brazil, has flower stalks up to 60 cm in length; the large flowers resemble butterflies and appear in late autumn. The flowers of *O. ornithorhynchum*, growing in loose sprays, resemble the beak of a bird. A good plant for beginners.

O. tigrinum has large fragrant flowers appearing in pendulous racemes in autumn and winter. It takes up little room.

O. varicosum is probably the most graceful and best known species. It is cultivated for cutting as well. The flower stem may grow to 1·50 m (4½ ft) and bear innumerable flowers. It usually flowers every two years, in winter.

One of the most interesting phenomena in the realm of orchids is *O. kramerianum*, whose three long sepals are shaped like a butterfly's antennae. A plant for amateurs, but not easy to grow.

Care: all the species mentioned are really hot house orchids. Instructions supplied with each plant should be followed.

Oncidium ornithorhynchum, fam. *Orchidaceae*

Oplismenus hirtellus, fam. *Gramineae*

Opuntia microdasys var. *albispina*, a white-thorned fig cactus, fam. *Cactaceae*

Oplismenus hirtellus

Seen from a distance this attractive tropical grass, used as a hanging plant, might be taken for a tradescantia. On closer inspection the difference is obvious, for here we have much more slender stems, divided by typical 'grass joints'; they are up to 50 cm (20 in) long. In the trade it is frequently called by its old name, *Panicum*.

Cultivars: it is not essential to grow it as a hanging plant. Young shoots planted in soil or grown in soilless cultivation, placed at the edge of a plant window, will rapidly root at a soil temperature of 18–20°C (65–70°F) and grow into graceful ground cover.

Growing season and dormancy: during the growing season from mid-March until mid-August, water freely. Feed very sparingly, to avoid the variegated or pink-bloomed foliage turning green. From the middle of August onwards decrease the water supply and gradually cease feeding. When the resting period starts, which generally coincides with the heating being switched on, avoid hanging the plant in a warm air stream if possible; find a cooler spot for it. Normal room temperature 22°C (72°F).

Light requirement: this is greatest in winter, so give the plant plenty of light. During the growing season a slightly shady position is preferable.

Repotting: if necessary before the new shoots appear, early February. It is hardly worth the trouble.

Propagation: it is better to grow new plants in spring, as with tradescantia: insert young top-shoots, about the length of your finger, in groups of ten or more in shallow pots or bowls with propagating soil.

Opuntia

Not so long ago it was predicted in trade circles that the end was in sight for the use of opuntia as a houseplant. The large species for garden and balcony use were becoming increasingly fashionable. These grow rapidly, but in our climate never flower. Meanwhile the prediction has been proved wrong.

The less vigorously growing small forms for indoor cultivation, which have less intimidating spines, are once more indispensible in the cactus assortment and one need not be a pronounced cactus lover to be able to appreciate these undemanding plants.

Origin: the whole American continent forms the native habitat of this genus of more than 200 species; both hardy and non-hardy species are found from Canada in the north to Patagonia in the far south. Several species were brought to other continents a long time ago and wild, non-herbaceous opuntias now occur in Australia, South Africa and southern Europe.

Appearance: the most widely distributed species, which at the same time are most suitable for indoor cultivation, are the well-known opuntias with flattened joints, combined in the strangest shapes. Among this group we find the vigorous species which in their original tropical habitat are valued as hedging plants or fruit-producing plants (fig-cactus!). Cylinder opuntias have an entirely different appearance: these are low-growing plants with round, fleshy branches which subsequently become

Opuntia bergeriana, fam. *Cactaceae*

woody. Many opuntias have interesting spine development; they all — including the fruits — have tufts of so-called glochids — barbed bristles; these may cause painful wounds, so be careful! Handle them in gloves and keep a pair of tweezers within reach.
Inflorescence: large flowers in clear colours, glossy as silk. In flattened opuntias they appear along the edges of the joints; in cylindrical opuntias they appear earlier in the season along the branches. The flowering season is usually from spring into summer.
Species for room cultivation: *O. basilaris*, without spines but rich in glochids. The reddish purple flowers do not appear in large numbers. *O. bergeriana* flowers readily; the flowers are orange-red; this species may grow into a tub-plant. *O. microdasys* embraces attractive small strains such as 'Minima', yellow. *O. rufida* is thornless but has numerous tufts of glochids; flowers freely, sulphur-yellow.
Care in summer: in an airy, warm and light situation it is an easy plant. The soil must be porous and rich. Water normally from the beginning of the growing period until August; then gradually decrease the amount until the beginning of the dormant period.

Until mid-July feed twice weekly with a special cactus fertiliser; thereafter decrease gradually until the resting period. If necessary repot in spring.
Care in winter: the dormant period should be spent in a well-lit, definitely dry, situation 6–8°C (43–46°F).
Propagation: from cuttings; the section should be allowed to dry out for a fortnight. Raising from seed is also possible.
Diseases: be on guard against mealy bug between the tufts of glochids.

Oroya gibbosa, fam. *Cactaceae*

Oroya

This magnificent globular cactus, of which only a few species are known, was found in the Andes mountains in central Peru at an altitude of over 4000 m (12 000 ft). Great favourites among collectors are *O. neoperuviana* and the main species *O. peruviana*. All oroya species are practically winter-hardy.

Appearance: as a pot plant it is about 7 cm ($2\frac{3}{4}$ in) high, up to 14 cm ($5\frac{1}{2}$ in) across. The body of the cactus is dark, almost blue-green with spiral ribs divided into hexagonal warts. The spines are initially pale brown and transparent, but later turn brown or black, often intermingled.

Inflorescence: pink to red, often with a yellow base; numerous.

Care: in view of its origin, oroya requires pebbly soil with a light acid reaction (*p*H 5–6). Grafted specimens are the easiest to grow. Keep cool and light in winter.

Pachyphytum

Like *Echeveria* this beautiful, thick-leaved plant is a succulent foliage-plant from Mexico, very suitable for room cultivation. In addition to the species there are all sorts of interesting hybrids, mostly the result of crossing with echeveria. These have been given the name ×*Pachyveria*.

The following three species are the most important. *P. bracteosum*, with 10–30 cm (4–12 in) tall stems, has bright red flowers in summer. *P. hookeri* is a shrub-like plant, 60 cm (2 ft) tall. *P. oviferum* forms short stems; the leaves are thick and fleshy, egg-shaped, covered in a white bloom; carmine-red flowers in May–June.

Care: as for echeveria. They are plants for the unheated greenhouse; in winter they should be kept cool but frost-free and in a good light. Propagation from cuttings or, if available, from seed.

Pachyphytum oviferum, fam. *Crassulaceae*

Pachypodium

Two plants which are distinguished almost solely by their tuft-like foliage are known as 'Madagascar palms' and are among the most popular succulents. Like the *Adenium* illustrated on p. 166 and the Oleander on p. 270 which provided its stock, it belongs to the Dogbane family.

Origin: *P. geayi* originates in arid regions of south-western Madagascar, *P. lamerei* produces its wonderful flowers only in its native habitat. Both species may develop into mighty trees, 5–8 m (16–26 feet) high. The thorny, succulent trunks may attain a diameter of over 1 m (3 feet).

Care. In our flower windows we can only grow young plants with a reversed growth pattern: they have their dormant season in the summer months, when they are watered very sparingly and not fed. In winter the plants should be kept in a very sunny place; the temperature should be a constant 18°C, even at night. At this time they are watered sparingly and given very little fertiliser.

Causes of disease. Ignorance of the above rules is probably the chief cause of damage to the leaves.

Preferably grown in hydroculture. Unfortunately the plants are very poisonous.

Pachypodium geayi, Madagascar palm, fam. *Apocynaceae*

Pachystachys lutea

There is now a plant available which greatly resembles the pink *Beloperone guttata*, and is in fact sometimes called the 'yellow beloperone'. It is by no means a new discovery, although, when a few years ago it was displayed at a horticultural show, most people had never seen it before.

Origin: in pre-1900 plant books, *Pachystachys lutea* is invariably mentioned. At that time it was strongly recommended as a houseplant. Later it appears to have been included in the genus *Jacobinia* and as such the plant, American in origin, has been introduced as a novelty.

Appearance: because of its vertical habit and stiffly erect 'ears', it does indeed resemble some *Jacobinia* species, which are, however, no longer marketed. In any case the close relationship between all these members of the *Acanthaceae* family is very obvious.

Old books already describe *Pachystachys lutea* as follows: semi-shrub with green branches; smooth, pointed, elongated oval leaves, vivid green with clear veining; abundantly flowering in dense, terminal ears; creamy white flowers surrounded by bright yellow bracts. Flowering continues from spring until winter; the bracts always remain on the plant for a long time.

Considerations: my experience with *P. lutea* is of relatively short duration, but it seems to me that its requirements lie between those of the fairly easy, robust *Beloperone guttata*, which is known to be suitable for use on the balcony and in the garden as well, and the slightly more demanding *Jacobinia* species, which are suitable for room cultivation, but have a shorter flowering season. *Pachystachys lutea* is therefore an acquisition for plant lovers who always have difficulty in growing the *Jacobinia* species, since these are actually greenhouse plants and require a moist atmosphere in winter. This pseudo-novelty must surely be better suited for use as a houseplant, otherwise it would not have been described so favourably in Karsch's *Vademecum Botanicum* published in 1894.

Care: I have more or less followed the growing instructions which apply to *Beloperone* (p. 138), except that *Pachystachys lutea* is unsuitable for the balcony or the garden. In my house it therefore remains indoors throughout the year; during the past year in an airy, north-east facing window, where the temperature in summer is normal and in winter about 15°C (58°F). And when I was not sure what to do next, I looked up the instructions for *Jacobinia carnea*, which you can find on p. 246.

Pachystachys lutea, fam. *Acanthaceae*

x Pachyveria

In botanical textbooks this member of the *Crassulaceae* family has a cross in front of its name to indicate that it is not a pure genus, but that it is the result of crossing. It is an 'intergeneric bastard', but none the less beautiful for all that.

x Pachyveria clavifolia, fam. *Crassulaceae*

Origin: as a rule they are the outcome of crossing *Echeveria* with *Pachyphytum* species. As x *Pachyveria* they are much-loved Mexican leaf-succulents.
Appearance: in general they resemble *Echeveria*, but the implantation of stemless or short-stemmed rosettes is derived from *Pachyphytum*. This is also the source of the plant's waxy bloom and the proportionately large reddish or almost white flowers growing on stems which at first curve, but later grow erect; see the photograph.
Species: only the most avid collectors of succulents will connect the plants with some of the numerous names, or will know that these are so-called generic hybrids resulting from crossing *Echeveria* and *Pachyphytum*. According to the professional literature the four most important hybrids occasionally available are: x *Pachyveria clavifolia,* x *P. pachyphytoides,* x *P. scheideckeri* and x *P. spathulata*.
Care: in general as for *Echeveria*; with the emphasis on a cool situation in good light in winter, practically without watering.
Propagation: especially from fresh seed, supplied by a dealer in exotic seeds; it will germinate in a fortnight.

Pandanus

I freely admit that this is one of my favourite plants. For many years I have always had one in my plant window; the last one had grown so enormous that it filled the entire case; I have had the present one for eight years and it is a showpiece which I would hate to lose.

Origin: the family of *Pandanaceae* includes nearly a thousand species, divided into 3 genera. Pandanus itself is a representative of a group of trees and shrubs most of which grow in the Indonesian archipelago, though they are found in Madagascar and tropical regions of Africa as well.
Appearance: they are all erect-growing and evergreen and at first sight look like all-green or variegated *Dracaena* plants, particularly when young. Nevertheless there is a great difference. To begin with the long, narrow leaves, up to 1–2 m (3–6 ft) long, depending on the species, grow in a spiral formation round the stem. After four or five years pandanus starts to lift itself from the pot by means of its stilt-like roots. This phenomenon is a characteristic of the plant and you should on no account attempt to cover the roots by repotting at a lower level. These stilt-shaped 'aerial roots', as botanists call them, serve to replace the stem which has meanwhile died off. Quite soon runners develop round, among and below the foliage midway along the stilt-roots; this, too, is a characteristic of pandanus. The runners can be left or used for propagating new plants. In our regions the plant does not flower.
Species: there are three species available:
1. *P. utilis* in its native Madagascar grows to 20 m (60 ft) in height. It has stiff dark grey-green leaves, to 10 cm (4 in) across and up to 150 cm (60 in) in length, densely covered along their edges and at the points by quite dangerous red spines. This is the most hardy species. It tolerates a winter temperature of 16°C (60°F), but not lower. If the temperature drops even for a few days, the plant will die. It may take five to six weeks before it begins to go downhill, but once it does, nothing can be done about it. Be careful, therefore, if you go away on holiday in winter, for large plants are quite valuable.
2. *P. sanderi*, a native of Timor and some of the other islands in the Indonesian archipelago. The leaves are 5–6 cm (2–2½ in) broad, 80–100 cm (32–40 in) long, striped with white or pale green. Young foliage is frequently white, the edging and terminal spines white or red. Minimum temperature in

winter 18–22°C (65–72°F).

3. *P. veitchii* comes from the islands in the Pacific. The leaves are longitudinally striped with yellow, with needle-fine green stripes in between. Softly spined; no terminal spines. Minimum winter temperature 18–22°C (65–72°F). This is the most important species in cultivation and tolerant of living-room atmosphere. It is, moreover, a very decorative plant, which can be kept for quite a long time, until it unfortunately grows too large.

Care: keep indoors throughout the year. In summer it should be given a position in good light but out of the sun at a minimum temperature of 20–25°C (70–78°F). Some degree of atmospheric humidity during the time the new shoots appear is desirable. Winter temperature: see above. Resting period from September to January.

Watering: during the first six months, when it is growing well, water liberally and spray from time to time. Only excessively hard water need be softened to a pH of 6–6.5.

Feeding: from March to July give a normal feed solution once a week or every fortnight.

Repotting: young plants every year; afterwards as required. It is easiest if you tie and wrap up the foliage while you do this. The new pot should be considerably larger.

Pandanus grows best in a mixture of (beech) leafmould, coarse peat and loam or clay turves. Such a compost cannot be bought, so you will have to make it yourself. If you find this too much trouble, use ordinary pre-packed potting compost. In view of the long stilt-roots it is best to use deep pots.

Propagation: *P. sanderi* and *P. veitchii*, as mentioned above, form runners. These should be at least 20 cm (8 in) long; when they are removed from the parent plant they will readily root in propagating soil in small pots, at a normal room temperature of 20–22°C (70–72°F).

P. utilis, which does not develop runners, can be raised only from fresh seed imported by a dealer in exotic seeds. Germinating temperature 25–30°C (78–86°F), it will take three to four weeks. Of course the other species can also be grown from seed, in so far as it is available.

Diseases: it is the experience of both professionals and private owners that pandanus does not suffer from diseases or pests. The only thing that may kill it is too cold a position.

Pandanus veitchii, fam. *Pandanaceae*

Paphiopedilum (syn. Cypripedium) — Venus'-slipper

Paphiopedilum means 'Venus'-slipper', for the Goddess of Beauty was called Paphia after the town of Paphos in Cyprus. Although the genus includes only about 60 species, the number of strains derived from this proud ancestor is so large that it would make many a learned genealogist blench.

Paphiopedilum fairieanum, Venus'-slipper, fam. *Orchidaceae*

Origin: in contrast to other orchids which originate mainly in the Americas, most of the species of the genus *Paphiopedilum* come from the Old World, from eastern Asia and the islands in the Pacific: from the Far East via the Philippines and Indonesia to New Guinea. The *Cypripedium* or lady's-slipper species often confused with these chiefly terrestrial-growing tropical plants, belong in the temperate regions of the Old and the New World, and include the wild orchids belonging to the genus which occur, among other places in central Europe.
Appearance: evergreen perennials, with clusters of roots or with rhizomes; no pseudobulbs. The leathery leaves develop from the base and are green, mottled or marbled, depending on the species. The stems vary between 15 and 50 cm (6 and 20 in), again depending on the strain, and emerge from between the leaves; each stem usually bears only one flower.
Inflorescence: as a rule this consists of the very broad, undivided dorsal sepal, with at either side petals pointing downwards, and the slipper-like lip, whose shape facilitates pollination. Behind it is the staminodium, broadened like a petal; this is an infertile stamen, often fleshy and shield-shaped. Every part of the flower may occur in any imaginable shape and colour, depending on species or strain. The flowering season of many well-known species and cultivars occurs in the winter months, but there are also forms which flower in other seasons, often for weeks or even months on end.
Species: some of the finest species suitable for room cultivation or for a simple plant window: *P. barbatum*, both petals edged with hairs; a native of Malacca. It has frequently been used for crossing. Leaves with dark green, chequered marking. Stem 25 cm (10 in) tall, flower 6–10 cm (2½–4 in) across; it flowers from May to July. *P. callosum* originates in Thailand and Vietnam. Leaves bluish green with chequered markings. Stem 35 cm (14 in) tall, flowers 10 cm (4 in) across; flowering season March to July. It has been crossed with the 'Sanderae' variety. *P. fairieanum* comes from Sikkim in western India. Pale green foliage. Stem 20 cm (8 in) tall, flower 6 cm (2½ in) across, flowering season July to October. A particularly beautiful and interesting species. *P. glaucophyllum*, a Venus' slipper with blue-green leaves, is a native of Java. Stem 40 cm (16 in); it bears several flowers, each 10 cm (4 in) across; flowering season almost throughout the year. *P. hirsutissimum*, a very roughly haired Venus' slipper from Assam. Green foliage. Flower-stalk 20 cm (8 in) tall, flower 12 cm (3¾ in) across; flowering season April–June, sometimes March–May. *P. insigne*, a magnificent plant growing on mossy slopes in the Himalayas at altitudes of up to 1800–2000 m (5400–6000 ft). Foliage green. Stem 20–25 cm (8–10 in) bearing a single flower, 8–10 cm (3¼–4 in) across. Flowering season November to March. This is the most frequently cultivated species and the one most suitable for beginners. It has frequently been crossed, resulting in well-known strains such as 'Aureum', 'Chantinii', 'Harefield Hall' and 'Sanderae'. Flowers from autumn into winter. An undemanding plant for the unheated greenhouse. On the plant the flowers last for several months; when cut they will keep for some weeks. *P. niveum*, with snowy white flowers, is a native of the islands to the west of Borneo and Malacca, where it grows on limestone rocks. Variegated foliage. Stem 15 cm (6 in), flower 6 cm (2½ in) across; flowering season April to June. *P. spicerianum*, a native of Assam. Foliage green. Stem 20 cm (8 in) bearing a single flower, 8 cm (3¼ in) across. Flowering season October to February. Much used for crossing. *P. villosum*, from southern Burma; an epiphyte. Foliage grass-green. Stem 25–30 cm (10–12 in), very large flower 12–30 cm (4¾–12 in) across; flowering season November to April. Many hybrids. The most important hybrids are: × *P. harrisianum*, flowering from October to January;

× *P. lathamianum*, winter-flowering;
× *P. leeanum*, also winter-flowering.
Care: there are three groups: orchids for the unheated greenhouse, for the temperate greenhouse and for the hot house. Of the species mentioned *P. hirsutissimum* and *P. insigne* and their hybrids are suitable for the unheated greenhouse; they over-winter at a temperature of 12–15°C (54–58°F). All the others, not having a real resting period, require moderate warmth in winter, not below 18°C (65°F) at night and a moist atmosphere throughout the year. In summer the temperature must not fall below 25°C (78°F) at night. Plenty of light throughout the year, but no direct sunlight. Adequate soil and air temperature are particularly necessary to maintain the health of species with variegated foliage. Warmth and moist air are especially important during the growing and flowering seasons. Errors in this connection cause great damage.
Watering: needless to say these orchids, too, must always be given water containing absolutely no lime, chlorine, fluorine or other harmful elements. There are a few exceptions; *P. niveum*, for instance, likes lime. In view of its native habitat, where it grows on limestone rocks this is logical.

In non-industrial regions rainwater may be used, but where this is polluted one must have recourse to completely demineralised water. It is advisable to keep a store of water in the greenhouse, so that it is always at the right temperature. The plants require practically no resting period; nevertheless you should water more sparingly in winter for the simple reason that at that time there is less evaporation. Normal watering at this season would result in the compost becoming soggy and this, together with the lowered temperature, would soon lead to root rot. Overhead spraying can be done throughout the year, except on very dark days.

Feeding: ordinary fertilisers may be used for terrestrial orchids. Liquids used in soilless cultivation are also very suitable. Feed only during the growing and flowering season.

Repotting: hot house species every year; species grown in an unheated greenhouse every second year. In both cases repotting should take place before the plant starts into growth, usually early in spring. Compost about the same as for epiphytic species, even for terrestrial paphiopedilum species. A suitable mixture is potting compost with the addition of finely chopped osmunda or polypodium fibre. An ample layer of crocks in the bottom of the pot. Do not compress the mixture too tightly.

Propagation: by division when repotting.
Diseases: greenfly and red spider mite are encouraged by too dry an atmosphere.

Paphiopedilum insigne 'Harefield Hall', fam. *Orchidaceae*

Parodia

Parodia microsperma, fam. *Cactaceae*

A large genus of beautiful globular cacti which, because of their strikingly fine spine development and the flowers produced even by quite young plants, are among the favourites of collectors.

Origin: most of the species cultivated here originate in the central regions of South America; they grow on arid mountain slopes and in hilly areas of northern Argentina as well as in Bolivia, Paraguay and southern Brazil.
Appearance: depending on the species, the spheres may achieve a diameter of 5–15 cm (2–6 in) or more; some become cylindrical in shape when older. The ribs are often spirally placed, warty or knobbly, densely covered in colourful spines, usually with long barbs. Flowers in shades of red between salmon pink and carmine, sometimes yellow, usually 4–5 cm ($1\frac{3}{4}$–2 in) across, often appearing in bunches on the crown; they may last up to ten days.
All parodia species are slow-growing and easily accommodated. To achieve more rapid growth they are frequently grafted. Some of the finest are: *P. chrysacanthion, P. maxima* – the largest species, spines up to 11 cm ($4\frac{1}{2}$ in) in length; *P. microsperma, P. nivosa, P. sanguiniflora, P. schwebsiana.*
Care: generally as for other small cacti (see *Mammillaria,* p. 254 and *Rebutia,* p. 312), but be particularly careful with watering, as the neck of the root is very sensitive to water. Keep at 8–12 °C (46–54 °F) in winter, light and cool.
Propagation: from seed.

Passiflora – Passion flower

There is a legend that around 1620 a Jesuit father had a vision in Peru: he saw the miracle of the passion flower. He saw the three pistils as the nails on the cross, the three-coloured circle round the anther as the crown of thorns, and the stemmed ovary as the Lord's goblet . . .

Origin: a number of species occur in the sunny headlands of Peru, but in all there are more than 400 species, most of which are also found in other South American countries. Only a few species originate in Asia and Australia, while a number of almost winter hardy species grow in sub-tropical areas of North America. Only about ten species are in cultivation here.
Appearance: climbing shrubs with runners often meters long, bearing tendrils; these are trained on frames or along bent wire. The stemmed leaves are green, smooth, a broad oval, 3-, 7-, or 9-lobed, depending on the species; as a rule the central lobe is longer than the others. There are also variegated hot house species, but these have a more inconspicuous inflorescence. The flowers appear without interruption from early summer until autumn; they are large and striking in shape and colouring. The colours are white or sky-blue, more rarely wine-red or violet. In favourable conditions pot-grown passion flowers may develop yellow or orange fruits.
Species: an attractive houseplant, rich in tradition, is *P. caerulea,* the heavenly blue passion flower, which has 5- to 7-lobed, evergreen leaves; the flowers are white with a blue circle. Hybrids are the ivory-white 'Constance Elliot' and 'Empress Eugenie', which will flower profusely as a young plant. Provided it is protected in winter, the latter will grow against a warm wall in a sheltered position out of doors as well. Frozen stems will rapidly recover and will then flower normally. *P. gracilis* and *P. incarnata,* known in the United States as 'Wild Passion Flower' and very popular as a garden plant to be grown against a wall, is best regarded as a foliage plant. It is hardly known here. *P. maculifolia* is a foliage plant with beautiful variegated leaves and small, creamy white flowers. Suitable only for a warm greenhouse. *P. quadrangularis* is an enormous climbing shrub. Flowers 8–12 cm ($3\frac{1}{4}$–$4\frac{3}{4}$ in) across, ▶

Passiflora caerulea, Passion Flower, ▷
fam. *Passifloraceae*

Passiflora racemosa, fam. *Passifloraceae*

with a corona consisting of 5 rings of colourful filaments. Edible fruits. Only suitable for a large hot house. *P. racemosa* has tough, 3-lobed foliage and bright red flowers growing in loose clusters. It is suitable for use as a house plant.

Care: from the onset of growth in March it should be placed in plenty of light in the sunniest position in a south-facing window. From late May until September it may be put on the balcony or in the garden. Plunge the entire pot; never put the plant directly into the soil. Tie to canes or train on bent wire. Large specimens should be planted in tubs. In August after the growing period, prepare the plant for dormancy by reducing the water supply without, however, allowing the soil-ball to dry out. Gradually reduce feeding and bring indoors before the first night frosts occur.

In winter the plant should be given an airy position in good light until February. The best temperature is 6–8°C (43–46°F); moderate warmth up to 12°C (54°F) is tolerated provided a little more water is given, but normal room temperature will sap the plant's health and affect its flowering. When the plant starts to put forth new shoots in March, gradually increase the water supply and place in a warmer position in preparation for the summer.

Pruning: towards the end of February, about three weeks before repotting, old stems should be cut back to 6–8 buds; this will encourage the development of freely flowering short sideshoots. Plants which continue to grow without ever producing flowers can be forced to flower by pruning in summer.

Watering: vigorously growing and profusely flowering plants may often have to be watered twice a day in warm summer weather. Otherwise follow the instructions given under Care. The water should not be too hard: pH 5·5–6.

Feeding: because of its enormous vigour the plant must be given a weekly nutrient solution from spring until August. Repotted specimens should not be fed until the roots are growing well.

Repotting: young plants every year; older specimens every two years; in March. Choose a pot adapted to the size of the soil-ball and to the space required by the runners. The compost should be very rich in humus and slightly loamy. Large specimens will tolerate a storage feed, like the oleander (p. 270).

Repotted plants must be put in a cool and shady position for eight to ten days.

Propagation: the easiest method is to remove the runners and pot them singly; this is done between May and July. A slower method, less likely to be successful, is by cuttings from half-ripe wood. Raising from seed is difficult and plants grown this way do not as a rule flower readily.

Diseases: rarely occur in a properly cared-for passiflora. Mealy bug will attack only if the plant is given too warm a position in winter and lacks fresh air. Incorrect treatment in summer may lead to red spider mite infestation. Preventive spraying after pruning in spring will provide a measure of protection.

Important: 1. If the leaves yellow and drop in spite of having successfully withstood the winter, there is no cause for anxiety, provided the plant does not become entirely bare. 2. Passion flowers will only flower in plenty of warmth and sunlight; however well cared for they may fail to flower in a cool and rainy summer. 3. Failure to flower while the stems continue to grow may also be the result of too light and peaty a compost. This can be remedied when the plant is next repotted.

Pavonia

Neither its name nor its appearance has anything to do with the Latin word *pavo*, meaning peacock. The plant is called after the early eighteenth century Spanish explorer and botanist Pavon. This magnificent winter-flowering plant deserves to be more popular.

P. multiflora, a native of Brazil, is the only one of the approximately 170 species which has made the jump to the European plantpot. In the United States a very similar species, *P. intermedia*, is cultivated as a houseplant as well.

Appearance: evergreen small shrub, sharing with many other *Malvaceae* a tendency to restrict itself to one stem. Its 15–20 cm (6–8 in) long leaves are slightly serrated along the edges; they are rough underneath.

Inflorescence: reddish purple sepals and corolla, which never opens. The flowers appear on short, rigidly erect stalks growing in the axils of the upper leaves. The prominent style is pink-veined and has blue, kidney-shaped pollen grains. Flowers from autumn to spring. This is a plant for hothouse lovers. As the flowers grow at the top and the plant must therefore not be pruned, the height which it may attain must be taken into account when placing it on the window-sill.

Propagation: by tip cuttings only, at an atmospheric and soil temperature of 30°C (86°F).

Pavonia multiflora, fam. *Malvaceae*

Pedilanthus

This plant, whose name is hardly known, is another succulent member of the spurge family. Of the 14 species, all natives of tropical Central America, we only know the cultivar *P. tithymaloides* 'Variegata', which has white-edged foliage.

Appearance: a typical feature of this plant is the zig-zag shape of the green shoots, to 1 cm ($\frac{1}{2}$ in) thick, forming bushy shrubs up to 2 m (6 ft) tall. The leaves of the species are green, slightly succulent, elongated oval to 12 cm ($4\frac{3}{4}$ in) long and 6 cm ($2\frac{1}{2}$ in) across. The white-edged leaves of the cultivar mentioned above are frequently covered with a carmine red bloom as well. In the autumn the foliage drops and the dormant season begins.

Inflorescence: the well-known spurge flowers appear in dense terminal umbels; they are surrounded by striking red bracts. Unfortunately the plant produces its magnificent inflorescence only at an advanced age when it has already grown too tall for the window-sill.

Experience: we know that at one time these plants with their variegated foliage and amusingly zig-zag branches were used in this country, and especially in the United States, as decoration in plant bowls and other arrangements. Perhaps this will become fashionable again one day. These young plants, 50–60 cm (20–24 in) tall, might tempt plant lovers to grow them separately. Two things should be remembered: firstly, every textbook mentions the fact that pedilanthus contains a particularly poisonous spurge liquid, and secondly too moist air is likely to cause mildew.

Pedilanthus tithymaloides, fam. *Euphorbiaceae*

Pelargonium — Geranium

'The geraniums in window boxes and on the window-sill belong to the genus of *Pelargonium*,' we read in a large plant encyclopaedia. This succinct statement is not as strange as it might appear, for there is in fact also a *Geranium* genus, the many species of which are called cranesbill in this country.

Origin: whereas cranesbill species are found mainly in temperate climates, the genus *Pelargonium* with its approximately 240 species occurs almost exclusively in South Africa. The first geranium of this genus was brought to Europe as early as 1710.
Cultivars: in the course of time four large groups have been created, of which three, including many strains, are constantly being further developed. Only one group, the geranium with scented foliage, contains a number of original species (see p. 290). These are just as suitable for room cultivation as the cultivars.

1. Pelargonium grandiflorum hybrids

These usually have one stem which becomes woody. The leaves are large, pale green, kidney-shaped, with dentate edges. The flowers are large or even very large, 5 cm (2 in) across. They appear in dense, sometimes looser, umbels in many different colours: white, pink, red to darkest blue-violet, always with multicoloured blotches, streaks or veins. The flowering season is from May (useful for Mother's Day!) to early autumn.
Care: keep them indoors throughout the year, though if necessary they may be put on the balcony in summer. They should have plenty of light in summer, with morning or evening sun; a lot of fresh air. In winter they should have a moderately warm or cool place, not exceeding 10–15°C (50–58°F), but certainly not in the cellar with the balcony geraniums.
Watering: during the growing and flowering season (March to August), water liberally; sparingly after removal to cooler position. Beware of excess water in the pot both in summer and in winter. The water should be slightly acid-reacting, pH 5–6.5.
Feeding: freely during the growing and flowering season (March to August), but do not overfeed.
Repotting: in late February, into slightly larger pots. Rich compost is advisable. Pre-packed potting compost is suitable but whichever is used, it should be slightly alkaline, in view of the pelargonium's relatively great

Pelargonium grandiflorum, hybrid, fam. *Geraniaceae*

liking for lime. The pots must always be very well draining, for stagnant water is something which the geranium definitely does not tolerate. It is also better not to use plastic pots.

Pruning: at the end of August, after flowering, cut back a little. More drastic pruning, if necessary, at the time of repotting.

Propagation: from cuttings obtained when pruning; vigorous shoot tips, which have not yet flowered, are cut 5 cm (2 in) below a pair of large leaves, cutting obliquely downwards; the leaves on the lower half of the cutting are removed. Insert the cuttings up to a leaf axil in small pots containing cuttings compost and place them out of the sun, but in good light. An indoor propagator with soil temperature of 10°C (50°F) is ideal. Plant out for the first time in October, maintaining the same temperature. Repot again towards the end of February and place in a slightly warmer position, 12°C (54°F). Too high a temperature, 15°C (58°F), will inhibit flowering. Cut back in spring to encourage bushy growth. Propagation in March is also possible; easier, but less likely to be successful. In this case cut back in September.

Raising plants from seed is also becoming popular. The grower prefers this method because plants increased by the vegetative method are prone to virus disease. However, only a few strains can be grown from seed, and these are by no means the most beautiful.

Diseases: every error in cultivation may result in an attack by greenfly or whitefly. The latter occurs especially if the plant is placed outside in summer, when it may be attacked by large swarms. The best method of prevention is by spraying the underside of the leaves. Foliage diseases caused by fungi or bacteria are unfortunately prevalent in geraniums; they cannot be cured in room cultivation. The same applies to grey mould, which attacks a plant in too moist a position or lacking fresh air.

Important: cultivated geraniums have only a limited life-span. They flower most profusely in the second or third year, after which they deteriorate. Keeping old plants is not worth the trouble.

2. Geraniums for bedding or window boxes

When speaking of geraniums one usually refers to the plants with a brown ring on their leaves, those that are officially called *Pelargonium zonale*. These are now cultivated in innumerable strains and colours. They have an erect, bushy habit. When bruised, the velvety leaves give off the typical geranium scent. Single or double flowers appear in every shade of red, from the palest pink and lilac to scarlet, usually in dome-shaped umbels. The flowering season is from the end of April often to late in October. The *P. peltatum* hybrids, or hanging geraniums, have creeping or trailing stems, more than half a metre (1½ ft) long, often jointed, and smooth and shiny, thick, ivy-shaped leaves. Double or single flowers in slightly ▶

Pelargonium zonale, the well-known, erect-growing hybrids which are also called geraniums.

more subdued shades. The flowering season is the same.

Care: both groups like to be placed out of doors in summer. In winter they need good light and fresh air, at a temperature of 6–8°C (43–46°F). Whereas the more robust *P. zonale* hybrids tolerate a sunny spot, the smooth-leaved hanging geraniums prefer semi-shade. Both will stand a position in a closed room for a short time only. Otherwise treatment is the same as for the cultivated pelargonium.

Propagation: from cuttings, in autumn or spring; *P. zonale* hybrids, and even F_1 hybrids, can now also be grown from seed.

3. Scented-leaved geraniums

These modestly flowering species with fragrant foliage are rarely available, but very attractive; they have pinnate leaves. They include: *P. capitatum*, with rose-scented foliage and little mauve flowers; *P. crispum*, lemon-scented, with crispate leaves and violet flowers; *P.* × *fragrans*, deliciously nutmeg scented foliage, flowers white with red veins; *P. graveolens*, the fragrant house geranium most often cultivated, flowers pink to purple; *P. odoratissimum*, the lemon geranium, flowers white to pink; *P. radens*, rose-scented leaves, small pale mauve flowers with dark veining. All these species have already been in cultivation in this country for about 200 years.

Care: as for hybrid geraniums, but to prevent lank growth, a sunny position in summer is essential. Cut back and propagate in spring only. Life-span only two to three years.

Pelargonium radens, fam. *Geraniaceae*

Pellaea

The name is said to be derived from the Greek word *pellos*, supposed to mean something like blackish and indicating the dark brown, almost black leaf-stems of some species. Of the approximately 60 species known, 6 are in cultivation. They are regarded as undemanding, but nevertheless attractive plants which will last for years.

Origin: unlike most ferns, which prefer shady and damp positions, pellaea is a xerophyte; that is to say that it grows in dry places. The plants are happy growing on rocks.

Appearance: basal stems with feathered or smooth-edged composite leaves grow from the creeping rhizomes. Clusters of spores are found on the underside of the leaves at the end of the veins, hidden by the curved edge of the foliage.

Species: *P. atropurpurea*, a dark, reddish purple fern, is an almost winter-hardy plant originating in the United States, where it grows everywhere from Vermont to Florida, Texas and Mexico. It likes limestone rocks and therefore not only tolerates, but requires fairly hard water. The leaf-stems are 10 cm (4 in) long, glossy brown-black in colour; the beautiful reddish-brown fronds are up to 25 cm (10 in) long. This is an undemanding plant for the cold greenhouse. *P. cordata* occurs from Mexico to Peru. It has straw-coloured leaf-stems, to 20 cm (8 in) in length, and fronds up to 30 cm (12 in), with small elongated triangular leaves. In spite of its tropical origin it is considered a fern for the cold greenhouse. *P. falcata*, with sickle-shaped fronds, grows in South Asia and tropical regions of Australia, in Tasmania and New Zealand. It looks like a larger version of *P. rotundifolia*. The latter originates in New Zealand and, being a dwarf species, it is the pellaea most suitable for room cultivation. The leaves are leathery and glossy, dark green *P. viridis* with long, red-black leaf-stems and doubly or trebly pinnate triangular leaves, is the most hardy species.

Care: the origin alone leads us to the conclusion that pellaea species differ in every aspect from other ferns. In summer the position must be warm, roomy, airy and in a good light, although some shade is tolerated. In winter, too, plenty of light and air is required, preferably at temperatures of between 12 and 15°C (54 and 58°F), for instance in a north-east facing

window without a heating radiator below, and shut off from the warm room by a curtain.
Watering: normally in summer, more sparingly in winter. Very sensitive to excessive atmospheric moisture. On the other hand drying out of the soil-ball is also harmful. As a rule it is unnecessary to soften the water, since these are lime-loving plants like *Sansevieria* and other *Liliaceae;* pH 7·0 is perfectly acceptable. Spraying is not desirable.
Feeding: normally in summer.
Repotting: as required, using ordinary pre-packed potting compost or special fern compost.
Propagation: from spores; a temperature of 18°C (65°F) is already sufficient, but it is not easy to carry out indoors.

Pellaea rotundifolia, fam. *Sinopteridaceae*

Peperomia

Once it had been established that these delightful plants were able to survive outside a technically controlled tropical plant window or hot house, their versatility and the variety of their forms turned them within the space of ten years from greatly admired and loved hot house plants into house friends.

Origin: they are all natives of the tropical rain forests of South America. More than 600 species are known and so many of these have been brought into cultivation that a peperomia collector will not quickly find himself limited in his choice. Moreover, they make very suitable objects for collection because of their small size. A couple of dozen of the most important representatives of the peperomia genus can quickly be assembled and such a varied group will give you a great deal of pleasure.
Appearance: in order to bring some clarity into this multiform group, they have been divided into species with variegated foliage, species with all-green foliage and species with creeping or trailing stems. Most of them are log-growing, perennial epiphytes with thick stems and fleshy leaves, the upper surface of which has a remarkable structure visible to the naked eye: if you cut through the leaf and see it in profile, you will notice that the epidermis is a thin membrane forming a kind of water reservoir covering the actual leaf tissue. Thin-leaved species do not have this structure. Flowers occur in only a few

Peperomia caperata, fam. *Piperaceae*

species. The majority are sold as foliage plants because of their interesting, frequently variegated, leaves.

Species: 1. Variegated species: *P. argyreia*, Brazil. Shield-shaped green and silver striped leaves, 8–12 cm ($3\frac{1}{4}$–$4\frac{3}{4}$ in) long, growing on red stems. *P. caperata*; this plant was imported from Brazil as late as 1954. Corrugated foliage, marbled in dark and pale green, growing on long red stems. In spring numerous flower spikes appear on red-blotched stalks. *P. clusiifolia* (sometimes sold as *P. obtusifolia*) has short-stemmed leaves, thick and leathery, slightly curved, green with a reddish sheen. There is also a yellow-variegated hybrid called 'Variegata'. This species has been known since 1793 and must be the oldest peperomia in cultivation. *P. griseoargentea* (trade names also *P. hederaefolia*, *P. pulchella*) has long, red and white striped leaf-stems, leaves to 6 cm ($2\frac{1}{4}$ in) across, heart- or shield-shaped, quilted between the main veins on the upper surface, and with a silvery sheen.

2. Green-leaved species: *P. arifolia*, and its variegated hybrid called 'Princess Astrid'; the white flower-spikes join to form the inflorescence. *P. fraseri* (formerly *P. resediflora*) has white flowers with a reseda scent, forming clusters of spikes. *P. serpens* (or *P. scandens*), with creeping runners; a fine trailer.

3. Species with creeping or trailing stems: *P. angulata* with thin, reddish shoots and dark green, yellow-veined leaves; *P. glabella* 'Variegata', now very popular in bottles or as ground-cover; it is densely branched and has green/white oval leaves; and *P. rotundifolia*. The latter has thin, creeping runners and brown/dark green marbled foliage (it used to be called *P. prostrata*).

Care: a position in good light in summer but out of the sun, as otherwise the leaves will change colour. For the variegated forms the temperature in winter should be 18–20°C (65–70°F), for the green species about 15°C (58°F). The most favourable position is as a decorative ground cover in tropical cases or plant windows; some can also be grown as epiphytes on treetrunks. Room cultivation will in the long run not be successful.

Note: Peperomia species with fleshy (succulent) leaves – especially the robust *P. clusiifolia* (obtusifolia) – are more tolerant of living-room temperature and dry air, as well as less light, than the delicate soft-leaved species and varieties, which are more suitable for a tropical window or indoor greenhouse.

Atmospheric humidity: high in summer, very much less in winter in order to avoid spotting of the leaves.
Watering: great care must be taken when watering peperomias; their succulent structure provides a pointer to this. Plastic pots are better avoided, as in such pots the compost becomes too wet rather than too dry. Always use soft water at room temperature and water sparingly, especially in winter.
Feeding: once every fortnight from March to July.
Repotting: as required into flat bowls. Light, humusy compost. Pre-packed potting compost is acceptable, but a mixture of rotted beech leaves, coarse sand and some rotted cow manure is better.
Propagation: green species from tip cuttings or leaf-cuttings; variegated species only from tip-cuttings. Seed is hardly ever available.
Diseases: when the soil is damp beware of slugs. If the situation in winter is too cool, the plant will rot.

A group of well known peperomias, from l. to r.: *Peperomia fraseri*, *P. obtusifolia* 'Greengold', *P. verticillata* and *P. caperata*

Phalaenopsis – Moth Orchid

Is it not remarkable? As long as a century ago the famous English horticulturist George Nicholson devoted no less than a page and a half of his monumental nine-volume work on gardening to the *Phalaenopsis* and mentioned no fewer than 34 species, from *P. amabilis* to *P. violacea*.

Development. His statement – that the *Phalaenopsis* is one of the finest orchid species in cultivation – is still valid today, for this truly magnificent stranger from the Far East is still regarded in the same way.

Except for true orchid experts or collectors, people generally speaking of the Phalaenopsis are generally referring to the white, pink or lilac flowering *P. amabilis* (the 'lovable'). This noble creature has come a long way before it reached its present position at the top. It was considered delicate and was grown in hot-houses for cutting. Then came the great break-through: hybrids were developed and the plants are now assessed in an entirely different manner. It is these hybrids which are referred to when the name is now mentioned.

The original species has disappeared from the market and other species or natural forms such as *P. aphrodite*, *P. cornucervi* ('Hartshorn Phalaenopsis'), *P. schilleriana*, *P. stuartiana* and even the magnificent *P. violacea* of the relatively large-flowered Borneo type can no longer, or not yet, keep up the pace.

Thanks to the fact that they have become much hardier in cultivation, modern *P. amabilis* hybrids are eminently suited to careful indoor propagation. They will thrive (and flower repeatedly) in hydroculture as long as we remember that the roots dislike too much water.

The artificially created flowering seasons of the various cultivars enable the orchid lover to have flowers throughout the year. It will, moreover, take several weeks, from the time the buds appear until the flowers fade. Afterwards it is important not to cut the panicles back too drastically. Each stalk should remain at least 12–15 cm (5–6 inches) long, since they often bear eyes which will produce new flowers.

Appearance. In its native habitat the *Phalaenopsis* grows as an epiphyte, preferably in the crowns of trees growing on the edge of forests at an altitude of 200–400 m (650–1300 feet), where it is protected from the sun and where there is a certain degree of humidity. The beautiful dark green leaves develop directly from the strong roots; there are no pseudo-bulbs. The body of the plant is surrounded by silver-grey aerial roots. All the above is characteristic of the life of this plant.

Care. This may be largely deduced from the plant's appearance, but it should be mentioned that when the plants are watered when in growth – of course always with tepid, soft water – the base of the leaves must not become wet. The plants should moreover not be watered daily, but at most twice a week, and in winter once or twice a month. They must not be fed with ordinary fertiliser, but should occasionally be given a special orchid feed – during the growing season only. All the above depends on the flowering season, which is believed by some experts to occur in the months July to November, by others between October and March. The truth actually lies midway, for flower development depends on the cultivar and on the way the plant has been prepared by the grower. Specific instructions will therefore be your best guide.

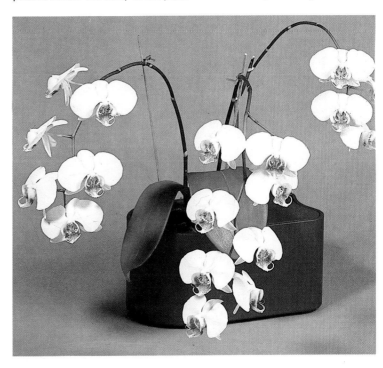

Phalaenopsis amabilis hybrid, moth orchid, fam. *Orchidaceae*

Philodendron

Of this plant, after the rubber plant surely the most popular houseplant, 275 species are known to date. They originate in the tropical rainforests of Mexico, Central America, Peru and Brazil. They also occur in Cuba and Jamaica. Most of them take up a good deal of space.

Appearance: shrub-like or climbing, sometimes stemless and herbaceous plants, with or without aerial roots. They are closely related to *Monstera* (p. 264). The evergreen, sometimes splendid velvety or white-dappled, leaves vary in form: even young and mature leaves of a single species show marked dissimilarities, making it difficult to describe them.

The characteristic inflorescence, resembling the arum-lily, has a white or coloured spathe and a plump spadix. Used as an indoor plant it rarely flowers. Older specimens grown in a greenhouse flower fairly regularly.

The well-known *Monstera deliciosa* is sometimes sold under its former name *Philodendron pertusum* (see p. 264, *Monstera*).

Usage: many of the species are suitable only for show collections and botanical gardens, as they are too large for the house and often even for the amateur greenhouse. Our selection therefore contains mainly the small-leaved, less vigorous varieties which are not restricted to the greenhouse.

1. Climbing or creeping plants with undivided foliage

P. erubescens, originating in Columbia, has 40 cm (16 in) long leaves, dark green on their upper surfaces. Flowers with red spathes and creamy white spadices. A very decorative and, moreover, hardy and lasting species, whose not too vigorous growth can be further restricted by pruning. Two well-known and popular strains are 'Green Emerald' and 'Red Emerald', whose respective colours are described by their names. Both are disease-resistant.

The native habitat of *P. ilsemannii* is unknown. It is one of the finest variegated trailers available. Its leaves are up to 40 cm (16 in) in length, 15 cm (6 in) wide, usually lanceolate, white/green marbled, the green often confined to a few blotches. In spite of its variegated foliage this is a strong, easily cultivated plant. *P. andreanum* originates in Colombia. Splendid, velvety olive-green leaves with white veins; 40–80 cm (16–32 in) in length, up to 30 cm (12 in) wide; they grow

Philodendron erubescens, fam. *Araceae*

Philodendron scandens, fam. *Araceae*

perpendicularly around the stem. They look beautiful growing up a moss-covered stump in the window. Requires an even temperature (not below 18°C (65°F) in winter) and moist atmosphere (illustration p. 297).
P. panduriforme has varying foliage, the lower leaves being oblong, the central leaves violin-shaped and the upper having two good sized basal-lobes or 'ears', somewhat reminiscent of *Syngonium*. A strong houseplant, tolerating the usually dry indoor conditions.
The well-known climbing *P. scandens* is a native of Jamaica and has been known in Europe since 1793. This plant was formerly called *P. cuspidatum*. It can be recognised by its heart-shaped leaves, small and firm when young, later doubling in size, up to 30 cm (12 in) in length and 20 cm (8 in) in width. A fine, undemanding trailer, equally suitable for training up walls. Puts up with little light. *P. verrucosum*, native habitat between Costa Rica and Colombia. Like *P. melanochrysum* it has large leaves with a velvety sheen, olive-green on their upper surface, with emerald-green edges, red-brown underneath. Its unusual leaf-stems, which may grow up to 50 cm (20 in), are covered in wart-like nodules. Demands a warm, damp atmosphere (up to 18°C (65°F) in winter).

2. Climbing species with divided foliage

P. elegans may be found everywhere in tropical areas of South America. The vigorous stem produces many aerial roots and 60 cm (2 ft) long, round leaf-stems, with deeply-incised, somewhat sickle-shaped leaves, 40–60 cm (16–24 in) long and 30–50 cm (12–20 in) wide. A fine, undemanding plant.
P. laciniatum, originating in Brazil, has somewhat rough leaf-stems, 40–50 cm (16–20 in) in length, and large, more or less triangular leaves, irregularly incised and lobed. A hardy variety.
P. squamiferum has the most remarkable leaf-stems of all: up to 60 cm (2 ft) in length, red, and covered from an early stage in projecting red scales, which later turn green. Flowers easily but requires warm conditions.

3. Non-climbing species with and without stems

P. bipinnatifidum, a native of Brazil. The straight stem bears a closely-growing crown of leaves; each falling leaf leaves an obvious scar on the stem. A hardy species, but usable only as a young plant, as later on it grows too large.
P. martianum, called after a 19th century German botanist, originates in southern Brazil. Its short, recumbent stem produces inflated leaf-stalks up to 40 cm (16 in) in length, with long, lance-shaped, somewhat leathery pale green leaves, which may grow to 35–65 cm (14–26 in) in length. Flowers easily, but soon grows too large for indoor use.
P. selloum shows great affinity with *P. bipinnatifidum* described previously, but is even larger. If you hold the leaves against the light you will

▶

Philodendron martianum, fam. Araceae

notice numerous parallel blotches or stripes. Suitable only for very large rooms, such as large entrance halls.
Aerial roots: not all species develop aerial roots, which serve mainly to absorb moisture from the atmosphere, but which may grow into the soil, where they turn into true roots. Aerial roots should be treated with care and removed only if absolutely necessary.
Care: all philodendrons detest direct sunlight and prefer a relatively dark situation. They should on no account be placed out of doors in summer. Winter temperature for green-leaved varieties and the hybrids occasionally available should be 14–16°C (57–60°F). Slightly higher soil temperature is desirable, but not essential. Variegated and velvet-leaved plants should be kept at a minimum temperature of 18°C (65°F). For these plants warm soil is important, for they dislike having cold feet.
Atmospheric moisture: essential for variegated and velvet-leaved species; desirable for green-leaved plants, especially at the start of the growing season (spring to August).
Watering: freely between spring and August, then more sparingly to stop the leaves growing. Luke-warm water, pH 5–6.5.
Feeding: from April to August, weekly for vigorous growers, fortnightly for the less rapidly growing kinds.
Repotting: as a rule the plants grow so fast that they will require repotting every spring. This is best done in a mixture of coarse leafmould (unsifted), loamy turf soil, some rotted cow manure and sharp sand. Polystyrene granules are often added to make the mixture warmer and more friable. However, the philodendron is such a strong plant that it will flourish even in ordinary pre-packed potting compost. The plant is also very suitable for soilless cultivation, provided sufficiently large pots are available.

The Luwasa system (with clay granules) should certainly be considered for such large plants. Since the plants may grow to such size, they will often be used as specimen plants in offices, etc. In such cases ordinary clay pots will not be sufficiently decorative. No matter, the philodendron may equally well be grown in a plastic cylinder or a glass fibre tub, provided the compost is never allowed to become too wet, for this will kill the plant. In other words: either provide very good drainage, or check the moisture frequently and water only when necessary.
Pruning: kinds with thin stems, in particular, should be cut down from time to time.
Propagation: from cuttings (from stem or growing tips), or by division. Young plants independently growing from the roots should be separated with their aerial roots; stronger plants should be air-layered. Can also be propagated by sowing fresh, freely germinating seed; exotic seed catalogues contain several species. Propagation of plants is a very satisfying occupation.
Diseases: if properly cared for, the plants will rarely be subject to disease or plant pests.
Important: brown leaf-tips indicate that the soil is too dry or too chalky.

Philodendron melanochrysum, fam. *Araceae*

Phlebodium

Only the very latest books on houseplants mention this plant by its present name *Phlebodium*, for until quite recently it was known as *Polypodium*, a member of the large family of the same name. In the chapter on plant families it has already been mentioned that this family has been, to some extent, divided among several smaller families.

P. aureum, the only species of its genus, is an immense South American tropical fern with yellowish, 50 cm (20 in) long leaf-stems and pendulous, feathery incised leaves, whose attraction lies in their striking blue colour. The blue sheen is possibly even more marked in a number of strains; as young plants these are frequently used in plant combinations. The best known are 'Glaucum Crispum' with crimped leaves, and 'Mandaianum' with wavy crimped fronds. A striking feature is the handsome brown-haired rootstock, which is usually seen only when the plant is repotted. It twines all over the pot.
Care: if you obtain a young plant you should remember that, if properly cared for in the living-room or in the plant window, it will soon grow too large. It is, however, very suitable for a greenhouse, where it can be planted in an elevated bed to trail freely. In spite of its tendency to hang down, phlebodium is not an epiphytic, but a terrestrial fern, which will grow even in ordinary potting compost, although it will do better in a coarse, crumbly special fern compost. A warm and shady position, humid air, temperature in winter 18–22°C (65–72°F). Water with warm, softened water (pH 4–4·5). Feed with liquid fertilizer. Propagation by division or from seed.

Phlebodium aureum 'Mandaianum', fam. *Polypodiaceae*

Phoenix – Date palm

Many plant lovers have successfully grown a date palm from a date stone and many a magnificent plant has resulted from these efforts, even though they did not develop into the useful tree with its delicious fruit grown in southern countries.

Origin: the 'real' date palm grows chiefly in Arabia and North Africa. Other representatives of the genus occur in tropical regions of Africa and Asia, especially in India. The date palm from the Canaries is the one that has long been used as a houseplant in pots or tubs.

Appearance: low-growing or tall palms, with or without a stem, the upper part of which bears the remaining leaf-stems. The leaves vary greatly in shape: some are stiff and erect, others gracefully arching. The inflorescence, in the shape of a spadix, appears only in older plants.

Species: *P. canariensis* comes from the Canary Islands. This is the hardiest species, most suitable for room cultivation. *P. dactylifera*, the 'real' date palm, grows more rapidly and vigorously. In its natural environment it grows to 20 m (60 ft) in height. *P. roebelinii*, from India; the most beautiful and graceful species, but the one most difficult to grow. A slow-growing dwarf form, suitable for the heated greenhouse.

Care: *P. canariensis*, the decorative palm for terrace and garden, prefers to stand in a sunny spot out of doors in summer. In winter its position should be frost-free, at 4°C (40°F) and in good light. Younger specimens of *P. dactylifera* should overwinter at a higher temperature, 8–10°C (46–50°F). *P. roebelenii* requires to be kept warm; even at night the temperature must not fall below 16°C (60°F). Moist air and a shady position.

Watering: freely in summer, but provide adequate drainage; sparingly in winter. Tepid water, pH 5–6.

Feeding: weekly from March to July, using a normal nutrient solution watered on to moist compost.

Repotting: initially every year, later as required. The plants occasionally have a tendency to develop stilt-roots. Medium heavy, humusy compost. Pre-packed potting compost may be used.

Propagation: from seed only; quite simple, but takes a long time. See also the chapter on Propagation (p. 73).

Diseases: infection by scale insect and mealy bugs is treated as in *Microcoelum* (p. 260). The main cause is too warm and humid a position in winter. Lack of fresh air frequently leads to foliage disease and mildew. Lack of light often causes small yellow spots.

Important: brown leaf tips may be due to various causes: 1. Calcification of the soil as a result of hard water. 2. Excess water due to overwatering, especially in winter. 3. Dried out soil ball when the plant is potbound. When removing the brown tips, leave a narrow strip of dead tissue. Whole leaves should always be removed with a sharp knife, leaving a clean cut.

Phoenix canariensis, fam. *Palmae*

Phyllitis — Hart's tongue

This magnificent evergreen and completely winter-hardy fern grows practically everywhere except in tropical heat and burning sun. It grows in a shady position in the rock garden, as well as in situations in the living-room where no other plant will flourish. It occurs in Europe, Asia Minor, North Africa, the northern part of the U.S.A. and in Japan.

P. scolopendrium is the sole species representing its genus; it belongs to the small family of *Aspleniaceae*. Numerous strains have been developed; in England, in particular, their cultivation is very successful. There are, for instance, strains with crinkly, incised or fringed foliage, completely deviating from the original species, whose leaves are usually 40 cm (16 in) long, 4 cm (1¾ in) across and smooth-edged. One of the finest is *P. s.* 'Undulata', with wavy-edged leaves. Other forms regularly available at the florists' are the enormous 'Capitatum', 'Crispum' (with crinkly leaf-edge), and 'Digitatum'. Only the species can be grown from spores. The hybrids are propagated from stems, to which a small section of rootstock must be attached; this is done in a heated propagating frame.

Phyllitis scolopendrium 'Undulata', fam. *Aspleniaceae*

Pilea

The common name of one of the species, Artillery plant (*P. microphylla*), refers to the fact that ripe seed is forcefully discharged. In favourable conditions self-sown young plants will be found at distances of up to a metre (3 ft) from the parent plant.

Origin: the botanical name means something like cap or hat. This interesting nettle-plant, whose genus includes about 200 species, is a so-called pantropical plant; in other words, it occurs in tropical areas all over the world.
Appearance: most of them are annual or perennial plants; they do not keep very well. It is therefore better to obtain new plants every year, either from cuttings or by growing the small self-sown plants. Not all species in cultivation are self-sown.

Pilea cadierei, fam. *Urticaceae*

Pilea spruceana 'Norfolk', fam. *Urticaceae*

Watering: maintain constant moisture in summer; keep drier in winter.
Feeding: weekly from March to August, using normal liquid fertiliser.
Repotting: not necessary if new plants are regularly grown from cuttings or seed. For young plants use a light and porous compost, e.g. prepacked potting compost.
Propagation: as mentioned under Repotting, from cuttings or by cultivating the self-sown plantlets. Very easy.
Diseases: hardly ever occur.

Species: the species most often available and at the same time most suitable for room cultivation is *P. cadierei*, which was discovered in the Vietnamese jungle as late as 1938 and, mainly as a result of political events, not introduced in this country until 1948. At the time plant lovers welcomed it as a special event, for it is not every day that such a novelty is introduced. It is a herbaceous plant, at its best when it has reached 20–25 cm (8–10 in); it does not remain in this state; as it develops ugly long shoots. These can be kept within bounds to some extent by pruning, but as said above, it is advisable to grow new plants every year, as young plants have a fine bushy shape. The leaves are 10–12 cm (4–5 in) long, oval, dark green, with four rows of silvery splashes between the veins; for this reason it is sometimes called the aluminium plant. Inconspicuous small flowers from spring until summer. *P. c.* 'Minima' is an attractive dwarf form, 12–15 cm (3¾–6 in) tall. *P. involucrata* is also sold under its old name *P. pubescens*; its unusual beauty speaks for itself. *P. microphylla* (in the trade often still called *P. muscosa*), a mossy species from tropical America, where it grows at altitudes of over 2000 m (6000 ft). This species is suitable for use as a greenhouse plant or as groundcover in the garden. *P. spruceana*, a native of Peru and Venezuela, a herb, 10 cm (4 in) in height, forming runners. Striking green nettle flowers. The seed is discharged as in *P. microphylla*. A beautiful hybrid of this species is 'Silvertree', with broad silver stripes on either side of the main vein and magnificent flowers.
Care: all the species mentioned require a shady or half-shady position and plenty of fresh air. Temperature in winter not below 10°C (50°F) and not above 18–20°C (65–70°F). Some degree of atmospheric humidity at the onset of the growing period is desirable; afterwards dry air is preferred.

Pilea involucrata, fam. *Urticaceae*

Piper — Pepper

As every housewife knows, not everything called pepper is in fact a condiment. The fruits of *Capsicum annuum*, the Spanish pepper, and the less sharp paprika, as well as those of *C. frutescens*, cayenne pepper, are pungent as pepper, but these plants belong to the family of *Solanaceae*.

Origin: *Piper nigrum*, the black pepper, from whose fruits, ripe or unripe, the black or white pepper-corns are won in all the 'pepper countries' of tropical Asia, is at the same time an attractive houseplant with great powers of adaptation. Advantages: it can climb against a wall or hang from a pot; it grows — slowly — even in a shady position; it tolerates normal room temperature, though during the dormant season in winter, this should be lower, 12–14°C (54–57°F). *P. ornatum* comes from Celebes. *P. porphyrophyllum*, another native of Indonesia, and *P. sylvaticum*, of unknown origin, are variegated plants of great beauty; to develop fully they belong in a tropical plant window with a high degree of humidity, in a shady position and a winter temperature of 16–18°C (60–65°F). Water fairly sparingly and use soft water. Feed once a fortnight from March to August. Repot as required, after dormancy, in pre-packed potting compost.

Propagation: only possible in a heated propagating frame.

Diseases: curled foliage indicates that the air is too dry. White pearls appearing on the lower surface of the leaves are a natural phenomenon.

Piper nigrum 'Black Pepper', fam. *Piperaceae*

Pisonia

When it appeared in garden shops a few years ago, there were enthusiasts who saw this attractive white-variegated foliage plant as a competitor for the white variegated ficus which is often rather difficult to grow. But as soon as the cultivar *F. elastica* 'Doescheri' became available, pisonia was somewhat forgotten.

Origin: there are about 30 known species of pisonia; they are called 'pantropical' plants, which means that they occur in all tropical areas of the Old and the New World, like, for instance *Asplenium nidus*, the bird's-nest fern, the *Commelinaceae* and the pepper plants. It should further be mentioned that *P. alba* which grows in Hawaii as well as in the Pacific islands and in Australia, is the sole representative of its genus in indoor cultivation and that it is closely related to *Bougainvillea*, described on p. 141. Both belong to the family of *Nyctaginaceae*.

Appearance: in its native habitat, *P. alba*, named after the Dutch physician Willem Piso, who died in 1648, forms shrubs or trees up to 6 m (18 ft) in height. As can be seen in the photograph, the leaves grow opposite, almost in a circle, the size of ficus leaves, but with more beautiful white marking, sometimes marbled.

Care: according to several plant books, *Pisonia* is a hot house plant. Personally I have grown it for many years simply in the window-sill, as a pot plant or in soilless cultivation, without recourse to a tropical window or a hot house. It was treated like ficus and flourished.

Repotting: in spring; use pre-packed

Pisonia alba, fam. *Nyctaginaceae*

potting compost or a loamy-humus mixture.

Propagation: soft cuttings or cuttings containing an 'eye' root readily in a mixture of sand and peat at a temperature of 25°C (78°F).

Pittosporum

Tourists who have spent their holiday in the mountain forests of Madeira and Tenerife often talk about a shrub conspicuous by its deliciously scented white flowers. Nobody knows why this shrub from grandma's day has in this country been practically forgotten as a houseplant.

Origin: the plant growing on the islands mentioned above is only one of the 150 species whose geographical distribution stretches from China and Japan to Australia and New Zealand. A number of species were brought into cultivation over here as much as 200 years ago.

Appearance: evergreen shrubs or small trees. Leaves leathery, smooth-edged or sharp dentate, usually placed in circles round the stem. Fragrant flowers, the colour varies with the species. The flowering season occurs in summer. The seeds in the leathery or woody seed capsules are suspended in a very sticky liquid.

Species: *P. coriaceum*, with leathery seed capsules, is the species found in Madeira and Tenerife. *P. crassifolium*, a thick-leaved species, is a native of New Zealand; the dark red flowers appear in terminal clusters. Another native of New Zealand is *P. tenuifolium*; this species, with its densely growing, pale green, wavy-edged leaves and dark red flowers is considered to be one of the most decorative. The best-loved species is *P. tobira*, which comes from the subtropical regions of China and Japan and has 10–12 cm (4–4¾ in) long, shiny dark green leaves. From May onwards it produces creamy white flowers in terminal clusters; according to one of my old flower books they perfume the entire room. A variegated hybrid called 'Variegata', very popular in the United States as well, of course does not flower so profusely.

Care: there is a reason for the fact that this once so famous plant disappeared from the florist's: pittosporum is actually a garden plant or a tub plant, overwintering at a temperature of 4–8°C (40–46°F) rather than a plant that can grow in the living-room throughout the year. Provide as much light as possible and treat the plant like callistemon, myrtle and oleander.

Propagation: from cuttings or from seed.

Pittosporum tobira, fam. *Pittosporaceae*

Platycerium — Staghorn fern

It is not so very long ago that this member of the family of *Polypodiaceae* was a much admired exotic stranger in our midst. It retains its exotic appearance, but has meanwhile developed into a greatly valued house plant which can give us pleasure for a long time.

Origin: the genus *Platycerium* includes 17 species, whose native habitat stretches from southern China via Malacca into Australia. Staghorn ferns also occur in the rain forests of Central and South America.

Appearance: they are all epiphytic plants, often resembling enormous nests high in the trees of the jungle, growing in the forks of branches or in other hollows filled with the remnants of dead plants. Nourishment is absorbed by ribbed scales as well as by the roots. All species have two kinds of leaves: 1. Broad and flat supporting leaves, layered like a pile of plates, growing pale green from the centre front, while the older leaves behind gradually turn brown and rot away at the bottom inside. In this way they produce their own humus, from which the roots can draw nourishment. The anchoring leaves are sterile, in other words, they do not develop spores. 2. The spore-bearing leaves emerge from the growing point; they vary greatly in shape, depending on the species, but are always narrow at the base, broadening considerably towards the end. In some species they grow to over a metre (3 ft) in length and protrude sharply. In young plants the leaves are usually covered with white bloom, which disappears at a more advanced age, especially near the bifurcation points. In room cultivation no spores will form.

Species: the most hardy, and therefore most important form is *P. bifurcatum bifurcatum* (in the trade often called by its old name *P. alcicorne*); a bi-forked staghorn from tropical regions of Australia, and its two cultivars 'Hillii' and 'Majus'. The Javanese *P. willinckii*, which has a particularly thick felty layer, may be considered for room cultivation, but requires some extra care. Once it is growing well, it soon becomes too large. All other species, including especially *P. wilhelminae-reginae*, belong in the warm environment of a very large plant window, or better still, in a hot house.

Care: because of the way in which the staghorn grows naturally, as an epiphyte in the crown of a tree, it is really senseless to plant it in a pot. Platycerium belongs in a hanging plant basket of the type used in the cultivation of tree-orchids; smaller specimens will grow on tree trunks. During the growing season it requires a high degree of humidity, and in winter, too, dry air in a heated room is not tolerated. When misting, try not to wet the leaves; do not spray after 3 p.m., for the plant must dry before evening. Temperature in winter 16–18°C (60–65°F); at night cooling to 15°C (58°F) will do no harm.

Watering: from March to August water generously with tepid water, pH 4–4.5. In winter water sparingly, but do not allow the plant to dry out; this can be fatal.

Feeding: from April to August give liquid, lime-free manure, for instance, the nutrient solution used in soilless cultivation. ▶

Platycerium bifurcatum, Staghorn Fern, fam. *Polypodiaceae*

Platycerium grande, Staghorn Fern, fam. *Polypodiaceae*

Repotting: young plants in pots are often difficult to repot. It is better to replace the top layer of compost at the beginning of the growing season by a special fern mixture (containing chopped fern roots).
Propagation: sowing and cultivation is a professional job.
Diseases: stubborn attacks by scale insect are encouraged by draughts and excessively dry air in too warm a position in winter.

Plectranthus

In the early nineteenth century the species *P. fruticosus* was thought to keep away moths because of the scent of its leaves. It has also been thought to be good against rheumatism. This species is a native of South Africa.

Origin: there are about 200 species, occurring in tropical regions all over the world.
Appearance: *P. fruticosus* is an evergreen shrub growing to about 1 m (3 ft) in height, with somewhat rough-haired foliage borne on reddish stalks. Innumerable erect spikes of pale blue flowers from October to late winter; the lip of the flower is usually speckled. In addition to this species with its modest requirements, simple in cultivation, we should mention the hanging or creeping *P. oertendahlii*, used as a hanging plant or as ground cover; it has white-marked foliage and white flower racemes, up to 15 cm (6 in) long.

Care: *P. fruticosus* requires a sunny situation on the window-sill, in a not too warm room where it can remain undisturbed throughout the year; in winter the most favourable temperature is 12–15°C (54–58°F). Dry air is tolerated. As it is winter-flowering, it does not have a resting period; nevertheless watering should be slightly reduced after flowering has ceased; the plant should be cut back drastically and kept at a temperature of 10–12°C (50–54°F) for some time. In summer feed weekly.
Repotting: older plants every two years in roomy pots. Use very rich compost.
Propagation: from soft cuttings.

Plectranthus fruticosus, fam. *Labiatae*

Plumbago — Leadwort

This tropical and subtropical genus, consisting of about 20 species, is called after *Plumbago europaea*, the juice of which was supposed to turn human blood a leaden grey in colour (from the Latin word *plumbum* = lead). The story about the liquid may be true, but who knows now?

Origin: only *P. auriculata* is worth mentioning as a house- or tub-plant. It is often still sold under its former name *P. capensis*, which indicates its South African origin.

Appearance: an erect growing, deciduous semi-shrub. By removing the side branches up to the desired height, it may also be grown as an attractive standard plant, although it may take up to three years before a crown is formed (on the subject of growing standard plants see also under *Abutilon*, p. 114). It has long stems and pendulous flowers, covered in small white scales, as are the undersides of the small, short-stemmed dark green leaves. The 5-petalled flowers are widely reflexed and have a slightly sticky throat. They grow in terminal clusters. The species has sky-blue flowers; there is a white cultivar called 'Alba'. In adequate warmth the plant will flower profusely from May until late autumn.

Care: until it grows too large, *P. auriculata* can remain indoors throughout the year, but in summer it prefers a warm, sheltered position out of doors, where it will have the 'cool foot' and 'warm crown' it likes. Being a tropical plant, it is not only very sensitive to frost, but will already be harmed by temperatures below 6–8°C (43–46°F). The resting period is in winter, and although during that time the plant is leafless, it must nevertheless be placed in a good light, not in a dark cellar. When it starts into growth again in spring, it should be kept as cool as possible, in order to hold back flowering to some extent.

Watering: water freely during the flowering period in summer; in winter keep only just moist.

Feeding: normally from spring until August.

Repotting: as required, before new shoots appear; the plant should first be cut back vigorously. Compost: prepacked potting compost or a mixture containing slow-acting fertiliser as for oleander (p. 270). If the plant is not repotted, it should nevertheless be cut back a little in spring.

The plants will grow even better if they are taken from their pots at the end of May and planted out in the garden in prepared soil. Repot in autumn; the roots and the crown may be pruned slightly.

Propagation: the simplest method of producing new plants is by taking cuttings of unripe shoots in late autumn or early spring. These will rapidly root at a soil temperature of 20–25°C (70–78°F). They will also develop roots if placed in a bottle of water, but this requires patience. If available, plants can be grown from seed as well.

Diseases: avoid drying out in summer and excess moisture in winter.

Plumbago auriculata, fam. Plumbaginaceae

Polyscias

The photograph is not a mistake — it really does show a plant belonging to the *Aralia* family. Most species have feathery foliage, like ferns. The plant illustrated is the most frequently grown cultivar *P. balfouriana* 'Peacockii', with leaves streaked and marbled with yellow.

The species *P. balfouriana* originates in New Caledonia; in a heated greenhouse it will grow into a dense, bushy plant. The stems do not grow woody and are pale green with dark stripes. Evergreen round leaves, irregularly blotched with white along the margin; they often grow in groups of three on 12–15 cm (3¾–6 in) long stems. Other species, such as *P. filicifolia*, *P. guilfoylei*, or *P. paniculata*, all natives of the Pacific area, have pinnate foliage. *P. balfouriana* is suitable for a plant window, conservatory or hot house; it requires warmth, a high degree of humidity and a shady position out of direct sun. Temperature in winter 16–18°C (60–65°F). General care approximately as for *Dizygotheca*, although it is slightly more particular.

Polyscias balfouriana 'Peacockii', fam. *Araliaceae*

Polystichum — Shield fern

It should be made clear that on the occasion of the most recent reclassification, *Polystichum* and *Cyrtomium*, described on p. 186, were classified as two separate genera in the new family of *Aspidiaceae*. However, this knowledge may be of little interest to the amateur plant lover.

Origin: it includes over 200 species spread all over the world, and thus offers an enormous choice of fine evergreen and completely winter-hardy ferns. The number of non-hardy species suitable for room and greenhouse cultivation, on the other hand, is much smaller; these are usually natives of eastern Asiatic regions between Japan and Ceylon.
Appearance: they are terrestrial ferns, developing dense groups of stemmed, erect growing fronds of varying shape from the rhizomes. The spore clusters are arranged on the veins on the underside of the leaves.

Two species deserve attention: 1. *P. auriculatum*, a fern from the mountain forests of India, better suited to the hot house, and 2. the undemanding *P. tsus-simense*, from China and Japan, which in northern countries is often regarded as a plant for the unheated greenhouse, although because of its small size and slow growth it is eminently suitable as a houseplant where space is lacking.

Very beautiful, but rarely available, are a number of cultivars of *P. setigerum*, whose much divided fronds almost resemble filigree: 'Plumusum Densum', a moss-green fern with plume-like fronds is possibly the finest; 'Proliferum' develops little plantlets along the narrow, finely divided fronds; 'Proliferum Wollastonii' is more vigorous and has broader leaves.

Care: as for *Pteris*. Water freely in summer (check the *p*H of the water) and make sure the plant never dries out. In winter water more sparingly and if possible do not let the temperature drop below 10°C (50°F). Feed only once a month. Repot as required when new growth appears in March.

Propagation: the species are grown from spores, in so far as these are available. Cultivars are only increased by division and from the plantlets mentioned above: sections of leaves with plantlets are placed in damp peat in summer; at a temperature of 18–22°C (65–72°F) roots will develop, from which the new plants will grow; in the autumn these are planted out separately in small pots.

Diseases: risk of rotting as a result of excess moisture.

Polystichum tsus-simense, fam. *Aspideaceae*

Primula

The name is derived from the Latin *primus*, first; these delightful little early flowering plants are usually thought of primarily as garden plants. However beautiful these harbingers of spring are in the garden, though, we should not forget the pot-grown primulas as houseplants.

Primula malacoides, fam. *Primulaceae*

Origin: with one or two exceptions, they are not indigenous plants, like the garden hybrids, nor are they immigrants from the tropics but originate mainly from temperate regions. The family of *Primulaceae* includes 28 genera with about 800 species; the genus *Primula* alone embraces about 500 species. Only three or four of these are pot plants.

Appearance: some are monocarpic, others perennial plants. The leaves, with or without stems, grow in basal rosettes. The arrangement and the shape of the flowers differ according to species and strain. Flowering occurs in spring.

Species: *P. malacoides*, a scented primula, with flowers growing at different levels. It was introduced in England from China as late as 1908 and was subsequently taken into cultivation elsewhere as well. At the time it was thought of only as a garden plant. Hybrids suitable for pot cultivation were developed later; the leading form is now the Swiss ▶

'Wädenswyler Typ'. Growers aim at a compact growth of the long-stemmed leaves, heart-shaped at the base, rounded at the top and slightly wavy along the edge, as well as at a dense arrangement of the multi-level inflorescence, which grows to 40 cm (16 in). The stems are covered in mealy farina. Colours range from pure white via bright red, salmon-pink, deep carmine red to bluish red. The flowering season depends on the time when the plants are brought into growth and may occur at any time from December to April. You will probably buy flowering plants. If dead flowers are immediately removed, flowering may continue for two to three months.

These primulas make few demands; they are sensitive only to excess heat, sunlight and drying out of the soilball. They require good light, out of the sun, slightly humid, fresh air and a temperature not exceeding 10–12°C (50–54°F). After flowering the plants should be fed until the autumn, and can be kept for another season.

P. vulgaris (trade name often *Primula acaulis*), the primrose, is considered to be the oldest and most important garden primula, but in the course of time has become a well-known pot plant as well. F_1 hybrids are available in the colours white, yellow with red-veined throats, and various shades from red to dark blue.

These primulas are frequently used in mixed bowls. They need a cool and airy, half-shady situation; keep them constantly moist and after flowering plant them in the garden. Do not try to keep them indoors. Flowering season from December to May.

P. praenitens (formerly *P. sinensis*) was brought to England from Canton in 1820 and since then many cultivars have been developed. There are large-flowered strains, but also hybrids with rose-shaped or with double flowers resembling narcissi. The flowers are funnel-shaped with an inflated calyx and are arranged in clusters or umbels. Flowering season February to May. Stemmed leaves, large, round, lobed and dentate; dark green. The entire plant is hairy. This Chinese primula is a strong and long-lasting pot-plant. It requires a position in good light, but not in the sun, in a moderate temperature, 15°C (58°F). Water carefully, for the plant is sensitive to excess moisture. During flowering give a little more water; otherwise keep moderately damp. There is no actual resting season. Be careful with feeding as well, for the plant is sensitive to salts: best to give half a dose every two weeks. Repot after flowering in not too large pots. The soil mixture must be light and humusy; e.g. normal potting compost mixed with a little extra sand. Propagation from seed – 800–1200 seeds to 1 g ($\frac{1}{28}$ oz) – is a professional job.

P. × *kewensis* was developed by chance in the greenhouses of Kew Gardens in 1898. As a rule the entire plant is covered in farina, but there are non-mealy strains as well. The leaves are elongated oval, irregularly toothed, and wrinkled. The dense fragrant flower heads grow on 30 cm (12 in) tall stalks. Their colour is a pure yellow, never seen in other primulas. The natural flowering season occurs from February to April; depending on the strain it may be earlier and more prolonged. This is one of the best pot primulas, also suitable for arrangements in bowls. Care as for *P. malacoides*. It is irritating to the skin of some people. *P. obconica*, famous of old, but also somewhat notorious because it causes irritation of the skin in allergic people; the plant came from China to England in about 1880. The whole plant is covered in softish hair. The long-stemmed leaves are up to 15 cm (10 in) long and 12 cm ($3\frac{3}{4}$ in) across, succulent and with lightly incised undulate margins. The flower-stems grow singly or in groups, up to 25 cm (10 in) tall. Flowers in

Primula vulgaris, fam. *Primulaceae*

Primula obconica, fam. *Primulaceae*

umbels up to 8 cm (3¼ in) across, mainly in shades of red, pink and lavender-blue. Flowering season at almost any time of the year; it can be brought into flower in any month, but strains overwintered by the grower and flowering in spring are most in demand. Their flowers last longest and are even suitable for cutting.

P. obconica requires good light, but a half-shady and cool position; in winter the temperature should not exceed 12–14°C (54–58°F). Water and feed as *P. praenitens*. Correct care, and the regular removal of dead flowers and stems induce vigorous growth which enables the plant to be kept for a subsequent season if planted in the garden in summer. Be sure to bring it indoors in good time, in September, before the first night frost occurs.

A disagreeable feature of this primula is the fact that the hairs contain primine, a substance which causes irritation of the skin in people with an allergy. In 1958 a strain called *P. obconica* 'Multiflora' appeared for sale; this lacked the unpleasant substance, but looked less attractive and is no longer grown.

Conclusion: all primulas require semi-shade, but more light in winter, all strains overwinter at a temperature of 10–12°C (50–54°F). Sensitive to drying out, but also to excess moisture. Water with tepid water, pH 5·5–6·5, *P. obconica* to pH 7·0. Water that is too calciferous will cause yellowing of the foliage. Excess warmth in winter will shorten the flowering season and encourage fading of the flowers.

Pseudoranthemum

Formerly this plant was called *Eranthemum* and it is frequently still sold as such. A number of species of the old genus still bear this name. Others were reclassified among different genera, for instance *Pseudoranthemum*. One of these is the variegated *P. kewense* (syn. *P. atropurpureum*).

Pseuderanthemum atropurpureum, fam. *Acanthaceae*

Origin: the plant probably originates in one of the Pacific islands and it is thought to grow wild in South America as well. The genus includes more than 50 species.

Appearance: six attractive foliage plants are in cultivation, some of them flower as well. They are shrubby and half-shrubby plants, freely branching and growing to a metre (3 ft) in height. The leaves are usually a broad oval or elliptical in shape, ending in a point, strikingly blotched, veined or marbled. The flower spikes are terminal or appear in the axils of the leaves and as a rule are white with reddish purple; the flowering season is early in the summer. The finest forms are *P. reticulatum*, which rarely flowers, *P. sinuatum*, flowering season June to August, and *P. tricolor*, with pink, purple and green leaves.

Care: these are plants for a warm and roomy flower window. They require plenty of light (but without direct sunlight), a high temperature and humidity. Temperature in winter 16–18°C (60–63°F); a cooler situation will cause damage.

Watering: keep moist, but avoid excess water. Water less during the resting period from October to January (pH 5–5·5). Smooth foliage; no risk of spotting when spraying.

Feeding: from April to August feed every three weeks with a weak solution, as the plant is sensitive to salts.

Repotting: in early spring, using shallow pots and ordinary potting compost.

Propagation: fairly simple, from cuttings in a heated propagator.

Diseases: errors in cultivation frequently lead to attacks by scale insect, red spider mite and thrips.

Pteris – Brake

Another common name for this plant is Ribbon Fern; the spores are arranged in a kind of band on the lower surface of the leaves. The name, derived from the Greek word *pteron*, meaning wing, indicates the shape of the fronds. The plant is internationally known for its beauty.

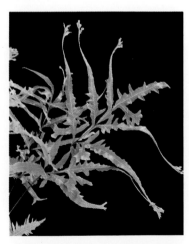

Pteris cretica 'Wimsettii', fam. *Pteridaceae*

Origin: the genus *Pteris* has been promoted to the family of *Pteridaceae* and is represented by about 300 species in tropical and subtropical regions all over the world. They are all terrestrial ferns. At least 50 species and strains are in cultivation; in countries with a favourable climate they are used out of doors, in cooler regions as houseplants.

Appearance: the fronds, varying greatly in shape, grow in dense clusters from the rhizomes, or from slender stalks, up to 30 cm (12 in) long and often straw-coloured or reddish brown. The spores appear in dark brown clusters or in uninterrupted bands on the underside of the leaves.

Species: the best known species, including many strains, some of them variegated, is *Pteris cretica*. The most widely used is undoubtedly *P. c.* 'Wimsettii'. Another popular form is 'Major', a very undemanding plant. 'Rivertoniana', a bushy fern with denser growing and finer feathering, is particularly beautiful. The white variegated 'Albolineata' has broad fronds with a greenish white central band. *P. ensiformis* is a native of tropical Asia, Australia and some of the islands in the Pacific; the species itself is rarely cultivated, but has produced two famous white variegated strains. For a long time florists frequently used 'Victoriae' in plant combinations, until it was more or less supplanted by 'Evergemiensis', a cultivar with glossy, silvery leaves, cultivated in Belgium. Since it is the progeny of a distinctly tropical fern, it cannot be unconditionally recommended for use as a houseplant. A more hardy, less rapidly growing form is *P. multifida* (also known as *P. serrulata*), a native of eastern Asia, and the beautiful strain 'Cristata'. *P. tremula* is a large and vigorous ornamental houseplant with fronds up to 1·20 m (4 ft) in length growing on chestnut brown stems, 30 cm (12 in) tall. An even hardier cultivar is *P. cretica* 'Major', while *P. umbrosa* will grow even in deepest shade. The fronds are dark green and glossy, to 60 cm (2 ft) in length; the stems, up to 50 cm (20 in), are reddish brown.

Care: these plants should be kept in an enclosed space throughout the year, and be taken outside to be sprayed only on warm summer days. They require a half-shady situation, definitely out of the sun. Growing season from March to July, dormant period from October to February. The most favourable winter temperature for the hardy, green species is 10–12°C (50–54°F), for the variegated forms 16–18°C (60–65°F). To maintain adequate humidity and prevent the plant drying out it is advisable to plunge the pots in damp peat fibre in a roomy outer container with a drainage layer of crocks and pebbles. If the plants are kept in a suitable environment in an unscreened plant window, these measures are unnecessary. It is important to provide the highest possible degree of atmospheric humidity when new fronds are growing, approximately from April to July. A normal atmosphere is tolerated, at lower temperatures, in winter, although the variegated forms, which require more warmth, will need slightly moister air during that period.

Watering: water liberally during the growing season, using soft, tepid water, pH 4·5–5·5. Like azaleas, orchids, etc., they do not tolerate limy water. Decrease the water supply in winter, depending on the temperature, but never allow ferns to dry out, or they will wither.

Feeding: all ferns are sensitive to lime and salts. It is therefore advisable to use a special lime- and salt-free liquid fertiliser.

Repotting: when necessary, in spring before the start of the growing season. Many pteris plants grow so rapidly that they must be repotted twice a year. Compost: equal parts of ordinary potting compost and peat fibre.

Propagation: this is best done from ripe spores caught in damp peat fibre when they drop from their capsules. They germinate at a temperature of about 25°C (78°F) after which they should be grown in a moderately warm greenhouse. A good catalogue will list spores of many pteris species and strains. Increase by division of the rhizome is not always successful.

Diseases: apart from fern slugs, 2–3 cm (1–1½ in) long, pests and diseases rarely occur.

Pteris multifida 'Cristata', fam. *Pteridaceae*

Punica granatum — Pomegranate

On seeing a flowering dwarf pomegranate or cuttings of the large standard form, few people remember that this plant has been known for many centuries. It was mentioned in the Song of Songs, and in the late Gothic era it was frequently used as a motif in art and architecture.

Punica granatum 'Nana', Dwarf Pomegranate, fam. *Punicaceae*

Origin: deciduous, spreading shrubs, to 2 m (6 ft) in height. The smaller *P. granatum* 'Nana' grows to 50–60 cm (20–24 in) and is often cultivated in standard form. The small, smooth leaves, pointed oval in shape, grow in pairs on red stems. The bright red flowers of the standard form are up to 3 cm ($1\frac{1}{4}$ in) across and have a fleshy margin along their calyces; they are single and stemless and appear in groups of 3–4 on strong side branches. There are also hybrids with double flowers, white, yellow, salmon-pink, streaked or blotched. Flowering occurs from summer to autumn. In our part of the world the fruits do not ripen, but drop with the leaves. The large punica does not flower until it is six to eight years old.
Use: the large pomegranate can be used as an ornamental plant in tubs, the dwarf form as a house plant, or in summer on the balcony or in the garden. These plants can grow to a great age.
Summer: in the second half of May the plant should be given a very sunny and sheltered position out of doors; the dwarf form is happiest if its pot is buried in the garden, but may also be kept in the living room, provided it has plenty of fresh air. Until the end of summer water and feed liberally. From September onwards withhold feed and gradually decrease the water supply.
Winter: the first night frost will make the plant shed its leaves; it will do this at approximately the same time if kept indoors. From now on it requires a resting season in a cool, frost-free position, 4–6°C (40–43°F). Water very sparingly but be sure not to let the soil-ball dry out. When the plant begins to put forth new shoots in March, it should be placed in good light in a slightly warmer position; the water supply should be increased and feeding gradually restarted.
Repotting: only when the plant is obviously outgrowing the pot; with advancing age repot less frequently. Rich compost containing garden compost.
Pruning: do not prune to shape. Lightly cutting back at the end of winter will encourage flowering.
Diseases: practically unknown.

Rebutia

All cactus collectors agree: with all due respect to *Mammillaria* and other cacti, rebutia bears the crown. As they are small plants, it is easy to establish a collection even if little space is available. Many species, moreover, flower twice a year.

Origin: rebutia species are natives of northern Argentina and adjoining areas of Bolivia. They often grow in the poorest conditions in rock clefts at altitudes of up to 3600 m (11,000 ft) with some hardy grasses as their only companions. However, this does not mean that as houseplants they can be starved and yet be expected to flower profusely. They owe their name to the French cactus grower Rebut, who died in Paris in 1898.
Appearance: the expert describes

Rebutia pygmea, fam. *Cactaceae*

Rebutia minuscula grandiflora, fam. Cactaceae

rebutia species as dwarf, globular or slightly flattened plants, spirally ribbed, like *Mammillaria*, and covered in soft, short spines. Unlike many other small cacti, they do not flower at the top, but develop their blooms near the base. These are usually scarlet, but may also be yellow, orange or violet-pink. Even one-year-old seedlings are capable of flowering.

Species: there are dozens of species and varieties, of which only the most important ones can be mentioned here. *R. chrysacantha*, from northern Argentina, grows to about 6 cm ($2\frac{1}{2}$ in) in height, 4–5 cm ($1\frac{1}{2}$–2 in) across; it has golden yellow spines. The clear red flowers may grow to a diameter of 4–5 cm ($1\frac{1}{2}$–2 in). *R. marsoneri*, another native of northern Argentina, is among the largest species and may achieve a diameter of 8 cm ($3\frac{1}{4}$ in); it is fairly densely covered in yellow to brownish spines, often more than 3 cm ($1\frac{1}{4}$ in) long. Magnificent yellow flowers, carmine red when in bud. *R. minuscula-grandiflora*, species originating in northern Argentina; strongly sprouting from the base; it has small white spines. The variety *grandiflora* has particularly large flowers. The form *violaciflora* has magnificent violet-pink flowers, usually appearing early in the year. *R. pygmea*, the dwarf rebutia, is illustrated in the photograph; this shows the delightful effect of a bowl filled with the smallest cacti. *R. senilis*, on the other hand, from northern Argentina, like *R. marsoneri* grows to what is for a rebutia the immense height of 8 cm ($3\frac{1}{4}$ in); the same size across; it is densely covered in long, snow-white, fine spines. The flowers are of a clear carmine red. A particularly multiform species is *R. xanthocarpa*, a native of Bolivia; this is one of the smallest species, only 3–4 cm ($1\frac{1}{4}$–$1\frac{1}{2}$ in) in height; fine white spines, 3–6 cm ($1\frac{1}{4}$–$2\frac{1}{2}$ in) long, and carmine flowers 2 cm ($\frac{3}{4}$ in) across.

Care: the same as for most other small cacti. *In summer:* an airy situation on the window-sill, in the cactus greenhouse, or in a glass-covered cactus frame out of doors. Fierce sunlight is harmful; in warm weather it requires a high degree of atmospheric humidity and should therefore frequently be sprayed. *In winter:* after a transitional period it must be allowed to rest from November until February. A light, cool position, 6–8°C (43–46°F); keep completely dry.

Watering: normally in summer; do not let the plant dry out. Do not water at all from November until February. In the transitional periods increase or decrease water gradually.

Feeding: from April to July with special, nitrogen-free cactus fertiliser.

Repotting: rarely necessary. The compost must be porous and rich, but at the same time sandy and mineral.

Propagation: by division, but especially from seed. The plant often seeds itself, but the seed will also be available from good seedsmen. It germinates in light and should therefore be thinly covered with some fine sand.

Grafting: a popular pastime with rebutia species, but really unnecessary, as their own roots are sufficiently strong.

Diseases: if the plant is left too dry in summer, red spider mite or mealy bug may occur. In winter there is a risk of rotting as a result of moisture and too warm a situation. These are the most usual causes of diseases.

Rechsteineria cardinalis

Professional journals warmly recommend this magnificent plant, named after the Swiss priest Rechsteiner, and it can be flowered easily if given the care applied to gloxinias (see p. 334).

The first flowering specimens often appear in the shops by Mother's Day. They require a moist atmosphere and a shady and warm position. Water with soft, warm water. If after flowering the plant is to be kept for the following year, watering should gradually be reduced, until in the autumn the plant has died off. During the winter the tubers are kept dry in their pots at 12–15°C (54–58°F). In February they are lifted, old roots are removed, and the tubers are planted in fresh compost in shallow pots or bowls. Keep in a temperature of 18–25°C (65–78°F) and gradually increase the water supply and start feeding. In addition to *R. cardinalis*, there are a number of fine new strains. *R. cardinalis* is frequently sold under its former name *Corytholoma cardinale*, or its even older name *Gesneria macrantha*. *R. leucotricha* is easily grown from seed; leaves are grey-green and the flowers rose and purple.

Rechsteineria cardinalis, fam. Gesneriaceae

Rhaphidophora

Changes in botanical nomenclature do not make things easier for us. When there was a genus *Scindapsus* we all knew where we were, but subsequently several species were reclassified and included in the genus *Rhaphidophora*. Its modern name is Epipremnum. *Scindapsus* still exists as well (p. 328).

Origin: more than a hundred species of this arum lily plant occur in the Indonesian archipelago. The majority are large lianas, hardly suitable for room or greenhouse cultivation, but occasionally found in botanical collections. Only *Rhapidophora aurea*, formerly known as *Scindapsus aureus*, in spite of its tongue-twisting name, continues its century-long career as one of our most modest climbing houseplants. When it was renamed it was first called *Raphidophora*, but linguists have since added another 'h'. It is therefore no printing error when we speak here of *Rhaphidophora*.

Appearance: it is a native of the Solomon Islands. For a long time it was taken to be a *Scindapsus*, until one day a specimen cultivated in the United States produced flowers. From the structure of these flowers botanists concluded that the plant was clearly not *Scindapsus*, but belonged to the genus *Rhaphidophora*. These botanical particulars are of no great interest to the amateur, but some system is indispensable and growers, florists and the general public will have to get used to the fact that the so-called 'Scindapsus', while still looking exactly the same, now has a new name. The rather tough, somewhat angular, yellowish green climbing stems branch freely and attach themselves by means of small aerial roots growing in varying lengths from the leaf-nodes. On close examination of the foliage we discover the interesting fact that the somewhat leathery, green, yellow-blotched or -streaked and cordate leaves are always more or less asymmetrical in shape. As is the case with the related *Monstera* and a number of *Philodendron* species, the leaves of young plants are usually much smaller than those of older specimens.

Apart from the species *Rhaphidophora aurea* there are a number of very striking strains, the most important being 'Marble Queen', whose almost pure white leaves indicate very slow growth.

Care: the plant should be kept indoors throughout the year in a position out of the sun. It can be used climbing against a stake or a frame, as wall cover in a plant window and in some cases, less popularly, as a trailing plant. In winter it is best kept at a temperature of 15–18°C (58–65°F), but the plants will tolerate being kept in a normally heated living room without an actual dormant period. However, they will then grow more slowly and produce smaller, less well developed leaves. The minimum winter temperature is between 12 and 15°C (54 and 58°F). The paler varieties require slightly more warmth than the species.

Air moisture: particularly desirable during the growing season from April to August.

Watering: keep constantly moist in summer, somewhat drier in winter. Water according to the prevailing temperature.

Feeding: from the beginning of the growing period until August feed once a week.

Repotting: when new growth appears, in shallow pots filled with light and humusy soil.

Propagation: from any short stem section with 1–2 nodes. An even simpler method is to cut off self-rooting runners of plants used as ground cover in a plant window, potting these separately.

Diseases: pests rarely appear. Too dry air results in leaf curl.

Rhaphidophora aurea, fam. *Araceae*

Rhaphidophora aurea 'Marble Queen', fam. *Araceae*

Rhipsalidopsis — Easter cactus

The Easter cactus is frequently mentioned in one and the same breath as *Zygocactus* or the Christmas cactus, but they are actually two different genera. It is true that they have many features in common, but nevertheless they prove their independence by obvious differences in inflorescence and flowering season.

Origin: Brazil is usually mentioned as the native habitat of both genera. Nevertheless there are many hundreds of miles between the eastern areas of Brazil around Rio de Janeiro, where *Zygocactus* occurs, and the tropical jungle of southern Brazil, where *Rhipsalidopsis* belongs, and there are great differences in the climates of these two regions.

Appearance: epiphytic small shrubs with flat or slightly angular offshoots, indented at regular intervals; they subsequently become round and crusty The internodes are 5 cm (2 in) long and 2 cm (1¾ in) across, lightly notched along the edges. Flowering season usually from April onwards and lasting for two months. All rhipsalidopsis species grow best if grafted on *Eriocereus jusbertii* or on *Pereskia*, the well-known stock cactus which, because of its lime-loving nature, counteracts many errors in watering.

Species: the most frequently cultivated species are *Rhipsalidopsis gaertneri*, the true Easter cactus, which has scarlet flowers, and *R.* × *graeseri*, whose flowers are orange-red outside and carmine red inside. *R. rosea*, a fine plant with many pink flowers, grows to only 25 cm (10 in); this plant is rarely available.

Care: some people say that the Easter cactus and the Christmas cactus require the same treatment, but rhipsalidopsis is more sensitive and demands more humid warmth than *Zygocactus*. Experts consider that, like tree orchids, they should be kept in a tropical plant window or heated greenhouse. However, if they die when grown in the living room, this is usually not due to lack of humidity, but to drying out of the soil-ball when they have not been inserted in a damp peat bed; or else to watering with hard water. More even than *Zygocactus*, rhipsalidopsis requires a lime-free soil mixture and softened water at pH 5. The plant is kept indoors throughout the year, in summer in a half-shady spot without direct sunlight; from January onwards in a light position at a temperature of 10–12°C (50–54°F). As soon as the buds are visible, the plant must be moved to a warmer situation, after which it should not be turned or moved again.

Watering: freely during the growing period from August to December. Sparingly until the buds form, then gradually more freely, while at the same time providing a higher degree of air moisture.

Feeding: two or three times during the growing period, with special cactus fertiliser.

Repotting: as required, after flowering. Use humusy, lime-free compost.

Propagation: ripe tip joints root easily at 22°C (72°F); the best time is May–June. *R. gaertneri* can also be grown from seed.

Troubles: red spider mite and mealy bug as a result of excessively dry air.

Important: very suitable for soilless cultivation.

Rhipsalidopsis gaertneri, Easter Cactus, fam. *Cactaceae*

Rhododendron — Azalea

It is difficult to get used to the fact that it has suddenly become incorrect to call this plant, known and loved for centuries, an 'azalea'. However, the botanical classification is inexorable on this point. Needless to say the plant's appearance and treatment have not changed!

Origin: the plant is a native of eastern Asia. In 1810 it was introduced to Europe as a garden plant and because of its great power of adaptation the newcomer was soon used for all kinds of experiments by English, Belgian and German growers. Innumerable, increasingly beautiful strains were developed and became internationally famous. In the United States one group of the old *Azalea indica* is still sold under the name 'Belgian indica'. The original species of the pot-grown azalea is *Rhododendron simsii*. A more recent newcomer is *R. obtusum*, a native of Japan. This plant has a twofold application: first as a houseplant, later in the garden.

Appearance: tree-shaped flowering

▶

Rhododendron obtusum, Azalea, fam. *Ericaceae*

Rhododendron simsii, Azalea, fam. *Ericaceae*

shrubs, also available in dwarf form, with brown shoots and greenish stems, like the leaf-stems, the leaf veins and the calyces covered in hairs. The small evergreen leaves are leathery, reverse egg-shaped and usually slightly pointed at both ends, dark green above, grey-green underneath. The colour range embraces all shades from white to deep red; any bluish tinge is regarded by growers to be a fault! There are single-, double-, small- and large-flowered kinds; never scented. Flowering occurs from Christmas to Mother's Day, depending on the strain and when the plant was started into growth.

Care: by considering the original environment of this inhabitant of cool, humid mountain forests with their lime-free, humusy soil, it is easy to avoid errors in treatment. As a rule the azalea is already in flower when it enters your home. These plants should never be brought straight into a normally heated living-room, but should first have a transitional period of two to three days in a half-shady spot at 10–12°C (50–54°F). If they can stay in this position, and adequate soil moisture is maintained, flowering will continue for several weeks and the leaves will not drop. In a warm room, and certainly in direct sunlight, they will soon cease to flower and loss of foliage is inevitable, particularly as diseases may occur, and you will be left with an unsightly bare shrub, fit only to be thrown away.

Keeping: all excessive conditions should be avoided: no dried out soil ball, no excess water in the pot, no draught. Dead flowers must be removed regularly, stem and all; ugly leaves should be cut off. The plant must be kept cool until the end of May and can then be plunged in a half-shady position in the garden in preparation for the following flowering season; balcony or window-sill are unsuitable. In September, before the first night frosts occur, the plant is brought indoors and placed in good light in a cool situation. When it is in full flower it can temporarily be kept in the living-room.

Watering: during the flowering and growing seasons until August water liberally; then gradually decrease the water supply. Use tepid, demineralised water (pH 4–4·5).

Feeding: after flowering until mid-July with a lime-free liquid fertiliser.

Repotting: every two to three years, after flowering. The compost should be pressed down firmly. Use special azalea compost or pre-packed acid potting compost mixed with extra peat.

Propagation: a professional job.

Diseases: errors in treatment will encourage pests. The only disease likely to affect indoor azaleas is azalea gall, when the leaves become thick, with a white bloom on them. Pick off such leaves as soon as seen and use a systemic fungicide.

Rhododendron simsii, Azalea, fam. *Ericaceae*

Rhoeo

This stately plant is very similar to a bromeliad or a *Dracaena*, but in actual fact *Rhoeo spathacea* is closely related to *Tradescantia*. This is a useful piece of information, since it indicates that it is a welcome addition to the range of houseplants: a beautiful and problem-free plant.

Origin: its origin is somewhat obscure; moreover, the genus has only one known species, namely *Rhoeo spathacea*. Rhoeo was discovered in Central America at some time before 1788 and in the course of 200 years it was found to grow wild in other tropical regions of the world as well.
Appearance: the basal, oblique leaves grow in loose rosettes; they are 20–35 cm (8–14 in) long and 5–7 cm (2–2¾ in) across, olive green above, as a rule dark red underneath. In the 'Vittata' form the leaves are usually striped longitudinally with pale yellow on the upper side. The inflorescence, consisting of numerous small white florets, are enclosed in shell-shaped bracts and grow close to the stem in the axils of the lower leaves. They appear in large numbers from late May until July, but last only a short time. Nevertheless the appearance of the plant and the form of the inflorescence are unusual as well as interesting. Occasionally *R. spathacea* is grown as a hanging plant.
Care: to be kept indoors in a half-shady or shady position throughout the year. In winter it should have plenty of light, but be kept out of direct sunlight, in a minimum temperature of 16–18°C (60–65°F). During the growing period from March to August atmospheric humidity is desirable.
Watering: freely in summer, with tepid water, *p*H 5–6·5. From August onwards water less. From October to January the plant should have a resting period and be kept dry.

Feeding: once a week from March to August with a normal feed solution.
Repotting: as required, in spring. Use roomy pots; prune the plant to encourage branching; or remove the side shoots, as a single-stemmed plant is more attractive. The soil should contain humus and be porous; pre-packed potting compost is suitable.
Propagation: sideshoots as well as shoot tips will root easily. When the shoots are at least 10 cm (4 in) long, they are removed and potted singly in sandy compost. As they root so readily, growth will be rapid and the new plants will soon have to be repotted in larger pots in normal pre-packed potting compost.

In addition seed won in early summer is usually viable; it is worth trying.
Diseases: lack of atmospheric humidity causes leaf curl. Excess moisture in winter leads to rotting of the neck of the roots, while drought will encourage red spider mite and thrips.

Rhoeo spathacea vittata, fam. *Commelinaceae*

Rhoicissus capensis

Naturally this distinctive, though certainly vine-like plant must be considered in the context of other *Cissus* species, extensively discussed on pp. 160–162. The plant is as yet a novelty and not widely available. Hopefully this situation will soon improve.

Origin: when a plant bears the specific name capensis, its origin is easily deduced. It is a native of the Cape province of South Africa, which has given us so many valuable houseplants.

Appearance: in the United States it is called 'Evergreen grapevine' and is described as a vigorous climber with globular tubers, brown-haired woody stems and long-stemmed, thick and leathery, metallic green foliage. The photograph shows that the leaves are almost circular or kidney-shaped, 16–18 cm ($6\frac{1}{2}$–$7\frac{1}{2}$ in) across, with a wavy, dentate margin; rust-red and felty on the lower surface and along the edge. Older plants will produce flowers, followed by blackish red berries. The size and shape of the foliage alone are sufficient to give *R. capensis* an attractive appearance. At the same time it gives an impression of vigour, indicative of great powers of resistance; it creates few problems in simple room cultivation. It will benefit the plant if the foliage is now and then carefully rinsed.

Care: see the treatment of *Cissus* species for the unheated greenhouse (*C. antarctica, C. rhombifolia*). However, *Rhoicissus capensis* is hardier than either of these: it may be placed in a sheltered position out of doors in summer. In the countries of southern Europe — the Italian and French Riviera — it is popular as a beautiful climber on house walls, balconies and pergolas. From this we may deduce that, unlike other *Cissus* species, *Rhoicissus capensis* is unsuitable for use as a hanging plant. Because of its robust stem and enormous leaves it requires to be grown vertically on strong supports. Nevertheless a few tendrils may be trained horizontally.

Rhoicissus capensis, fam. *Vitaceae*

Rochea

The French botanist La Roche (d. 1813), after whom the plant was named, was a descendant of the ancient noble family of literary fame. At one time it was unkindly said that the thick, stiff leaves of rochea resemble goat's feet, but this comparison is no longer valid.

Origin: its four main species still occur in South Africa, but the finest and best known species, *R. falcata* (the one with the goat's-foot-shaped leaves), has long since been re-classified under the genus *Crassula*.

Species: we now have four beautiful, freely-flowering and even fragrant succulents: *R. coccinea*, a semi-shrub, almost a garden plant, 30–60 cm (12–24 in) tall; it flowers from May–June onwards. An even more vigorous cultivar, 'Gräsers Rote', produces its magnificent flowers slightly later. The green or brownish stems, arising from the soil, are closely covered in leaves up to 2 cm ($\frac{3}{4}$ in) in length, growing crosswise. The less fragrant *R. jasminea* produces its flowers, first white, later red, a few at a time.

R. versicolor develops an umbel composed of numerous florets, white, pink, red, yellow, or even speckled and scented into the bargain.

Care: as for *Crassula*, see p. 177. Dead flower-stems should be removed to make room for new shoots. Subsequent flowering will be encouraged if these are cut back to 12–15 cm ($4\frac{3}{4}$–6 in). Winter temperature 6–8° C (43–46°F). In recent years there has been a marked drop in the numbers of rochea plants cultivated. I do not know the reason: is it because the plant is not thought to be attractive, or because it is difficult to keep?

Rochea coccinea, fam. *Crassulaceae*

Rosa – Rose

Naturally only miniature roses are sold as houseplants. When on exceptional occasions a garden rose is pruned to 'pot size', it will flower for only one season, after which it will demand its liberty. Miniature roses, on the other hand, can be kept indoors.

Introduction: although miniature roses are of some value as houseplants, they are not long-lasting. The rose is a deciduous woody plant and this makes room cultivation a difficult matter; the number of species available is therefore not large.

Species: *R. roulettii* does not appear under this name in any official list, but every rose lover knows that *R. chinensis minima*, which is in fact its correct name, is the progenitor of all miniature and pot roses. Height 20–25 cm (8–10 in). Small, pale pink flowers; there are both single and double-flowered varieties. Other dwarf roses with obsolete names are *R. lawrenceana*, and *R. × noisettiana*, now classified respectively under the *R. chinensis* group and the *R. noisettiana* hybrids. These further include all modern pot roses such as 'Baby Masquerade', yellow and red; the American 'Little Buckaroo', pale red; 'Coralin', coral red; 'Mother's Day', deep red; 'Pour toi', white; 'Rosina', yellow; as well as the famous pair 'Dwarf King', blood red, and 'Dwarf Queen', pink. We should not omit to mention the Dutch miniature roses, which, though growing to only 20–30 cm (8–12 in), flower with incredible abundance. These strains were given both English and German names. The finest are: 'Happy'/'Alberich', scarlet, semi-double; 'Sleepy'/'Balduin', pink, double; 'Doc'/'Degenhardt', pink, semi-double, large-flowered. Miniature roses flower from March to October. To obtain flowers in winter the plants must be forced. Faded flowers should be removed immediately, to avoid hip formation.

Care: good light and a sunny situation, but screen them from too bright sunlight. If plenty of fresh air and some humidity is provided, a living-room atmosphere will be tolerated. After the foliage has dropped, the plant requires a resting period in a cool position, 6–8°C (43–46°F); from February onwards it should be placed in a warmer spot, 12–15°C (54–58°F); once growth restarts the plant should gradually be acclimatised to living-room temperatures.

Watering: normally during the growing season; during dormancy the soil should be kept moist.

Feeding: from February to August every one to two weeks.

Repotting: before new growth appears, in only slightly larger pots; ordinary potting compost.

Pruning: before the new shoots appear, cut back by half or one-third; old wood and spindly branches should be removed.

Propagation: though possible, propagation from cuttings in July–August is a very complicated process. Easily grown from seed; the plants will flower six months after sowing. Mixtures of seeds are available.

Diseases: mildew and greenfly should be treated as prescribed.

Rosa chinensis 'Rosina', fam. *Rosaceae*

Saintpaulia — African violet

Some flowers have a remarkable history. The African violet, which is now receiving increasing attention, was a German discovery, but it has achieved most popularity in the United States where it has for many years headed the list of best loved plants, and where innumerable new cultivars have been produced.

Origin: the native habitat of the saintpaulia is in the East African Usambara mountain jungle, where it was discovered in 1893 by Walter von Saint Paul-Illaire, district governor of the former German colony of East Africa. Deaconesses of the German mission brought the plant to Europe and were the first to cultivate it as a houseplant.

Appearance: *S. ionantha* is an herbaceous, low-growing and rosette-shaped perennial, 20–25 cm (8–10 in) across; in soilless cultivation it grows larger. The elongated oval leaves are deep dark green, fleshy, often with reddish brown veining; lightly dentate along the edge and, like the short fleshy stems, densely covered in short hairs. The flowers are held in loose clusters on their stems, just above the foliage; except in the darkest months they appear throughout the year. In addition to the unexcelled violet-blue original species there are more than a hundred variants: white, pink, reddish purple, single or double, with crispate or wavy margins and varying leaf shapes.

Care: keep in an enclosed space, a shady, cool situation in summer. A position near a north-east facing window is very suitable. If possible the plant should be left in the same spot in winter; good light, but out of direct sunlight. A constant temperature, not below or above 16–18°C (60–65°F); it dislikes hot, dry air. The most favourable position is a north-east facing plant window with bottom heat.

Watering: with tepid, soft water, pH 4·5. The leaves and the heart of the plant should never get wet; it is therefore advisable to water from the bottom. During the growing and flowering season constant moisture should be maintained. In the middle of winter, when the plant has a short rest, water more sparingly.

Feeding: except during the resting period, give a weak solution every fortnight.

Repotting: only when the plant is obviously outgrowing the pot. Use pans rather than pots.

Propagation: from leaf cuttings in a sandy medium or in water (see also the section on Propagation, p. 71). Young plants may flower after six months.

Important: prolonged bright sunlight will cause yellow blotches, drops of water leave brown spots on the foliage. Soft stems, making the leaves collapse, are caused by water in the heart of the plant or by excess water in the saucer. Temperature fluctuations, dry air or insufficient light will halt flowering.

Properly cared for, the plants will last for years.

Saintpaulia ionantha, African Violet, fam. *Gesneriaceae*

Sansevieria – Mother-in-law's tongue

This plant was named after a nobleman: Raimondo de Sango, prince of Sanseviero, 1710–1771; he was born and died in Naples. It is popularly called mother-in-law's tongue, and is frequently treated in a way that would rapidly kill other plants in similar circumstances.

Origin: there are more than 70 species, most of them natives of tropical regions of Africa and South Africa, some also occur in Asia. Because of its hemp-like fibres, the West African *S. trifasciata* is still used as a crop plant.

Appearance: a maximum of 6 stemless, fleshy leaves grow from a creeping rootstock, often thicker than a finger; they are rarely more than 7 cm (2¾ in) broad, but may grow to 1½ m (4½ in) in height. Their growing habit is stiffly erect; they are dark green with irregular grey-green cross-banding and terminate in a sharp green point. Well cared for plants will in spring produce fragrant greenish white flower spikes growing on a slender stem not more than two-thirds of the length of the leaves.

Species: there are several fine cultivars of *S. trifasciata*. The best known and most beautiful is 'Laurentii', in which the leaves are edged by golden-yellow bands along their entire length. In the less well known 'Craigii', the bands are creamier in colour. 'Hahnii' differs in habit from other cultivars: it grows in a low rosette; the marking of the leaves is similar to that of the species. In general it resembles a small bromeliad. There is also a 'Golden Hahnii', with gold-edged leaves. All these plants are very suitable for soiless cultivation.

Care: indoors throughout the year. The plants are suitable for the windowsill, for a succulent case, in shops and in hotels, for they are entirely insensitive to dry or close air. Although it looks more attractive in the shade, this sun-loving plant from the steppes should be placed in the best possible light. To avoid damage, the winter temperature should not drop below 15°C (58°F).

Watering: normally in summer; keep very dry in winter. They are chalk-loving plants and ordinary tapwater may be used.

Feeding: only in summer, with nitrogen-free fertiliser (cactus fertiliser).

Repotting: only when the rootstock crowds the pot and emerges from the compost, and the pot almost breaks. Use coarse, pebbly compost, no peat. The soil reaction should be slightly alkaline, *p*H 7·5. Lack of flowering is frequently due to the use of incorrect compost.

Propagation: this is best done by division of the rootstock. Propagation from leaf cuttings will succeed only in adequate soil temperature and will always produce the original species, even if taken from a leaf of a cultivar. The cuttings are subject to rotting.

Important: the plant's chief enemy is excess water in the pot. This will cause the neck of the root to rot and the leaves to collapse and die. Too cold a situation will have similar results.

Sansevieria trifasciata and cultivars, fam. *Agavaceae*

Sauromatum venosum

This is one of the most remarkable dry-flowering plants on the market, and the magnificent leaf may moreover be enjoyed for many years, during which time the bulbs multiply. In temperate climates the bulbs will even flower again after spending the winter out of doors.

Origin: there are six species, occurring in tropical Africa, in India and in Sumatra. Only *Sauromatum venosum* (syn. *Arum venosum*), a native of Burma, is known as a dry-flowering plant, a miracle flower, blooming without soil.

Appearance: the slightly flattened bulb, almost the size of your fist, with a curved growing point at the top, is bought at the end of December or later, and is placed without any soil or water, on the window-sill of a warm room. The typically arum-lily-shaped flower smells like a carrion flower, but is of a very unusual dark red colour; the inside is brown-spotted on a paler background. It continues to flower till spring, when it should be put in a roomy pot and covered with a 3 cm ($1\frac{1}{4}$ in) layer of compost, or be planted in the garden. It will then develop a foot-shaped leaf on a 50 cm (20 in) stem, so deeply incised that it gives the impression of an entire crown of leaves. In autumn the leaf will fade. The bulbs are left in the pot until the end of the dormant season, overwintering at 8–10°C (46–50°F). At the end of February it is taken up and left to flower without soil or water as before. It should be watered and fed only while the leaf is growing, until August. A large number of offsets will be produced, especially in garden cultivation; these will flower after two to three years.

Sauromatum venosum, fam. *Araceae*

Saxifraga

Saxifraga, the chief genus of its family, includes about 350 species, most of which occur in the high mountains of moderate regions and have become known as plants for the rock garden. Only one is a true houseplant; it has been cultivated in living-rooms for 200 years.

Saxifraga cotyledon, fam. *Saxifragaceae*

Origin: *S. stolonifera*, is a trailing plant originating in China and Japan (illustration p. 326); it was brought to Europe in 1771. *S. cotyledon*, a rosette-forming plant, grows wild among the primeval rocks in the mountains of Switzerland, Norway, Iceland and the Pyrenees, it is a winter-hardy garden plant, but because of its magnificent flowers is sold as a pot plant for use indoors and on the balcony in June.

Appearance: the literal meaning of the name *S. stolonifera* is 'runner forming stone breaker'. The slender, trailing red runners, up to half a metre ($1\frac{1}{2}$ ft) long, bear attractive young plantlets at their tips. A large number of runners together give the impression of an untidy beard. Because of these small plantlets the plant is sometimes called Mother of Thousands. The yellow-speckled small white flowers appear from May to August; the sepals are reflexed in two directions. Apart from this, *S. stolonifera* is a hardy plant with basal leaf rosettes. The leaves grow on red stalks and are rounded kidney-shaped, slightly wavy and lobed; the upper surface is dark green with white veins, the reverse red and speckled. The flowers grow in panicles, 20 cm (8 in) tall, often red-tipped when in bud. In addition to

the species, which in a wild climate can spend the winter out of doors provided it is adequately protected, there is a white variegated cultivar: *S. stolonifera* 'Tricolor'. The latter requires more warmth, is slow growing and flowers less profusely; nevertheless its beautiful foliage make it equally attractive. Young leaves often have traces of red as well.

Care: the parent plant requires a great deal of light in a half-shady position, whether standing or hanging. Lack of light will make the leaves fade and turn limp. These plants dislike being hemmed in by others and require plenty of fresh air. During the resting period from October to February the most favourable temperature is 8–12°C (46–54°F), although higher temperatures will be tolerated if absolutely necessary. 'Tricolor' likes normal room temperature and will flourish in living room atmosphere provided this is slightly damp.

Watering: from the time the first shoots appear until August, keep constantly moist; then decrease the water supply in conformity with the temperature. 'Tricolor' grown in a heated room naturally needs more moisture than the all-green species kept in a cool or moderately warm position. The water need not be softened.

Feeding: fortnightly during the growing season, until August.

Repotting: every spring in small pots. Adequate drainage layer and a porous compost, rich and humusy. Pre-packed potting compost is satisfactory.

Propagation: by removing the plantlets in summer; they are potted in groups of three.

Diseases: too warm a situation in winter and lack of moisture in the air may encourage greenfly.

Important: the plant should always hang freely and be protected against draught. Not suitable for use on the balcony.

Saxifraga stolonifera 'Tricolor', fam. *Saxifragaceae*

Schefflera

As it provides so many opportunities for decoration, while being fairly slow growing, schefflera has gained in importance in recent years, not only as part of the choice for large plant communities in soilless cultivation, but also as a specimen plant. The rules pertaining to temperature are particularly important.

Origin: of the approximately 150 species occurring in many tropical countries, only the Australian *Schefflera actinophylla* is cultivated over here. Its botanical name commemorates Jacob Christian Scheffler, an 18th century Danish botanist and a friend of Linnaeus.

Appearance: in its native habitat schefflera is an enormous tree, up to 40 m (120 ft) tall; in the Australian spring it is decked in scarlet flowers. In cultivation it will never achieve this; even at an advanced age and when grown in a tub it will remain a sparsely branching shrub which, like the related *Dizygotheca*, will in the course of many years form small trunks. In young plants the long-stemmed leathery green foliage is initially divided into 3 or 5 leaves only, but as the plant grows older it divides further. The maximum division is thought to be 16 sections, 30 cm (12 in) long.

Care: in summer it should have plenty of light and air, but not in direct sunlight or in a draught. From late May onwards the plant may be put on the balcony or in the garden. Don't forget to bring it back indoors before the first night frost. Dormant season from October to February. Plenty of light in winter; temperature not below 12°C (54°F) or the foliage will drop, and not above 16–18°C (60–65°F) since this will cause the plant to grow lanky and be subject to scale insect. The plant's health depends entirely on whether or not these rules are adhered to. Further treatment as for *Fatsia japonica* (p. 208).

Propagation: only from freshly imported seed sown in a heated seedbed. The most beautiful effect is achieved by combining three plants in a pot or tub.

Schefflera actinophylla, fam. *Araliaceae*

Scindapsus – Devil's ivy

The remarkable fate of scindapsus, of which the best known species was reclassified in the genus *Rhaphidophora*, has been described on pp. 314–315. The plants discussed on this page are a particularly beautiful species and one of its cultivars, suitable only for a heated greenhouse.

S. pictus originates in the same Malay regions as its former relations. Its habit is similar, but the dark green foliage is white-flecked. The cultivar *S. p.* 'Argyraeus' has slightly smaller leaves, small, clear silver spots and a silver margin. It is more graceful than the species but also, like the cultivars of *Rhaphidophora*, more demanding. It belongs in a tropical window, or at any rate in a plant window with bottom heat; it thrives on moss-covered tree-trunks or epiphyte trees.

Scindapsus pictus 'Argyraeus', fam. *Araceae*

Scirpus cernuus – Club rush

It is thought that a rush-like plant with the name scirpus was known in Roman times. We now realise that many hardy and non-hardy species of this member of the sedge family occur all over the world. Only one of them, *S. cernuus* is grown as a houseplant.

Appearance: initially this 20–25 cm (8–10 in) tall, fresh green grass grows erect, forming dense clumps. At a later stage the tender, threadlike blades curve, resulting in a beautiful hanging plant. In summer small grass flowers appear in broad spikes at the tips of the blades.
Care: in general as for the umbrella plant (p. 183). The chief difference is the fact that *S. cernuus* cannot be placed out of doors; it requires a half-shady position indoors throughout the year. In a warm environment it will continue to grow in winter. Moist air and plenty of water in the saucer are essential to avoid the tips of the blades turning brown.
Propagation: from seed in spring, or by division when repotting in January–February.
Important: very suitable for soilless cultivation.

Scirpus cernuus, Club Rush, fam. *Cyperaceae*

Sedum

Tradition has it that the Romans knew this plant and planted it on walls and on the roofs of their houses as a protection against thunderstorms. The fact that our good old *S. acre* has the popular name Wall Pepper fits into this story, but the same tradition exists for the houseleek (*Sempervivum*).

Origin: sedum occurs in all temperate regions of the world. Mexican foliage succulents are used as houseplants.
Species: there are so many beautiful forms of the non-hardy species suitable for room cultivation that it is easy to establish a fine collection. appeared. In Germany *S. pachyphyllum* is called 'Schnapsnase' ('red nose'); this describes its appearance; it is vigorously erect-growing and has circular, red-tipped leaves. *S. rubrotinctum* grows to 20 cm (8 in) and branches close to the soil; the leaves are cylindrical, the flowers yellow; a autumn; they like to overwinter at 8–10°C (46–50°F) but will tolerate a higher temperature and fairly dry living-room air.
Watering: in summer normally; during its resting season in winter it should be kept almost dry, particularly in a cool room.

Sedum rubrotinctum, fam. *Crassulaceae*

S. bellum, in which the foliage is almost entirely hidden by creamy white flowers, is highly recommended as a winter-flowering semi-trailing plant; it is suitable for bowls or hanging pots. Another magnificent hanging plant is *S. morganianum*, in which the trailing or creeping shoots grow to half a metre (1½ ft) in length and are surrounded by cylindrical, succulent pale green leaves. In my house this plant has been growing in a south-facing window for two years; the rose-red flowers have not yet very distinctive plant. *S. sicboldii*, a native of Japan, produces pink flowers in autumn and is more of a balcony or garden plant. It is winter-hardy, though it fades in late autumn; the leaves of the species have a bluish bloom; the cultivar 'Mediovariegatum', created 90 years ago, bears an irregular yellow blotch on each of its scarlet-edged green leaves.
Care: all tropical indoor forms of sedum require a great deal of light; they tolerate a position, either standing or hanging, in a south-facing window or can be placed in an airy situation out of doors, provided they are brought indoors in good time in the

Sedum sieboldii 'Mediovariegatum', fam. *Crassulaceae*

Feeding: not advisable.
Repotting: when required, in only slightly larger, shallow pots. Use compost with extra sand.
Propagation: from stem- or leaf-cuttings which must be left to dry for a few days before being inserted obliquely in propagating compost (1 part sand, 1 part peat). Can also be grown from seed, if available.
Diseases: rarely occur. Most damage is caused by excess moisture.

Selaginella – Creeping moss

The family of *Selaginaceae* consists of only one genus; on the other hand there are over 700 species. Most of them belong to the tropical rain forest and it is these that have produced some of the magnificent hot house plants, growing to 50–60 cm (20–24 in). A number of lower-growing species originate in temperate regions.

S. apoda (centre front), only a few inches high, is yellowish green and clump-forming; it occurs across the entire continent of North America, from Canada to Texas. When brought into cultivation it soon became popular as ground cover in an unheated plant window or in a conservatory: long lasting, undemanding and easily increased. *S. martensii* is a native of Mexico; it grows to 30 cm (12 in) and is therefore very suitable as a pot plant. Minimum temperature in winter 14–16°C (57–60°F). The variegated form *S. martensii* 'Watsoniana' is grown on a large scale. Selaginellas need shade and in summer a moist and fresh atmosphere; they must be watered with lime-free water and dry out very easily. Feed with a very weak solution.

Right: *Selaginella martensii;* above: *Selaginella martensii* 'Watsoniana'; centre front: *Selaginella apoda,* fam. *Selaginaceae*

Selenicereus grandiflorus
– Night cactus

Whenever it is reported in the local press that a 'Queen of the night' is about to flower, either in a botanical garden or in a private home, it causes a small sensation among plant lovers. However, you could witness this nocturnal miracle in your own home.

Apart from a large measure of patience – it takes years before the plant will flower – it does not require a great deal of expertise. In the Argentine and in Texas, its native habitat, the night cactus is a semi-epiphytic climbing shrub; the 2·5 m (7½ ft) thick ribbed stems, to 5 m (15 ft) in length, are covered in yellow thorns. The fragrant flower, 20–30 cm (10–12 in) long, is 20 cm (10 in) wide and lasts for only one night.

Care: a warm position during the growing period in summer, but not too much sun; plenty of light and a temperature of 10–12°C (50–54°F) in winter; spray frequently until the ▶

Selenicereus grandiflorus, Night-flowering cactus, fam. *Cactaceae*

plant flowers. In summer feed regularly with special cactus fertiliser. The pot should be roomy and the soil light, and rich in humus (pre-packed potting compost mixed with extra sand). When repotting, wrap the entire plant in a piece of plastic or cloth; it is possible to repot older plants as well. It is best to tie the long stems to curved bamboo supports; if the plant is to grow up the wall of a greenhouse use straight sticks.

Propagation: easily grown from cuttings, either in soil or by the soilless method. Possible also from seed; apart from *S. grandiflorus* and *S. pteranthus*, seeds are available of the slightly smaller new strains developed from crossings with *Epiphyllum*. These flower at a much earlier age than the original species.

Senecio herreianus, fam. *Compositae*
▽

Senecio

Senecio, with its more than 2000 species, is the largest genus of the family of *Compositae*. It is therefore not astonishing that in addition to the well-known herbaceous plants, shrubs and trees there is also a group of succulents, occurring all over the world, many of them of remarkable form — a surprise to any collector.

Origin: the majority are natives of Madagascar, where they form dense carpets in the tropical highlands at altitudes between 1500 and 2000 m (4500 and 6000 ft). Other species originate in South and South-west Africa.

Appearance: there are creeping and trailing species, whose cylindrical runners are covered in globular leaves. If these remarkable leaves, greatly resembling prickly berries, contain a transparent longitudinal 'window' on their upper surface, the plant is a South-west African *S. herreianus*. If, on the other hand, the leaves are a pointed globe in shape, like lemons, and are covered in a grey green waxy bloom, with transparent veins, we know it is *S. citriformis* from the Cape Province. *S. haworthii* has been popular among experts for more than 150 years; it is a freely branching, bushy plant with thick, cylindrical runners covered in felty, white-haired leaves, also cylindrical in shape. All these senecio species may flower; as a rule the flower-heads are small and inconspicuous.

Care: similar to that of *Crassula*. Place in a very sunny position. Keep dry in winter.

Propagation: from cuttings and from seed.

Senecio cruentus hybrids — Cineraria

They were formerly called *Cineraria hybrida*. These much loved plants flower at the end of winter and decorate our window-sills around Easter, but they are also used in the garden and in boxes on the balcony.

Origin: *S. cruentus*, the original species, comes from the Canary Islands, where it grows as a perennial, 40–60 cm (16–24 in) in height, in the frost-free climate of damp, cool mountain forests. About 200 years ago the first seeds of the plants were brought to England; at that time their inflorescence was quite modest. They were brought into cultivation; France and other countries joined in their development, which continues to this day.

Appearance: to the achievements of these growers of different nationalities we owe the many beautiful strains in which the flowers, which occur in many colours, appear to emerge like an old fashioned round posy from the tender foliage. Professionals divide the enormous number of cultivars into five different groups: the medium tall large-flowered varieties, the medium and small-flowered, the double varieties and the tubular-flowered forms. A recent novelty is the dwarf form 'Scarlet', only 15 cm (16 in) tall. The chief colours are: bright blue, cornflower-blue, red, copper, crimson, carmine pink, soft pink, white, copper with a white ring, crimson with a white ring, blue with a white ring. It is remarkable to note how the former often unattractively lurid and artificial looking colours have been replaced by more subtle shades. The illustration on this page provides an example.

Care: the natural environment in which the plant originated and the conditions it demands as a result, are only too frequently forgotten. The best advice is undoubtedly to give the cineraria a cool position in light shade and to water freely, using tepid water. This treatment provides the best guarantee for prolonged flowering without infestation by aphids, inevitable in a warm living-room. If you are able to keep the plants in a temperature of 8–10°C (46–50°F), their flowers will give you pleasure for four to six weeks. After flowering they are practically worthless, since they never flower for a second time.

Propagation: a job for the expert.

Senecio cruentus, hybrid, Cineraria, fam. *Compositae*

Setcreasea

If you like *Tradescantia*, you will probably enjoy growing a setcreasea as well. The two plants are very similar, but the setcreasea is slightly more attractive, as in a good light its unusual colour makes it stand out more than the normally all-green *Tradescantia*.

Origin: its native habitat is in Mexico. Of the few species occurring there, *S. purpurea* was discovered as a houseplant as late as 1955. Other setcreasea species, for instance *S. striata*, have meanwhile been re-classified under the genus *Callisia* (see p. 146). *S. purpurea*, called Purple Heart in the United States, now represents both species and genus.

Appearance: correctly treated, the entire plant will be deep violet-purple in colour. Initially the juicy stems are erect growing, but later they become more recumbent and trailing in habit, making it very suitable for use as a hanging plant. The blue-frosted leaves can grow to 3–4 cm (1½–1¾ in) across and over 15 cm (6 in) long. The close-packed small blue-violet flowers are enclosed by two bracts.

Care: the truly magnificent colour of the foliage is dependent on the plant's situation. It requires plenty of light, out of direct sunlight. In winter the temperature should not drop below 18–20°C (65–70°F). In cooler sur-roundings the plant will wilt and lose its fine colour. Older plants become unsightly and new specimens should therefore be grown from cuttings every two years. The cuttings will root readily if some bottom heat can be provided. Further treatment as for *Tradescantia*. Very suitable for soil-less cultivation.

◁ *Setcreasea purpurea,* fam. *Commelinaceae*

Siderasis fuscata, fam. *Commelinaceae*
▽

Siderasis fuscata

The generic name *Siderasis* is derived from the Greek; its approximate meaning is 'rust-coloured fur'. The specific name *fuscata* also refers to the brown colour of the hairs in which the entire plant is covered.

It is a native of Brazil and has been known as a houseplant for only a short time. It is a most attractive small plant, somewhat similar to *Rhoeo*. Its outward appearance gives little indication of its relationship with *Tradescantia*. The beautifully marked leaves are up to 20 cm (8 in) long and 8 cm (3¼ in) wide, green above with white streaks down the centre; the under-side is red. The lovely flowers range from sky-blue to reddish. This is a hot house plant; like the *Maranta* it should be given a damp and shady position. Propagation only by division of mature plants, never from seed or cuttings.

Sinningia hybrids — Gloxinia

Both the botanical and the popular name are based on personal names. Wilhelm Sinning, 1794–1874, was curator of the Botanical Garden in Bonn; Benjamin Peter Gloxin was a Strasbourg physician. Together they named one of our best loved plants, one that flowers during summer and autumn.

Origin: its country of origin is Brazil, where about 20 modestly flowering tuberous species occur, typical inhabitants of the tropical rain forest.

Appearance: above ground all its parts are velvety; the stems are succulent. The foliage grows horizontally or curves over the edge of the pot, depending on the strain. Large bell-shaped flowers, widely reflexed at the edge. There are numerous colour variations, especially in the velvety red and purple shades. Flowering occurs during summer and autumn.

Care: the most favourable situation which should remain unchanged, is in good light, protected from direct sunlight. The temperature should be fresh and constant, but draught must be avoided. Adequate humidity, but if spraying do not do so directly on to the plant, as this will stain the flowers.

Watering: use lukewarm water until the end of the flowering period; if you want to keep the plant, use water with a pH of 4–4·5.

Feeding: if the plant is kept through the winter, give a diluted liquid fertiliser until it is well in flower. The nutrient solution used in soilless cultivation is very suitable.

Keeping: after flowering gradually give less water. Too rapid withering of the foliage and drying out of the tuber will affect the success of overwintering; this should take place in a warm, 16°C (60°F), and dry position. Towards the end of February lift the tubers and remove the compost and the old roots. Place the tubers in the same or slightly larger pots containing very humusy compost and cover them with a 2-m ($\frac{3}{4}$-in) layer. Water sparingly and start the plants into growth at 18–20°C (65–70°F). For further treatment see under Care. Leave only one to two shoots on each tuber; remove the rest.

Propagation: from leaf cuttings or shoots. Increase from seed should be left to the grower.

Diseases: too dry air will cause leaf curl, will check growth and encourage greenfly. If the soil is too damp or the water too cold, the stem will rot and the leaves will collapse.

Sinningia, hybrid, Gloxinia, fam. *Gesneriaceae*

Skimmia

Skimmia belongs to the family of *Rutaceae*, and the plants of this genus originate in eastern Asia. As its specific name indicates, *S. japonica*, the species which first became known over here, is a native of Japan, skimmia is from the Japanese *skimmi*. Its natural habitat is in frost-free regions in cool mountain forests.

Actually skimmia belongs on the balcony in winter and in a sheltered position in the garden in summer. However, it has become the fashion to market the young plants with their burden of scarlet berries together with the *Capsicum annuum* described on p. 150, around Christmas, when they make inexpensive presents. If you continue to grow the little evergreen shrub, you may be rewarded in April by the sight of the attractive small, fragrant white flowers and later by the scarlet berries which keep until the following autumn. There are no special rules for treatment.

Skimmia japonica, fam. *Rutaceae*

Smithiantha

Nowadays it is not unusual to name cultivars after queens and female filmstars, but to call an entire botanical genus after a woman is an unusual event. The lady in question was called Mathilda Smith; she was a botanical artist attached to Kew Gardens.

Smithiantha, hybrid, fam. *Gesneriaceae*

Origin: three species are known to us, all originating in the tropical mountain forests of Mexico. However, the species as such are now rarely cultivated, since finer and more valuable hybrids have been developed. Among these are 'Pink Domino', 'Orange King', 'Swan Lake' and 'Vespers'; there is also a semi-dwarf form called 'Deschene's Strain'.

Appearance: in the professional literature, smithiantha is justly described both as a fine foliage plant and as a beautiful flowering plant. They are all succulent herbaceous plants, growing from scaly rhizomes; they have softly haired, velvety green or reddish brown leaves.

Inflorescence: the flowers appear from June to October, depending on the time when they are started into growth; as a rule they are numerous and keep reasonably well. The colours are yellow, pink or red, paler in the centre and with darker tips. The inflorescence consists of erect growing loose racemes.

Care: of course it cannot be expected that these exotic beauties will be satisfied with simple room cultivation; but with a little extra care they can nevertheless be used indoors, provided you do not buy hot house grown plants already in flower, but grow new plants from rhizomes yourself.

Bringing into growth: after the plant has faded in autumn, the rhizomes should be left to overwinter in the pot, at a temperature of 10–12°C (50–54°F). Towards the end of winter they are lifted, planted in groups of three or four in humusy soil and covered with a 2 cm (¾ in) layer. Ordinary potting compost or a garden compost-based mixture may be used. The pots should now be kept warm until the flowers appear; a soil temperature of $\pm 22°C$ (72°F) is essential. Give only lukewarm water and spray overhead in order to satisfy the plant's need of moist air.

Feeding: when the roots are about the length of your finger, feed every

two weeks until the plant flowers.
Important: the plant will definitely not tolerate direct sunlight. In a temperature-controlled plant window or in a small greenhouse everything will be much easier.

Solanum pseudocapsicum, Christmas Cherry, fam. *Solanaceae*

Solanum – Christmas Cherry

Let me start by correcting a misapprehension. The Christmas Cherry, whatever its origin, is usually grown as an annual. It is regarded as a 'disposable' plant, worthless when the flowers and fruits have done their duty. But this is incorrect, for the Christmas Cherry will continue to grow!

Origin: two species are available. *S. capsicastrum,* from South America, has small red fruits; a variegated-leaved form called 'Variegatum' is sold on a large scale. A second species is *S. pseudocapsicum,* a native of Madeira; two strains are available in particular: 'New Patterson', with orange-red fruits, and 'Goldball', which has golden-yellow, fairly large spherical fruits.

Appearance: all Christmas cherries enter your home as fairly spreading shrubs, but if properly cared for they can grow into most attractive standard plants, more than a metre (3 ft) in height. They are evergreen by nature; if the foliage should drop they will put forth new shoots in March after having been vigorously pruned.

Flowers and fruits: the typical nightshade flowers resemble those of the potato; they are white and appear from May to June. The cheerful fruits are developed until the autumn and provided the plant is given the cool situation it requires, 8–10°C (46–50°F), they will last for a long time. It must be admitted that in a centrally heated house the plant will be difficult to keep through the winter.

Care: it is not difficult to care for. From the time the first shoots appear in March–April, the plant should be given an increasingly warm position; it requires a sunny situation. Water and feed freely during the summer; in the autumn decrease the water supply and gradually stop feeding. If it is kept in the garden in summer, be sure to bring it indoors in good time, for the plant is very sensitive to frost. Dormant period from October to February. Be careful not to let the plant dry out.

Repotting: before the appearance of new growth, in pre-packed potting compost. Give the plant only a slightly larger pot; careful trimming of the root system is better.

Propagation: a professional job.

Diseases: too warm a position and draughts lead to plant lice.

Sonerila margaritacea 'Argentea', fam. *Melastomataceae*

Sonerila

As we pointed out in the chapters on the little *Bertolonia* (p. 138) and on the *Medinilla* (p. 258), these are exotics of the first order. Like these two, the sonerila, another member of the *Melastomataceae,* and a native of the Indonesian archipelago, should only be grown in a hot house.

Like *Bertolonia,* the small semi-shrub *S. margaritacea,* with its strikingly beautiful foliage and profuse flowering in the autumn, requires a shady situation as well as plenty of warmth and very humid air. The plant is not often available; the one most frequently seen is the cultivar 'Argentea', whose leaves are even more silvery than those of the speckled and blotched species.

Sparmannia — House lime

Like many other plants, the lime tree is regarded as typically English, but this is a mistake, for the Lime family includes nearly 50 genera with over 400 species, most of them trees or shrubs occurring in the tropics. The summer- and the winter-linden trees of the temperate zones of Europe are exceptions.

Origin: of the 7 species occurring in tropical regions of Africa and in Madagascar, only *S. africana* or African hemp, has found acceptance as a houseplant.

Appearance: one species grows to 4 m (12 ft) in its native habitat, but there is also a dwarf form reaching only 80 cm (32 in). Both are evergreen shrubs, whose lower branches rapidly become brown and woody. The new stems and the large, cordate-oval, often sharply indented leaves are pale green while at the same time covered in felty hairs. The white flowers, appearing in late winter or early spring — rarely in summer — have unusual yellow and red stamens. As a rule the dwarf form flowers more profusely than the large original species, which in turn does better than the double-flowered cultivar 'Flore Pleno'. In addition there are house limes which never flower and should be regarded as fine foliage plants.

Care: the house lime prefers to be kept in the living-room throughout the year; it is not very happy when placed on the balcony or in the garden for a time. During the growing period, which partly coincides with the flowering season, it should have a sunny situation, well ventilated and not too warm. It must be placed away from other plants and is therefore unsuitable for the plant window. The growing period continues until August, and after a transitional period, is followed by the dormant season from October to December, at 6–10°C (43–50°F).

Watering: freely during the growing period; keep drier in the dormant season, but avoid inadequate drainage as well as drying out of the soil.

Feeding: weekly while the plant is growing.

Repotting: vigorous plants should as a rule be repotted into a larger pot or tub every year. Ordinary pre-packed potting compost.

Propagation: in summer from fairly mature shoots, preferably from a plant that normally flowers well.

Diseases: yellowing and dropping foliage indicates lack of nourishment and other errors in cultivation. Plants that have grown bare should be cut back. Be on the lookout for greenfly and thrips.

Sparmannia africana, House Lime, fam. *Tiliaceae*

Spathiphyllum — Peace lily

Among arum lily plants the one described here, and more particularly the freely flowering species *S. floribundum* from Colombia, is one of the most attractive as well as the most modest. It lasts for years, faithfully flowering summer after summer; the long-lasting flowers are even suitable for cutting.

Habit and care: this evergreen herbaceous plant grows to only 30–40 cm (12–16 in). It is satisfied with a shady or half-shady position in normal room temperature, which in winter should not drop below 16–18°C (60–65°F); during the resting period from October to January it is not sensitive to dry air, while in summer it may occasionally be sprayed, though only during bud formation. During that period the soil should be kept constantly moist and the plant should not be fed more than once a fortnight or the leaf tips may turn brown.
Repotting: when the plant begins to put forth new shoots in February–March; the pots should not be too large.
Propagation: by division at the time of repotting. Easily raised from seed as well. Seed of various species is available.

Spathiphyllum floribundum, fam. *Araceae*

Stapelia — Carrion flower

Some people call this plant the Starfish Flower; others, more forthright, call a spade a spade and refer to it as the Carrion Flower, because of its unpleasant smell. If you are not put off by this, you could build up a beautiful collection of these interesting succulents.

Cultivation is simple. The most important point is to satisfy the plants' winter requirements: a cool room, if possible not exceeding 10°C (50°F); plenty of light and practically no water. Flowering occurs in summer. The largest flowers, 25–35 cm (10–14 in) across, are produced by *S. gigantea*; those of *S. grandiflora* are 10–15 cm (4–6 in) in diameter and *S. variegata* has the smallest flowers, 5–8 cm (2–3¼ in) across.
Propagation: easily grown from seed; there is a large choice.

Stapelia grandiflora, Carrion Flower, fam. *Asclepiadaceae*

Stephanotis floribunda
— Madagascar jasmine

This magnificent, robust plant has for a great many years been one of the most acceptable gifts to plant lovers on Mother's Day. Now that the flowering season can be adjusted by the 'long-day treatment', there are even more opportunities to buy stephanotis as a gift sure to be welcome.

Origin: its original habitat is in Madagascar, where *S. floribunda* grows in the cooler regions as an evergreen trailing shrub. In Europe it has been known for more than 150 years as an undemanding, long-lasting plant for moderately warm rooms; in the United States, too, the Madagascar Jasmine is known and loved. In northern countries and in England the flowers are used for cutting, and are even included in bridal bouquets.

Appearance: the stems are tough and flexible, up to 5 m (15 ft) in length. Foliage firm and leathery, dark green and glossy, 5–9 cm (2–3¾ in) long and 4–5 cm (1¾–2 in) broad, growing opposite on short stems. The deliciously scented, star-shaped white flowers, up to 5 cm (2 in) across and 4 cm (1¾ in) deep, appear in loose umbels from the axils of the leaves. The natural flowering season occurs from May till August.

Use: in the greenhouse or in the conservatory it can cover entire walls. It can also be used as a vigorous house plant, to be trained along a firmly fixed frame or bent wire.

Care: it requires plenty of light, space and fresh air out of direct sunlight. Do not place the plant behind a curtain or in a crowded plant window. The worst mistake that can be made is too much warmth in winter; the correct temperature is 12–14°C (54–57°F), otherwise the plant will grow spindly and be attacked by pests, which will reduce its vigour and hold back flowering. Dropping of the flowerbuds or the leaves will occur if the plant is moved or turned: mark its position. Dried out compost, temperature fluctuations and draught are three factors detrimental to growth. Growing season from March to September, dormancy from October to January.

Watering: when the plant starts into growth, water freely until mid-August, using tepid water, pH 5·5–6·5. Then gradually decrease the water supply and during the resting period keep only just damp, but beware of drying out (see under Care). Spray frequently during the growing season.

Feeding: in view of its often impetuous growth, the plant should be generously fed every fortnight until mid-August; subsequently at longer intervals. From September to February do not feed, then start again gradually.

Repotting: young plants every year, older plants at longer intervals, depending on root development. The best time for repotting is in February–March. Use a humusy, rich and porous compost and provide an adequate drainage layer of crocks and pebbles. Pre-packed potting compost mixed with extra sand is satisfactory.

Propagation: from cuttings of fairly mature stems in spring and summer, in a heated propagating bed at 25–30°C (78–86°F). The cuttings will root in four to five weeks.

Diseases: too high a temperature in winter will encourage mealy bug and scale insect.

Stephanotis floribunda, fam. *Asclepiadaceae*

Streptocarpus – Cape primrose

This charming plant, with flowers resembling small orchids, is the Cape Primrose. In Germany it is called *Drehfrucht* (spiral fruit), which refers to the fact that in the seed-bearing species and their hybrids the seed capsules are coiled. To date the large-flowered 'Wiesmoor' hybrids are considered to be the finest.

Origin: it is a native of South Africa; its natural habitat is in the tropical rain forest. The magnificent hybrids are much-loved houseplants.

Appearance: herbaceous biennials, the same size as the primula, with elongated oval, slightly wrinkled and practically stemless leaves, curving over the rim of the pot.

Inflorescence: large, funnel-shaped flowers growing in groups of 2–3 on stems about 25 cm (10 in) tall. They occur in the most magnificent range of colours, from white via pink, soft lilac, pale blue to bright red, blue and velvety dark purple, often with a throat veined in a contrasting colour and with a fringed edge. The latest cultivar is the English 'Constant Nymph', whose pure lavender-blue flowers no longer develop seed, but which by way of compensation flowers almost throughout the year and exceeds the normal two-year lifespan (ill. p. 14).

Care: whether the streptocarpus hybrids are marketed as autumn-flowering or as spring-flowering plants depends on the time of the year when the grower took the cuttings. On the whole, treatment is the same as for the African violet: keep indoors throughout the year in plenty of light and good ventilation and protect the plant against sunlight. Because of the horizontal growth of the foliage, streptocarpus is also very suitable for use as flowering ground cover in a simple plant window. Winter temperature 18–20°C (65–78°F) (both soil and air). Temperatures below 16–18°C (60–65°F) will damage the plant.

Watering: freely in summer; in winter more or less sparingly, depending on the situation. Always use tepid water, *p*H 4·5. In summer some degree of air moisture is desirable; in winter this should depend on whether the plant is flowering or resting.

Feeding: weekly in summer; never pour the fertiliser on to dry compost.

Repotting: as required, in February–March, in rich, humus soil; prepacked potting compost.

Propagation: divide the leaf along the central vein and insert the cut edge obliquely in damp sand to a depth of 1 cm (½ in). In the course of time numerous tiny plants will appear.

Streptocarpus, hybrid 'Wiesmoor', Cape Primrose, fam. *Gesneriaceae*

Syngonium — Goosefoot plant

Most people think that this is some sort of *Philodendron*, and they are in fact not as wide as all that of the mark, for arum lily plants such as *Philodendron*, *Monstera* and *Scindapsus* are closely related to the trailing and climbing *Syngonium* plants.

Syngonium vellozianum, fam. *Araceae*

Origin: the genus originates in Central and South America, where 20 species occur, of which only a few are in cultivation. I really could not tell you which of these are the plants which have for many years been growing in my plant window, some in soilless cultivation.

Appearance: to distinguish it from other arum lily plants it is important to know that all syngoniums contain a milky liquid and are further characterised by the fact that the leaves, which in the course of time change in shape until they finally are 3-, 5-, or even 8-lobed, root very readily. They are creeping or climbing and can therefore be used as hanging or climbing plants, or on moss-covered tree trunks or epiphyte trees. In spring or summer older plants produce greenish, arum lily-shaped flowers, of which the spathe is reddish purple inside. The most suitable forms for room cultivation are: *S. vellozianum* (formerly called *S. auritum*), with thick green foliage, and *S. podophyllum* and its cultivars, some of them variegated. Care as for *Philodendron*.

Tetrastigma

The popular name Chestnut Vine refers to the size of the leaves and hence is an indication of the size of the plant. You will be able to accommodate it only if you have a floor-level plant window or a well-lit, moderately warm landing at your disposal.

Tetrastigma voinierianum, fam. *Vitaceae*

Origin: there are about 100 species, occurring from tropical Asia, via New Guinea, to the north-eastern part of Australia. Only one, *T. voinierianum*, from the region of Tonkin, is in cultivation here.

Appearance: in accordance with its habit, it belongs to the group of evergreen lianas or trailing plants. It has quintuple dentate foliage, dark green above, reddish brown and felty beneath, growing on 5–7 cm (2–2¾ in) long stems; each leaf can grow to 25 cm in length. Planted in a greenhouse, it will within a year easily grow tendrils several metres (feet) long. In cultivation it practically never flowers.

Care: good light or a little shade in winter as well as in summer; never direct sunlight. In winter the temperature should not drop below 12–15°C (54–58°F). Dry air is tolerated.

Watering: water generously in summer, more sparingly in winter. Some lime in the water is desirable (*p*H 7·5).

Feeding: only in summer, weekly with a normal strength solution.

Repotting: as required in a roomy pot or tub. Humusy, very rich and porous soil mixture.

Important: use strong canes and tie the runners in such a way that the foliage will be able to develop properly.

Diseases: yellowing of the young leaves indicates lack of nourishment. Drought frequently encourages red spider mite infestation.

Thunbergia

Thunbergia alata, Black-eyed Susan, fam. *Acanthaceae*

In most European countries the only thunbergia species known is *T. alata*, Black-eyed Susan, usually grown as an annual balcony plant. The botanist, on the other hand, is familiar with others, of which some profusely flowering species have been introduced as houseplants in the United States.

Origin: thunbergia forms part of the tropical flora of Asia, Africa and Madagascar and includes about 200 species.

Appearance: botanists distinguish three different groups. The first includes erect growing, as a rule magnificently flowering tropical shrubs, which unfortunately grow too large for the living-room, such as *T. erecta* (hot house) and *T. natalensis* (warm house).

The second group embraces vigorous trailing plants, with equally beautiful and prolonged flowering. One of these is *T. lutea*, with tendrils not exceeding 2–3 m (6–9 ft); even quite young plants will flower profusely and the plant's perpendicular habit can be preserved by regular pruning. In the United States it has an excellent reputation. Finally, the third group of perennial, twining plants includes *T. alata*, Black-eyed Susan, in this country usually grown in hanging pots on the balcony or climbing up a cane on the window-sill. It frequently lasts for only one summer.

Inflorescence: Susan, whose dark eyes encircled by golden-yellow petals will, in a good season, open from mid-June until mid-September, has a number of close relations of which we know practically nothing. There are, for instance, variants such as 'Alba', white with a dark centre, 'Aurantiaca', orange with a blackish red centre, 'Fryeri', pale yellow with white, 'Lutea', pure yellow.

Care: to grow it yourself, you sow the seeds in March in groups of two or three in small pots containing seed compost. Pot on once, and after the young plants have hardened off, place them out of doors at the end of May. It is easier to buy young plants from the florists'. A sunny position and rich compost, which may contain some lime. The plant should be brought indoors in good time and allowed to overwinter in a good light at a temperature of 8–10°C (46–50°F). Cut back rigorously in February and place in a warmer position to obtain strong, early flowering plants.

Tillandsia

This plant owes its name to Elias Tillands (1640–1693), Professor of Botany at the University of Turku in Finland. Because of their varied forms and their beauty, tillandsia species are undoubtedly the bromeliads most valued by plant lovers, and tillandsia collections are no longer a rarity.

Origin: the genus *Tillandsia* includes about 500 species and is thus among the largest; at the same time it has the widest geographical distribution. It occurs from the southern United States via Central America to deep into South America and grows in tropical rain forests as well as in deserts and subtropical steppes.

Appearance: these varying habitats explain the difference in types: in addition to epiphytic growing tree-tillandsia species there are also terrestrial forms from the steppes in coastal areas and from the rocky plateaux of the Andes. Strange as it may sound, these plants grow from the low-lying regions near the sea to altitudes of up to 4000 m (12 000 ft). The epiphytic species include large and small rosettes, with narrow, broad or grassy foliage, with an erect growing inflorescence in the shape of broad spikes. The barely developed

Tillandsia lindenii, fam. *Bromeliaceae*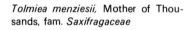

root system is another indication of their epiphytic growth: they are very suitable for epiphyte trunks in a tropical plant window. There are even species which have no roots at all and whose nourishment is absorbed entirely by means of the scales on their stems. A good example of the latter is *T. usneoides*, the Spanish moss; in times of drought it is paper-dry, but in the rainy season it swells and develops into an immensely long plant hanging from trees.

T. lindeniana has leaves to 30 cm (12 in) long, growing in a rosette. The bracts are green, streaked with pink to lilac; the flower is of a purplish colour. The flowering season lasts for several weeks.

Use: epiphytic species can be kept only in a well-lit tropical plant window with a high degree of humidity. Place the smaller forms and the small plants of *T. usneoides* on a treetrunk or drape them on branches. During the growing season they should be sprayed every fortnight with a nitrogen-free foliage fertiliser.

Tolmiea menziesii, Mother of Thousands, fam. *Saxifragaceae*
▽

Tolmiea menziesii — Mother of thousands

The popular names of this plant are an indication of its appearance. In the U.S A. it is called 'piggyback plant'; the Germans say 'hen and chicken', and in Austria and the Netherlands its popular name is 'child-on-mother's-lap'.

All these names are due to the fact that this fresh green *Saxifraga*, only 20–30 cm (8–12 in) tall, forms tiny rooted plantlets at the base of the leaves, which resemble the leaves of the house-lime. As is the case in similar plants, these plantlets drop when they are ready and continue to lead an independent existence. The inflorescence consists of a 15–20 cm (6–8 in) long raceme of inconspicuous greenish brown small flowers with orange tips.

Care: as for *Saxifraga stolonifera*, p. 325.

Tradescantia

This plant was named after two Flemish gardeners: John Tradescant, gardener to Charles I of England, d. 1637, and his son John Tradescant, who started as a gardener and later became a botanist and explorer; he died in 1662. Tradescantia has been known in Europe since that time.

Origin: there are more than 30 species. The winterhardy species, much valued as garden plants, arrived here from North America as early as about 1600. The non-hardy species were discovered 200 years later in tropical areas of South America; the first was *T. crassula*, found in Brazil in 1825.

Appearance: long-lasting, hanging or creeping herbaceous plants, with long, succulent stems divided by nodes. The leaves, also very succulent, are as a rule pointed-oval in shape, with longitudinal veins and at the base surround the leaf-joint. There are green, variegated and even true succulent species. The flowers are terminal growing or appear from axils of the upper leaves and are frequently surrounded by bracts resembling the foliage; the flowering season is late in winter or in spring.

Species: *T. albiflora*, the single, green-leaved species, is in demand everywhere as a pot plant or hanging plant; its name is a little misleading, for the small white flowers appear only rarely. In its hybrid 'Albovittata', the leaves are longitudinally striped with white. *T. blossfeldiana*, growing to only 20 cm (8 in) in height, is a creeping perennial from Argentina; softly haired. The upper surface of the leaves is brownish, the underside red. The flowers are pink with white; profuse. It needs a great deal of warmth. *T. crassula* is erect growing, to 40 cm (16 in); large, glossy fringed leaves, large white flowers; a fine plant. *T. fluminensis* is very similar to *T. albiflora*, but has short-stalked leaves, rather more bluish green above, purplish red underneath. Flowers readily; many small white flowers. A cultivar called 'Albovittata' has white-striped foliage. *T. navicularis* is a slow-growing succulent species with lilac foliage and pink flowers.

Care: throughout the year in a well-lit position. Normal living-room temperature is acceptable. Temperatures below 10–12°C (50–54°F) will cause damage. Tradescantia requires constantly moist soil, but if kept on the dry side, the leaves of 'Albovittata' will have a pink tinge. Feed from March to August. They grow rapidly and it is best to produce new plants from cuttings every year; these will root readily. Drought will frequently lead to an attack by greenfly.

Tradescantia blossfeldiana, fam. *Commelinaceae*

Tradescantia albiflora 'Albo-vittata', fam. *Commelinaceae*

Trichocereus

This genus includes mainly columnar species, but also some growing in the shape of a candelabra or bushily branched. They are interesting for two reasons: they are a valuable subject for collecting, producing magnificent specimens, and in addition they are very suitable for use as stock in grafting.

Origin: there are about 30 trichocereus species growing in the Andes. Those cultivated here originate mainly in northern and western Argentina.

Appearance: variously shaped ribs with strong to very strong spine formation, creamy white to brown in colour. Even young plants produce flowers of enormous size; diurnal flowers are golden-yellow, orange, pink or red; nocturnal ones white and fragrant.

Species: *T. candicans* has grass-green to bluish offshoots; diameter to 15 cm (6 in), height to 80 cm (32 in). Flowers at night. *T. grandiflorus* lives in small groups. Usually does not exceed 20 cm (8 in), but at a height of only 10 cm (4 in) produces large, red diurnal flowers. *T. pachanoi* grows in the Andes at an altitude of up to 3000 m (9000 ft) and is therefore winter hardy. Branched like a tree, up to 6 m (18 ft) in height. Very suitable for stock in grafting. Night-flowering. *T. purpureopilosus* forms small groups; it has 20–25 cm (8–10 in) tall columns, 6 cm (2½ in) across. Creamy white spines, red at the base. Flowers at night with blooms up to 20 cm (8 in) in length and 15 cm (6 in) across; they appear already on plants only 10–15 cm (4–6 in) tall. *T. schickendantzii*, vigorous plant with soft flesh and therefore suitable to be used as stock for grafting, as is *T. spachianus*. Grows to only 25 cm (10 in). Well worth growing for the sake of its annually appearing nocturnal flowers. *T. spachianus* is the best known and most often used graft-base. It can grow up to 2 m (6 ft) in height, glossy green with dangerously sharp spines. Magnificent flowers appearing at night; these are 20 cm (8 in) long and 15 cm (6 in) across.

Care: position in summer very sunny, if possible out of doors; in winter it must be given a cool situation and be kept completely dry. On the whole undemanding plants, satisfied with almost any kind of compost. In order to restrict the too vigorous species to some extent, they can be kept in pots that are too small for them. Small, less vigorous species naturally grow better in good compost and roomy pots.

Watering: freely in summer.

Feeding: regularly in summer, using nitrogen-free cactus fertiliser.

Repotting: when necessary in normal potting soil.

Trichocereus spachianus, fam. Cactaceae

Propagation: simple to grow from cuttings. Decapitated sections of stem will readily develop shoots. A large selection of seed is available from merchants in exotic seeds. Easily cultivated.

Trichocereus candicans, fam. Cactaceae

Tulipa — Tulip

The tulip is not really a true houseplant and in spite of various aids these bulbs are more difficult to raise than hyacinth bulbs. Nevertheless, enthusiasts refuse to give up this pleasure, even though they could buy flowering tulips from any florists'.

Origin: in the course of many centuries the tulip has had an eventful career. It was the emblem of the rulers of ancient Turkey and many a Turkish poet sang of the wild tulip. In the 16th century it was a curiosity in the gardens of the Fuggers family in Augsburg, and became a symbol of the baroque. It was the cause of the Tulip Mania in Holland, which led to speculation on an incredible scale. About 1730 the tulip was again the most popular flower of the Turkish court. It is also worth mentioning that in 1945 the Netherlands despatched a million tulip bulbs to various countries in gratitude for their liberation.

Appearance: it would be impossible to describe all the bulbs separately, but it is interesting to note that some strains, such as 'Keizerskroon', and the various parrot tulips, will shortly have enjoyed the favour of flower lovers for four whole centuries. It would be pointless to mention special strains in this book, since the selection changes from year to year.

Growing: in order to maintain the condition of the bulbs, it is necessary to use pots with drainage holes, filled with a 2–3 cm ($\frac{3}{4}$–$1\frac{1}{4}$ in) layer of fine grit or sand, and light, humusy compost. In September–October a number of bulbs are placed in each pot in such a way that their noses are 4 cm ($1\frac{3}{4}$ in) below the soil. The pots are then buried at a depth of 20 cm (8 in) in a container of damp peat fibre and placed in a cool cellar or on a north-facing balcony. Provided they are protected with a layer of straw, they may also be buried at a depth of 20 cm (8 in) in the garden. Lift the pots in mid-February, place them in good light in a cool room and water them. Only when the flower-bud can be clearly felt outside the bulb should the pots be placed in a warm windowsill, 20°C (70°F), in good light. Water normally. The flowers should appear after four to five weeks. Remove dead flowers; avoid seed formation. When the entire plant has died down, plant the bulbs in the garden.

Tulipa gesneriana, Tulip, fam. *Liliaceae*

Vanda caerulea, fam. *Orchidaceae*

Vanda

This plant has an original Indian orchid name and is considered to be the uncrowned queen of the family: *V. caerulea*, the heavenly blue orchid. It is the most important representative of the genus *Vanda*, which consists of about 50 species, some of which have produced valuable hybrids.

Origin: vanda is a native of India, the Philippines and New Guinea. *V. caerulea* was found in 1849 growing as an epiphyte at an altitude of 1500 m (4500 ft) in the foothills of the eastern Himalayas.

Appearance: the main stem is 60–80 cm (24–32 in) erect growing and closely covered with evergreen leathery, somewhat fleshy short leaves. The flower-stem grows to one side, to a length of up to 45 cm (1½ ft) and bears 7–15 widely spaced pale blue fragrant flowers, 7–10 cm (3–4 in) wide; the lip is dark violet in colour. Flowering season from September to November. Some of the species are very suitable for room cultivation. *V. c.* 'Hennisiana' is a particularly fine cultivar, suitable also for cutting; it is unfortunately rarely cultivated.

Care: in summer it is best treated as a hanging plant in a special orchid basket; semi-shade and little sun; temperature 22°C (72°F). Overwintering at 18–20°C (65–70°F). At night the temperature must not fall below 12°C (54°F). The plant feels most at home in a tropical plant window, since it is very sensitive to fluctuations in temperature and during the growing season requires a very humid atmosphere.

Watering: normally in summer, using tepid, softened water, pH 4–5·0, also for spraying. Keep drier during the resting period.

Feeding: very sparingly during the growing period; lime-free feed.

Repotting: as required; in spring. The roots are very fragile and easily rot where broken. Use a fibre compost containing polypodium and sphagnum moss.

Propagation: by division where possible. Growing from seed is an expert's job.

Diseases: to combat the typical orchid diseases, follow instructions.

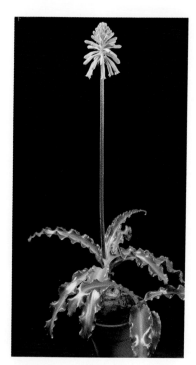

Veltheimia capensis, fam. *Liliaceae*

Veltheimia

The plant owes its botanical name to the 18th century German botanist Veltheim. It is not unlike a small red-hot poker, another member of the lily family.

Origin: of the five species occurring in South Africa, only *V. capensis* is in cultivation in this country. It is also listed by nurserymen under the name of *V. viridifolia*.

Appearance, inflorescence and habit: this beautiful bulbous plant would be an attractive winter-flowering acquisition, if only its colour and blooms would last in a heated living-room instead of requiring a temperature of 10–12°C (50–54°F). Unfortunately a higher temperature spoils the plant and veltheimia is therefore suitable for the unheated greenhouse and not much in demand as a houseplant.

After a dormant period in summer – from May until August it is left somewhere in a corner – the bulb is potted in fresh compost at the beginning of September.

The somewhat crinkly leaves appear fairly soon and grow to 20–30 cm (8–12 in) long and 10–12 cm (4–4¾ in) across. Towards Christmas the arrow-shaped, soft pink or salmon-pink flower raceme appears at the tip of a tall, brown-flecked stem; in a cool position it will last for two to three months. Older specimens with more than one flower stem are particularly beautiful. From March onwards the foliage will wilt. The resting period starts in May. During the growing and flowering seasons the plant should be watered sparingly and rarely be fed. It is very tolerant of dry air. Propagation from offsets.

Vriesea

This plant was named after the Dutch botanist Prof. W. H. de Vriese (1807–1862) and has for several decades been the uncrowned queen of all internationally cultivated bromeliads. Innumerable hybrids have been developed from the various species. A valuable plant.

Vriesea psittacina, fam. *Bromeliaceae*

Origin: they occur in Central and South America, especially in Brazil.
Appearance: medium-sized or large, usually funnel-shaped rosettes, growing singly or in groups. There are two groups:

1. Green-leaved species, as a rule with a fine and striking inflorescence. These include the beautiful small *V. psittacina* and the vriesea hybrid 'Rex'. The largest green-leaved species, rarely seen in show collections, is *V. imperialis*, a native of Brazil, with leaves 1·50 m (4½ ft) in length. It is said that it takes more than twenty years to develop its inflorescence, which is several metres tall.

2. Variegated species, such as the well-known *V. splendens* with its famous strains 'Flaming Sword' and 'Major'. The loosely arranged leaves of the species are about 50 cm (20 in) in length, lightly curved, bright green with striking dark brownish red cross-banding; they end in a sharp point. The most striking variegated species is *V. gigantea* (formerly *V. tessellata*). The leaves are more than half a metre (1½ ft) long, dark green and glossy, with yellowish chequerboard marking on the upper surface, reddish violet beneath. As a houseplant it rarely flowers; the flower-stem attains up to two metres. One of the finest of all

vriesea species is *V. hieroglyphica*. Its leaves, more than 80 cm (32 in) long and up to 12 cm (3¾ in) across, are marked on the upper side with dark green, on the lower surface with almost black, irregular cross-banding against a green background; these markings resemble hieroglyphs. The vigorous leaf rosettes grow to a metre (3 ft) in height. Even small, immature plants show the interesting marking. Like *V. gigantea*, this is a beautiful foliage plant.

Inflorescence: usually the flowers appear on a stem of varying length, either in a flattened spike (*V. splendens* 'Flaming Sword') or in branched humidity and soil and air temperatures which even at night do not drop below 18–20°C (65–70°F). Only in these conditions will the colourful bracts last as long as the parent plant which, in favourable circumstances, will develop offsets. Vriesea forms which rarely or never flower have an entirely different life rhythm. If no inflorescence develops in the centre (which would inevitably lead to their demise), they may be regarded as long-lasting foliage plants. This applies also to *V. imperialis* which flowers only after twenty years.

Among the vriesea hybrids now available there are a number of strains the water from the vase, leaving a little only if the plant stands in a warm room. Adjust the humidity in accordance with general rules.

Feeding: as for *Aechmea* (p. 118).

Repotting: not applicable to the flowering cultivars. When necessary, in special bromeliad compost with sphagnum moss.

Propagation: from offsets or from seed. The latter requires expertise and will be successful only in a heated greenhouse.

Diseases: pests hardly ever appear. Plants are more likely to be damaged by rotting or by the sun.

Vriesea splendens 'Flaming Sword', fam. *Bromeliaceae*

Vriesea, hybrid 'Rex', fam. *Bromeliaceae*

racemes. As in the case of *Aechmea*, the colourful bracts remain on the plant for a long time. The yellow, red/yellow, greenish, sometimes white, flowers do not last very long. Even the inherently readily-flowering vriesea cultivars in cultivation need an average of four years before they flower; hence their high price.

Use: they are best suited to the heated flower window. Vriesea plants with tender, green or colourfully marked foliage in particular require a shady situation with a high degree of which with a little extra care will survive on the window-sill. Their reduced sensitivity to living-room atmosphere depends on the foliage: as a general rule, all strains with more grey-green, plain foliage are better suited for room cultivation.

Watering and spraying: the same as for *Aechmea* (p. 118), *Guzmania* (p. 221) and *Neoregalia* (p. 267), although an even higher degree of humidity is required. Water only with lukewarm, tepid water (pH 4–4.5). During the growing season pour the water into the vase and keep the compost just moist. After flowering, when the resting season starts, pour

Yucca

Yucca aloifolia, fam. *Agavaceae*

Y. aloifolia, whose foliage resembles that of the aloe, is considered to be the finest of about 40 species; the illustration convincingly shows its exotic and handsome form. In this instance 'exotic' is not synonymous with demanding. Few plants are more modest and at the same time more readily-flowering than the yucca.

While the winter-hardy, sessile forms are more suitable for use in the garden, the non-winter hardy, trunk-forming types are to be preferred as house plants and tub-plants. There are two magnificent variegated strains of *Y. aloifolia*: 'Quadricolor' and 'Tricolor'. Another popular indoor form is *Y. elephantipes*, also grown in hydroculture. All yucca species are natives of the southern states of North America and of Central America.

Situation in summer: preferably in the lightest and warmest possible position out of doors, where it will flower in July. In winter it definitely requires a cool position at about 6°C (43°F) in a cool room.

Watering: generously in summer, but ensure good drainage; keep dry during the dormant season. Dry air is tolerated.
Feeding: weekly from April to August.
Propagation: from seed and from offsets.
Diseases: hardly ever occur.

Zantedeschia aethiopica
– Calla, Arum lily

Francesco Zantedeschi was an Italian who lived from 1773–1846. His fame both as a botanist and as a chemist was such that the calla was named after him; *Z. aethiopica* and *Z. elliottiana* previously bore the botanical names *Calla aethiopica* and *C. elliottiana* respectively.

Zantedeschia aethiopica, Calla Lily, fam. *Araceae*

Origin: it originates in South Africa where it occurs in marshes which dry out in summer. This knowledge is useful in determining its correct treatment.
Appearance: it grows to 80 cm (32 in) in height and after flowering dies down partially or entirely. The broad arrow-shaped succulent green leaves grow on perfectly straight stems from strong, fleshy tubers. The flower-stem appears at the end of winter and projects from the leaf stems. The inflorescence consists of the typical arum lily spadix, in indoor forms surrounded by a large white spathe. Some strains are suitable for cutting; there are also garden species with yellow or red flowers.
Care: to achieve regularly recurring flowering, it is essential that the resting period from May to July is strictly adhered to. In accordance with its natural environment in marshland drying out in summer, the calla should be kept so dry during that period that nearly all the foliage drops. You would do best to put the plant out of sight after flowering and forget it for a time. When it starts to sprout it should be placed on a light to sunny balcony or plunged, pot and all, in the garden.

In October the plant is brought indoors once more and is kept cool, 8–10°C (46–50°F) until December; 12–15°C (54–58°F) will be tolerated without affecting the bud formation. Keep warm from December until the end of the flowering period.
Watering: in summer, after the dormant season, in particular water freely; in winter reduce the water supply. During the growing season spray frequently.
Feeding: weekly during the growing and flowering season.
Repotting: in October, when the plant is brought indoors. Plant at the same level as before.
Propagation: by division of the rootstock.
Diseases: drought in winter and lack of fresh air will encourage greenfly.

Zebrina

If you like growing tradescantia in your living-room, you will also want to own the closely related zebrina. The few species belonging to this genus all originate in Mexico and South America and in room cultivation require a great deal of light and warmth. The magnificent *Z. pendula* is the best known species.

This plant can be used not only as a hanging plant, but will also supply fine ground cover in a plant window. When grown as a hanging plant, the stems, with their colourfully marked leaves, are up to half a metre (1½ ft) in length. The finest strain is 'Quadricolor', which has rose-red/white/green striped leaves.

Care: general care as for *Tradescantia* but water with lime-free water. The compost should be slightly lighter and more sandy. Feed very sparingly. In winter keep indoors, not below 15°C (58°F).

Zebrina pendula 'Quadricolor', fam. *Commelinaceae*

Zygocactus – Christmas cactus

Generations of plant lovers have known this plant by its former botanical name *Epiphyllum*. If you find it difficult to remember the new name, just continue to call it Christmas cactus and remember that in the field of jointed and leaf cacti there is a well-nigh Babylonian confusion of tongues.

The native country of the zygocactus is Brazil, where it grows as an epiphyte in the tropical rain forest. Only hybrids are now in cultivation; they grow best in compost rich in humus and mixed with sphagnum. They require a shady situation, constantly humid and warm air and moist compost, especially in winter. To encourage bud formation, the plants should be placed in a cooler position in July and August; water and feed should be withheld at this time. The leaves formed in early summer are thus allowed to ripen and buds are formed. In autumn provide a moderate temperature, 15°C (58°F) and water very sparingly, but until the buds appear spray regularly and gradually increase the water supply. Until the plant has ceased flowering it should not be moved or turned (make a mark). After flowering the plant loses its gloss; this indicates the first rest period, which lasts for two months.

Zygocactus truncatus, Christmas cactus, fam. *Cactaceae*

5. FOR QUICK REFERENCE

Technical terms

Although we have used as little technical jargon as possible in this book, it nevertheless contains some terms which may not be immediately clear to the layman. The following short glossary should be helpful. * Indicates that the word so marked is defined under a separate entry in the list.

Acidity. Also called soil reaction. Measured by ascertaining the concentration of hydrogen ions and expressed by the values on the pH scale*. See also p. 57. The appropriate acidity varies from plant to plant. Deviations can cause great damage.

Assimilation. In the widest sense of the word this means the change of inorganic matter into organic matter. This process takes place in the leaves of plants under the influence of light. The chief elements produced are carbohydrates, vegetable fats and protein. The term includes photosynthesis*.

Bromeliad compost. Most bromeliads require a light and acid compost consisting of chopped fern roots, coarse peat and sphagnum moss. Ordinary potting compost or rotted cow manure is added to provide nourishment.

Callus. Tissue formed by the plant to cover wounds. Roots may emerge from a callus; this phenomenon is seen when cuttings are taken from woody plants.

Chlorophyll. Plays a part in photosynthesis*, acting as a kind of catalyst. See also under variegation*.

Chlorosis. The leaves fail to turn green or become yellow for lack of chlorophyll. The most likely causes are excess lime in the compost when the plant is not repotted in time, too much fertiliser, or incorrect watering. The plant is unable to absorb the iron present in the compost. Occurs especially in acid-loving plants watered with hard water. Remedy: use softer water, repot the plant, water with iron sequestrols.

Comb formation. See cristate*.

Crest formation. See cristate*.

Cristate. Crest or cockscomb forming. This is a deviation occurring especially in cacti; it is caused by disturbance of the growing point. The plant takes on bizarre and monstrous shapes. The scientific term is fasciation.

Demineralisator. Plastic tube filled with polystyrene granules. Water passed through this apparatus is purified since the granules retain all polluted elements.

Distilled water. Chemically pure water, with a natural pH of 7·0. Formerly this was obtained by means of distillation; the water was made to evaporate. Can now easily be produced with the aid of a demineralisator*.

Epiphytes. Plants living on trees, rooted in the remains of rotted foliage and wood. Occur mostly in tropical rain forests. Not to be confused with parasitic plants, such as the mistletoe, whose roots penetrate the wood of the host tree.

Fasciation. See cristate*.

F₁ hybrid. Product of first crossing of two species. The F₁ hybrid itself does not produce seed which comes true and crossing must therefore constantly be repeated. See also under hybrid*.

Heel. Spur attached to a cutting obtained by not detaching the cutting cleanly, but by tearing off the end section. You will thus incorporate a piece of bark, the heel, which will help rooting.

Hybrid. New plant grown from seed via (artificial) pollination, having properties different from those of the parent plants. See also F₁ hybrid*.

Internode. Section of stem between two nodes*

Leafmould. This consists of rotted foliage and is made by piling leaves, especially beech leaves in a heap. After a year or so it is sieved and used in composts.

Long day plants. Plants which will flower only if exposed to daylight longer than normal. With the aid of artificial daylight lamps these plants can be brought into flower earlier and their flowering may be prolonged (begonia, African violet).

Mutation. Spontaneous change in hereditary characteristics. Cause obscure.

Node. The joint on a stem from which leaves usually develop. Also the part from just below which roots may emerge when taking cuttings. In tradescantia, etc., the nodes are particularly obvious.

Osmunda. The rotted black roots of the Royal fern, *Osmunda regalis*, are used in special orchid compost.

Peat fibre. The second layer of peat moors, consisting of decomposed bog-mosses (sphagnum)*. Use in composts.

Perlite. An alumino-silicate of volcanic origin, which is expanded by heating, supplied in 2–3 mm ($\frac{1}{8}$ in) lightweight granules. Chemically inert. Mix with compost to improve porosity, especially in cactus cultivation.

Photosynthesis. Part of the assimilation process*, in which plants convert carbon dioxide absorbed from the air and nourishment absorbed by the roots into sugar. Oxygen is a by-product of photosynthesis.

Polypodium. Reddish brown roots and rootstock of the oak fern *Polypodium vulgare*. Use in composts for orchids, etc.

Polystyrene granules. Round, white very lightweight granules used to improve drainage of certain composts; sometimes replaces peat.

Sand. See sharp sand.

Sharp sand. A fairly coarse, lime-free quartz or river sand. Use in composts to improve porosity. If pure sand is not available it is possible to use charcoal, polystyrene granules* or perlite* instead.

Short day plants. Plants which will develop flowerbuds only if they are exposed to daylight for fewer hours than usual (generally about ten). By giving extra darkness these plants can be made to flower earlier than they would do normally. Example: chrysanthemum, *Euphorbia pulcherrima* (poinsettia), kalanchoë.

Soil reaction. See acidity*.

Spadix. Fleshy flower spike.

Spathe. Bract surrounding the spadix in arum lily plants (*Araceae*).

Stopping. Removing the tips of shoots of young plants by pinching out or cutting, in order to encourage the formation of side growth.

Turf loam. It is best to use clay turves. Cut the turves and stack them, turf side down, until they have broken down, usually about a year later. After sieving use in composts.

Variegation. In white or yellow variegated leaves chlorophyll is, to a greater or lesser degree, missing in some patches. Usually hereditary and perpetuated by crossing in order to improve the 'ornamental value'. Variegated plants require more light than all-green forms. Most variegated plants can be propagated only by the vegetative method; sexual propagation as a rule produces all-green plants.

Xerophytes. Plants able to live in very dry regions. They have adapted themselves to their environment in such a way that transpiration is practically non-existent.

Light requirements and winter temperature

The healthy growth and flowering of most houseplants is closely linked to the correct conditions of overwintering. The most important factors are the degree of light and the temperature of the environment. Below, a number of houseplants have been classified according to the amount of light they require; in each case the most favourable winter temperature is mentioned as well.

In the autumn this information will assist you in collecting together plants with similar requirements. To avoid plants being damaged during dormancy, you should pay particular attention to the minimum temperature. Maximum temperature is much less critical. However, a number of plants, especially cacti and succulents, depend for their subsequent flowering on a low temperature in winter.

Plants requiring little light. They will survive even at distances of 2–3 m (6–9 ft) from a window.
Aspidistra, cool or warm
Chlorophytum, cool or warm
Cyrtomium, 10–12°C (50–54°F)
Howea, 14–16°C (57–60°F)
Monstera, not below 12°C (54°F)
Philodendron, not below 12°C (54°F)
Pilea, warm
Rhaphidophora, not below 12°C (54°F)
Scindapsus, warm
Tradescantia, all-green, cool or warm

Plants suitable for a north-facing window. Tolerate a great deal of daylight, but no direct sunlight.
Cyclamen, 10–12°C (50–54°F) until in full flower
Dracaena fragrans 'Rothiana', cool or warm
Erica, 10–12°C (50–54°F)
Nephrolepis, 12–16°C (54–60–F)
Polystichum, 6–8°C (43–46°F)
Primula, 10–12°C (50–54°F)
Pteris, 10–12°C (50–54°F)
Rhododendron, 10–12°C (50–54°F) until in full flower
Saintpaulia, 16–20°C (60–70°F)

Plants for light shade. Overwintering near a north-east or north-west facing window receiving little sun.
Adiantum, 12–16°C (54–60°F)
Anthurium, 18–21°C (65–71°F)
Araucaria, 8–12°C (46–54°F)
Begonia rex, 14–18°C (57–65°F)
Begonia, shrubby forms, 12–14°C (54–57°F)
Calceolaria, 10–12°C (50–54°F)
Chamaedorea, not below 12°C (54°F)
Cordyline, 10–12°C (50–54°F)
Dieffenbachia, 15–18°C (58–65°F)
× *Fatshedera,* 8–12°C (46–54°F)
Hibiscus, 12–14°C (54–57°F)
Maranta, not below 18°C (65°F)
Pandanus, 16–18°C (60–65°F)
Pelargonium grandiflorum, 8–12°C (46–54°F)
Peperomia, 16–20°C (60–70°F)
Phoenix roebelinii, not below 15°C (58°F)
Rhoicissus, 8–12°C (46–54°F)
Streptocarpus, 10–12°C (50–54°F)

Plants for moderately good light. Window facing east, south-east, west or south-west. Direct sunlight is not tolerated between 11 and 3 o'clock approximately.
Aechmea rhodocyanea, not below 12–15°C (54–58°F)
Aloë variegata, this is the only species tolerating warmth
Asparagus setaceus, 8–10°C (46–50°F)
Asparagus densiflorus, preferably 12–15°C (54–58°F)
Beloperone, 12–16°C (54–60°F)
Billbergia nutans, preferably 10–16°C (50–60°F)
Cacti, between 6 and 12°C (43 and 54°F), depending on the species
Callistemon, 6–8°C (43–46°F)
Ceropegia, 12–16°C (54–60°F)
Cissus antarctica, cool or warm
Clivia, 15–18°C (58–65°F)
Cyperus, cool or warm
Euphorbia milii, from 10°C (50°F) to warm
Euphorbia pulcherrima, 15–18°C (58–65°F)
Haemanthus albiflos, 15–18°C (58–65°F)
Hedera, not below 8–12°C (46–54°F)
Hoya carnosa, 12–15°C (54–58°F)
Microcoelum weddellianum, not below 15°C (58°F)
Myrtus, 6–12°C (43–54°F)
Plectranthus, 8–15°C (46–58°F)
Plumbago, 10–15°C (50–58°F)
Rhipsalidopsis, dormancy at about 10°C (50°F)
Rochea, 12–15°C (54–58°F)
Saxifraga stolonifera, not below 16°C (60°F)
Sparmannia, 8–12°C (46–54°F)
Stapelia, 12–15°C (54–58°F)
Stephanotis floribunda, 12–15°C (54–58°F)
Succulents, usually between 6 and 12°C (43 and 54°F), depending on species
Tradescantia, variegated species, cool or warm
Vallota, 15–18°C (58–65°F)
Vriesea, not below 12–15°C (54–58°F)
Zygocactus, 10–15°C (50–58°F)

Plants requiring plenty of light. Window facing south-east or south-west, or even south, provided the plants can be protected from strong

sunlight.
Abutilon, 10–12°C (50–54°F)
Amaryllis, not below 12–15°C (54–58°F)
Impatiens wallerana, not below 12–14°C (54–57°F)
Kalanchoë (Bryophyllum), 10–12°C (50–54°F)
Passiflora, 6–8°C (43–46°F)
Sansevieria, not below 15°C (58°F)

Borderline cases. A number of plants like a great deal of light in winter, while tolerating little or no sun.
Bougainvillea, 6–8°C (43–46°F)
Cordyline terminalis, not below 18°C (65°F)
Dracaena draco, not below 10°C (50°F)
Ficus species, not below 12–14°C (54–57°F)
Narcissus, cool until flowering
Rosa, 6–8°C (43–46°F)

Plants liking a light and cool situation in winter. Temperatures between 4 and 6°C (40 and 43°F).
Agave
Citrus
Hydrangea

Winter-flowering plants.
Abutilon species; not all
Aechmea rhodocyanea
Amaryllis
Ardisia
Begonia species; not all
Calceolaria
Camellia
Convallaria
Erica, the non-winter-hardy species
Euphorbia pulcherrima
Hyacinthus
Kalanchoë
Lobivia species; not all
Narcissus
Orchid species; some
Plectranthus species; some

Primula malacoides
Primula obconica
Primula praenitens
Sauromatum venosum
Sedum bellum
Tulipa
Vriesea species
Zantedeschia
Zygocactus

Plants which can overwinter in the cellar. Adequate ventilation is important. Temperatures between 4 and 6°C (40 and 43°F).
Fuchsia
Grevillea robusta
Nerium oleander
Phoenix canariensis
Punica granatum
Yucca gloriosa

Plants for the plant window

As you will have read on p. 30 and following, a plant window is a special facility for houseplants, varying between a simple, enlarged window-sill with a plant tray, and a glass case partitioned off from the room, in which conditions are automatically controlled. Various examples may be seen in the photographs.

The plants to be grown in such a plant window must be selected with care, since it is essential that their requirements should be similar. Only in this way can such an arrangement be maintained in good condition for many years. To introduce some colour, you might enliven the collection with flowering plants of a more or less temporary nature. By placing these in front of or behind other plants you can give them more or less light as required.
When installing such a window you should further keep in mind that watering should be easy. By the correct choice of pots (plastic or clay)
and by placing some plants at a higher level, you will be able to organise matters in such a way that each plant gets its due.
Simple arrangement. Suitable for a shady position, without glass partition on the room side, and without heating.
Climbers may be grown on the walls; green-leaved and variegated ivy species are particularly suitable, as well as various philodendrons.
Hanging pots may contain: *Chlorophytum, Scindapsus,* trailing *Philodendron,* green-leaved *Tradescantia.*
Taller plants to the right and left:
Aspidistra, Philodendron, Begonia. Various ferns, for instance *Nephrolepis,* might be placed at a slightly higher level to do them justice.
The bottom of the plant window might be covered with *Peperomia* species and *Pilea.* If you have a peat-filled container, you could also use the green *Tradescantia* which will spread rapidly.
Flowering plants may be added throughout the year, for instance *Cyclamen,* slipper plant, summer- and winter-flowering begonias, *Primula, Nertera* and bulbous plants in flower.

Larger plant windows. Light shade,

with or without glass partition on the inside and without heating.
Against the wall: *Cissus antarctica, C. rhombifolia, C. striata;* ivy and *Philodendron* might also be considered. If the temperature is adequate, *Ficus repens,* which can at the same time serve as ground cover.
For hanging pots: *Asparagus densiflorus, Ceropegia, Chlorophytum, Scindapsus, Saxifraga stolonifera,* variegated *Tradescantia,* and also *Platycerium.* To the left and right you might place specimen plants, for instance *Dracaena* species, *Sansevieria, Cyperus* species (remember they require plenty of water); also *Billbergia nutans, Haemanthus albiflos,* some of the taller *Begonia* forms. If the ventilation is adequate, *Ficus* species as well. At the bottom you might accommodate the same plants as mentioned for a 'simple arrangement' provided they can be shaded from very strong sunlight.
Simple bromeliad species, such as *Aechmea rhodocyanea,* as well as a number of hardy orchids and of course ferns, may be attached to an epiphyte tree placed in such a window (see also 'The plant window in practice', from p. 85 onwards). Flowering plants which must be replaced from time to time: *Senecio, Sinningia, Kalanchoë, Streptocarpus, Saintpaulia,* as well as *Abutilon* and *Beloperone.*

Enclosed plant windows, with inside glass partition, thermostatically controlled heating, humidifier with hygrostat, ventilation and screening facilities. Average temperature in summer, when there is adequate ventilation: 22 to 25°C (72 to 78°F) or more; night temperature not below 15 to 18°C (58 to 65°F). The average winter temperature should be adjusted in accordance with the requirements of the various plants. Provided the daytime temperature is adequate and moderate humidity can be provided, plants from the tropical rain forest can survive at a minimum night temperature of 12–14°C (54–57°F) in winter. In cases of doubt, refer to the instructions given for individual plants and make notes.
To check conditions you will require: a soil thermometer, an air thermometer, a hygrometer and possibly a third thermometer placed close to the outside window.
The choice of plants should be carefully adapted to the light conditions. An enormous variety of plants can be grown in such a window, provided, of course, they are available. In addition to the following there are numerous other possibilities, to mention only the enormous assortment of cacti and orchids.
Climbers: tropical *Cissus* species, *Clerodendron, Stephanotis floribunda, Ficus pumila,* various species of *Philodendron.*
As hanging plants in special pots or orchid baskets: *Columnea* species, *Epiphyllum, Tillandsia usneoides, Platycerium,* epiphytic growing orchids.
To the epiphyte tree you might attach: bromeliad species such as *Aechmea, Cryptanthus, Vriesea,* tropical ferns and orchids. *Ficus pumila* may be grown up the trunk; *Scindapsus* is also very suitable for this purpose.
Erect growing plants: *Aglaonema, Anthurium, Asplenium, Cordyline, Codiaeum, Dieffenbachia, Dracaena, Ficus, Philodendron,* all these in various forms. Typical terrestrial growing bromeliads and orchids are best confined to the bottom of the window. For ground cover you might consider: *Adiantum,* tropical creeping ferns such as *Pellaea, Fittonia, Maranta, Peperomia, Selaginella, Cryptanthus.*

Illustrations

Nearly all the colour photographs in this book were taken by Rob Herwig specially for this purpose, in the Netherlands, Germany, Switzerland and Scandinavia. They were then selected in collaboration with Margot Schubert. In addition a number of photographs were provided by the following:

Colour photographs: W. Blaicher, Eppelheim, 303; E. Drave, Munich, 173; H. Eisenbeiss, Munich, 187; Florabild, Stuttgart, 223; bottom, 294; D. Hinrichs, Munich, 125; Hortus-Verlag, Bad Godesberg, 205, 224, 243, 322, 331 top, 340; Dr. H. Jesse, Cologne, 177; O. Klees, Eppelheim, 233; K. Krieger, Herdecke, 27, 28, 35; G. Lindner, Reinbek, 25, 26; Prof. Dr. W. Rauh, Heidelberg, 157, 249 top, 265, 337; W. Schacht, Frasdorf, 167; H. Smith, Westcliff-on-Sea, 126 bottom.

Drawings: W. Eckhardt, Munich, 50; H. Hoffmann, Germering, 53, 57; all others by G. D. Burckart, Heidelberg.

6. INDEX

Bold page numbers after generic names refer to the plant descriptions in the section 'Houseplants in Colour'.

A
Abutilon 55, 107, **114**, 229
  megapotamicum 114, 230
    'Aureum' 114
  striatum 114
    'Souvenir de Bonn' 114
    'Thompsonii' 114
Abutilon hybrids 114
Acalypha 105, **115**
  hispida 115
  wilkesiana hybrids 115
Acanthaceae 100
Acanthus 100
Acanthus family 100
Achimenes 106, **115**
Achimenes hybrid 'Rose' 115
Achimenes hybrids 115
Acidity 355
Acid loving plants 61
Adaptation, powers of 12
Adenium obesum **116**
Adiantaceae 100
Adiantum 100, **116**
  capillus-veneris 84
  cuneatum 116
  pedatum 84
  raddianum 84, 116
    'Brilliant Else' 116
    'Decorum' 116
    'Fritz Luethli' 116
Aechmea 103, **117**
  chantinii 118, 268
  fulgens 118
  miniata 118
  rhodocyanea 117
Aeschynanthus 106, **118**
  speciosus 118
African violet 23, 71, 323
Agavaceae 100
Agave 100, **119**
  americana 100, 119

  filifera 119
  parrasana 119
  schidigera 119
  victoriae-reginae 119
Agave family 100
Aglaonema 101, **120**
  commutatum 84, 120
    'Pseudobracteatum' 120
  costatum 84, 120
  modestum 120
  pictum 120
Air 49
Air layering 75
Aizoaceae 22, 101
Allamanda 101, **121**
  cathartica 121
    'Grandiflora' 121
    'Nobilis' 121
    'Schottii' 121
Alloplectus vitatus 24
Aloë 107, **122**
  arborescens 122
  variegata 122
Amaranthus 244
Amaranthaceae 101
Amaryllidaceae 100, 101
Amaryllis 46, 77, 101, **231**
  belladonna 231, 232
Amaryllis hybrids 232
Ampelopsis 112, **122**
  brevipedunculata 122
    'Elegans' 122
Ananas 103, **123**
  comosus 123
    'Variegatus' 123
Anthurium 46, 101, **124**
  andreanum 124
  hybrids 124
  crystallinum 124
  scandens 124
    'Violaceum' 124

  scherzerianum 84, 124
    hybrids 124, 125
Anthurium mixture 124
Apocynaceae 101
Aphelandra 100, **126**
  squarrosa louisae 'Fritz Prinsler' 126
Aphelandra hybrid 'Dania' 126
Aporocactus 55, 103, **126**
  flagelliformis 126
Araceae 101
Araliaceae 102
Araucaria 55, 102, **127**
  excelsa 127
  heterophylla 102, 127
Araucariaceae 102
Ardisia 108, **128**
  crenata
Areca **128**
  tiandra 128
Argyroderma 22, 101, **129**
  festiculare 128
Artificial light 43, 47, 48, 54
  or daylight 54
Artificial substance 62
Artillery plant 299, 300
Arum venosum 171, 325
Arum family 101
Asclepiadaceae 102
Asclepias 102
  curassavica 236
Asparagus 106, **129**
  asparagoides 129
  crispus 130
  densiflorus 129
  falcatus 129
  plumosus 130
  setaceus 130
  sprengeri 129
Asparagus fern 129
Aspidiaceae 102

Aspidistra 17, 106, **130**
   elatior 130
Aspleniaceae 102
Asplenium 102, **131**
   adiantum-nigrum 84
   nidus **131**
   obovatum 84
   ruta-muraria 84, **131**
   trichomanes 84
Assimilation 49, 353
Astilbe 111
Astrophytum 103, **131**
   asterias 131, 206
   myriostigma 131
   ornatum 131
Azalea 51, 52, 55, 57, 317, 318
Azalea indica 317
Azalea pot 63

B
Bacteroids 128
Balcony plants 56
Balsaminaceae 102
Balsam family 102
Basalt splinters 45
Bead plant 272
Begonia 103, **132**
   acutifolia 133
   albo-picta 134
   'Cleopatra' 135
   corallina hybrid 'Lucerna' 133
         'Madame Charat' 133
         hybrids 133
   credneri 134
   crispula 137
   diadema 136
   eliator hybrids 133
   foliosa 134
   fuchsioides 136
   'Gloire de Lorraine' 132, 133
   hispida cucullifera 137
   hydrocotylefolia 137
   imperialis 137
      hybrid 'Hildegard Epple' 137
      'Marbachtaler' 137
   limmingheiana 134
   Lorraine hybrid 'Gloire de Lorraine' 132, 133
   masculata 136
   manicata 137
      'Aureomaculata' 137
      'Crispa' 137
   masoniana 137
   metallica 134, 136
   rex 136
      hybrids 132, 136
      leaf cuttings 72
   scharffii 136
   semperflorens hybrids 133
   serratipetala 136
Begonia family 102

Begonias, flowering 133
Begonias shrubby 133
Begoniaceae 102
Bell flower 149
Bell flower family 104
Beloperone 80, 100, **138**
   guttata 138, 279
Bertolonia 107, **138**, 336
   maculata 138
   marmorata 138
Bignoniaceae 103
Bignonia family 103
Billbergia 55, 103, **139**
   nutans 139
   hybrid 139
Bird's-nest fern 130
Bishop's-cap 131
Black-eyed Susan 342
Black pepper 301
Blechnaceae 103
Blechnum 103, **140**
   gibbum 29, 140
   spicant 84
Blood, dried 62
Bonemeal 62
Bottle gardens 26, 81, 82, 84
Bougainvillea 40, 108, **141**, 301
   buttiana 'Mrs. Butt' 141
   glabra 'Alexandra' 141
   spectabilis 141
Bowiea 29, 106, **142**
   volubilis 142
Brasilicactus 272
   haselbergii 273
Brassia 108, **142**, 262
   maculata 142
   verrucosa 142
Bromeliad 75, 117, 139, 180, 221, 267, 319, 342, 348
Bromeliad soil 353
Bromeliaceae 103
Bromeliad family 103
Broom 186
Browallia 112, **143**
   grandiflora 143
   speciosa 143
   viscosa 143
Brunfelsia 112, **143**
   calycina 143
   pauciflora var. *calycina* 143
      'Floribunda' 143
Bryophyllum 105, 247, 248, 249
   calycinum 249
   crenatum 249
   daigremontianum 249
   tubiflorum 249
Busy Lizzie 243
Buxus 242

C
Cactaceae 103
Cacti 22, 52

Cactus family 103
Cactus spurge 206
Caladium 101, **144**
   bicolor 144
      hybrids 144
   humboldtii 144
   picturatum 144
   schomburgkii 144
      hybrids 144
Calanthe 108, **145**
   vestita 145
      hybrid 'William Murray' 145
Calathea 107, **145**, 181, 257
   bachemiana 145
   insignis 145
   lancifolia 145
   makoyana 145
Calceolaria 111, **146**
Calceolaria hybrids 146
Calcium poisoning 56
Calla 350
Calla aethiopica 350
   elliottiana 350
Callisia 104, **146**, 333
   elegans 146
   repens 146
Callistemon 108, **147**, 225
   citrinus 147
   lanceolatus 147
Callus 71, 353
Camellia 51, 52, 56, 148
Camellia 57, 112, **148**, 217
   japonica 148, 165
      'Chandleri Elegans' 148
   sinensis 148, 165
Campanula 104, **149**
   fragilis 104, 149
   isophylla 104, 149
      'Alba' 149
      'Mayi' 149
   pyramidalis 149
Campanulaceae 104
Campsis 103
Canna 104, **150**
   indica 150
      hybrid 'Alberich' 150
      'Lucifer' 150
      'Perkeo' 150
      'Puck' 150
Canna family 104
Cannaceae 104
Cape jasmine 217
Cape primrose 340
Capsicum 112, **150**
   annuum 150, 301, 335
      'Gnom' 150
   frutescens 301
Carbonate hardness 58
Carbon dioxide 49
Carex 105, **151**
   brunnea 151
      'Variegata' 151

Carex elegantissima 151
Carissa 101, **152**
  spectabilis 152
Carrion flower 338
Catalpa 103
Catharanthus 101, **152**
  roseus 152
Cattleya 108, **153**, 200
  trianae 153
Celastraceae 104
Celosia 244
Cephalium 259
Cephalocereus 103, **154**
  senilis 154
    'Aureus' 154
    'Tetetzo' 154
Cereus 103, **155**, 157
  chalybaeus 155
  jamacaru 155
  peruvianus 156
    'Monstrosus' 155, 156
Ceropegia 102, **156**
  distincta ssp. *haygarthii* 156
  fusca 156
  linearis 156
    ssp. *woodii* 156
  sandersonii 156
  woodii 156
Chamaecereus 103, **157**
  silvestrii 157
Chamaedorea 109, **158**
  elegans 40, 158
    hybrid 'Bella' 158
  tenella 158
Chamaerops 109
  humilis 19, 109
Chenille plant 115
Chinese evergreen 120
Chinese rose 229, 230
Chlorophyll 353
Chlorophytum 17, 56, 90, 106, **159**
  comosum 159
    'Variegatum' 159
Chlorosis 353
Christmas cactus 52, 55, 352
Christmas cherry 336
Chrysanthemum 104, **160**
  indicum 160
    hybrids 160
Chrysanthemum, garden 160
Cinchona 111
Cineraria 332
Cineraria hybrida 332
Cissus 71, 74, 112, 122, **160**, 321
  antarctica 160, 161, 162
  bainesii 161
  crameriana 162
  discolor 161
  gongylodes 161
  juttae 161
  rhombifolia 161
    hybrid 'Mandaiana' 161
  striata 90, 161
    hybrid 162
  quadrangularis 161
Cissus species for plant- and tropical-windows 161
  for room cultivation 160
  succulent 161
Citrus 111, **163**
  aurantium myrtifolia 163
  limon 163
  microcarpa 163
  mitis 'Calamondin' 163
  sinensis 163
  taitensis 'Otaheite' 163
Clarkia 108
Clay granules 92
Clay turf loam 61
Cleistocactus 103, **164**
  albisetus 164
  smaragdiflorus 164
  strausii 164
    'Fricii' 164
  wendlandiorum 164
Clerodendrum 112, **164**
  fragrans 164
  philippinum 164
  splendens 164
  thomsoniae 164
Cleyera 112, **165**
  japonica 165
    'Tricolor' 165
Climbing plants 25
Clivia 40, 50, 52, 75, 101, **166**
  miniata 23, 166, 167
  nobilis 166
Cockscomb family 101
Coconut palm 29, 260
Codiaeum 105, **168**
  variegatum pictum 39, 168
Coelogyne 108, **169**, 200
  cristata 169
  dayana 169
  fimbriata 169
  massangeana 169
Coffea 111, **170**
  arabica 170
    'Nana' 170
Coffee plant 170
Colchicine 171
Colchicum 106, **171**
  autumnale 171
    hybrid 'Autumn Queen' 171
    'Lilac Wonder' 171
    'Princess Astrid' 171
    'The Giant' 171
    'Waterlily' 171
Cold, damage from 99
Coleus 70, 106, **171**
  blumei hybrids 171, 172
  fredericii 172
  pumilus 172
  rehneltianus 172
  thyrsoideus 172
Collecting 20
Columnar cactus 17, 55, 74, 155
Columnea 106, 118, **173**
  gloriosa 173
    hybrid 'Purpurea' 173
  hirta 173
  kewensis 173
    'Stavanger' 173
  microphylla 173
  schiedeana 173
Columnea hybrids 173
Comb formation 354
Commelinaceae 104
Compositae 104
Compost 60, 61, 79
  dried out 79
Conophytum 101, **174**
  ficiforme 174
Contract plants 36
Convallaria 106, **174**
  majalis 175
Cordyline 75, 100, **175**, 193, 195
  australis 176
  congesta 176
  stricta 176
  terminalis 176
    'Tricolor' 176
Cornaceae 104
Cornus florida 104
Coryphantha 103, **176**
  cornifera 176
  elephantidens 176
Corytholoma cardinale 314
Crassula 104, **177**
  arborescens, 14, 177, 178
  argentea 177
  falcata 178
  lycopodioides 178
  obliqua 177
  portulacea 177
  rupestris 178
  socialis 178
  teres 178
Crassulaceae 105
Creeping moss 330
Crest formation 354
Cristate 254, 353
Crocus 106, **179**
Crossandra 100, **179**
  infundibuliformis 179, 180
    'Mona Wallhed' 179
Croton 39, 168
Crown of thorns 205
Cryptanthus 103, **180**
  beuckeri 180
  bivittatus 180
  bromelioides 180
    'Tricolor' 180
  fosterianus 180
  zonatus 180
Ctenanthe 107, **181**, 257

oppenheimiana 181
  'Variegata' 181
Cyclamen 51, 56, 110, **182**
  persicum 182
  'Thusnelda' 14
Cymbidium 108, **183**
  giganteum 183
  hookerianum 183
Cymbidium mini-hybrid 'Galiodin' 183
Cymbidium mini-hybrids 183
Cyperaceae 105
Cyperus 105, 151, **183**
  alternifolius 183, 184, 185
    'Variegatus' 184
  argenteostriatus 184, 185
  diffusus 184, 185
  haspan 26, 185
Cyphostemma 162
Cypripedium 282
Cyrtomium 102, **186**, 306
  falcatum 186
    'Rochfordianum' 186
Cytisus 106, **186**
  × *racemosus* 186

D
Daffodil family 101
Daisy family 104
Date palm 73, 298
Datura 112
Demineralisator 58, 354
Dendrobium 108, **187**, 200
  crysanthum 187
  densiflorum 187
  nobile 187
  phalaenopsis 187
  pierardii 187
  wardianum 187
Deutzia 111
Devil's ivy 328
Dictamnus 111
Didymochlaena 102, **188**
  trunculata 188
Dieffenbachia 29, 41, 101, 120, **188**
  'Arvida' 190
  × *bausei* 190
  bowmannii 189, 190
  maculata 190
  picta 190
    'Jenmannii' 190
  seguine 190
    'Janet Weidner' 190
    'Mary Weidner' 190
    'Rudolph Roehrs' 190
Dipladenia 101, **191**
  boliviensis 191
  sanderi 191
    'Rosea' 191
Dipladenia hybrid 'Amoena' 191
  'Rosacea' 191
  'Rubinia' 191

Diseases, prevention and cure 96
Disposable plants, lovers of 11
Distilled water 58, 354
Dizygotheca 102, **192**, 326
  elegantissima 192
  veitchii 192
    'Gracillima' 192
Dogbane family 101
Dogwood family 104
Dracaena 17, 62, 75, 100, 175, **193**, 280, 319
  deremensis 193, 194
    'Bausei' 193
    'Warneckii' 16, 193
  draco 193, 194, 195
  fragrans 193, 194
    'Lindenii' 193
    'Massangeana' 193
    'Rothiana' 193
    'Victoriae' 193
  godseffiana 193, 195
    'Florida Beauty' 193
    'Kelleri' 193
  goldieana 193
  hookeriana 194
    'Latifolia' 194
    'Rothiana' 194
    'Variegata' 194
  margianata 10, 194
  reflexa 194
    'Song of India' 194
  sanderiana 194
  stricta 176
Drainage 64, 66
Dry air 59
Duchesnea 111, **196**
  indica 196

E
Easter cactus 52, 55, 316
Echeveria 105, **196**, 220, 278, 280
  agavoides 196, 197
    'Cristata' 197
  carnicolor 196
  coccinea 197
  elegans 197
  gibbiflora 197
  harmsii 197
  nodulosa 197
  secunda 'Pumila' 197
  setosa 196, 197
Echinocactus 103, **198**, 272
  grandis 198
  grusonii 94, 198
  horizonthalonius 198
  ingens 198
Echinocereus 103, **199**
  delaetii 199
  knippelianus 199
  pectinatus 199
  pulchellus 199
Echinopsis 103, **199**

  calochlora 199
  eyriesii 199
  kermesina 199
Elatostema 291
Environment 16
Epidendrum 108, **200**
  ciliare 200
  fragrans 200
  'Rainbow' hybrid 200
  vitellinum 200
  × *Epidrobium* 200
Epiphyllum 55, 103, **201**, 331, 352
Epiphyllum hybrid 201
Epiphyte tree 87, 88, 89
Epiphytes 18, 354
Epipremnum 101, 314
Episcia 106, **202**
  cupreata 202
  dianthiflora 202
  fulgida 202
  reptans 202
Eranthemum 242, 310
Erica 105, **202**
  gracilis 202
  hiemalis 202
  ventricosa 202
  × *willmorei* hybrids 202
Ericaceae 105
Eriocereus jusbertii 316
Espostoa 103, **203**
  lanata 203
Euphorbia 105, **204**
  caput-medusae 206
  globosa 206
  grandicornis 206
  lactea 206
  milii 205
    hybrids 205
  obesa 206
  pseudocactus 206
  pulcherrima 54, 203
    'Eckespoint Pink' 203
    'Eckespoint Red' 203
    'Oakleaf' 203
    'Paul Mikkelsen' 203
  splendens 205
  submammillaris 206
Euphorbiaceae 105
Eurya 165
Evergreen grapevine 321
Exacum 105, **206**
  affine 105, 206, 207
    'Midget' 206

F
F_1 hybrid 354
False castor oil plant 203
Fasciation 354
× *Fatshedera* 102, **207**
  lizei 207
    'Variegata' 207
Fatsia 55, 102, **208**

Fatsia japonica 207, 208
   'Albomarginata' 208
   'Moseri' 208
   'Reticulata' 208
   'Variegata' 208
Faucaria 101, **209**
   felina 209
   lupina 209
   tigrina 209
   tuberculosa 209
Feeding 43, 60, 94
Fenestraria 219
Fern roots 61
Ferocactus 103, **209**
   hamatacanthus 209
   latispinus 209
   stainesii 209
Fertiliser 62
Fibrous roots 63
Ficoidaceae 101
Ficus 14, 17, 22, 71, 75, 79, 107, **210**, 301
   aspera 210
     'Parcelli' 210
   bengalensis 210
   benjamina 42, 210
   carica 107, 210
   cerasiformis 210
   cyathistipula 210
   'Decora' 211
   deltoidea 107, 210
   edulis 210
   elastica 210, 211, 214
     'Decora' 211
     'Decora Tricolor' 211, 212
     'Doescheri' 301
     'Schrijvereana' 20, 212
     'Variegata' 211, 212
   lyrata 37, 212
   montana 212
   pandurata 212
   pumila 212, 214
     'Minima' 212
     'Serpylliolia' 212
     'Variegata' 212
   quercifolia 212
   radicans 212, 214
     'Variegata' 213
   religiosa 212
   retusa 212
   rubiginosa 212
     'Variegata' 212, 213
Ficus, creeping and climbing 212
   evergreen 210
Figwort family 111
Filling, soilless cultivation 92
Finger Aralia 192
Fittonia 100, **215**
   gigantea 215
   verschaffeltii 215
     'Argyroneura' 215
     'Pearcei' 215

Fixed rules 11
Flame nettle 171, 172
Flamingo plant 124
Flowering plants 38
Flowerpot 63
Fluorescent tube 48
Foliage fertiliser 62
Fuchsia 56, 80, 108, **216**
   procumbens 216
   triphylla hybrids 216
Fuchsia hybrid 'Pink Galore' 216
Fungi 99

G
Gardenia 52, 111, **217**
   jasminoides 217
     'Veitchii' 217
Gasteria 107, **218**
   armstrongii 218
   candicans 218
   pulchra 218
   verrucosa 218
Gentian family 105
Gentianaceae 105
Geraniaceae 106
Geranium 288
Geranium 56, 75, 80, 288
   for window boxes 289
   garden 289
   with fragrant foliage 290
Geranium family 106
Gesneria macrantha 314
Gesneriaceae 106
Glass case, enclosed 28
Glass domes 81
Glassy appearance 96
Globular cactus 17, 55
Gloriosa 107, **218**
   rothschildiana 218, 219
   simplex 218
   superba 218
Glory lily 218, 219
Glottiphyllum 22, 101, **219**
   oligocarpum 219
Gloxinia 17, 334
Gloxinia family 106
Godetia 108
Goosefoot plant 341
Grafting 74
Graptopetalum 105, **220**
   amethystinum 220
   filiferum 220
   paraguayense 220
   weinbergii 220
Grass family 106
Greenfly 99
Greenhouse, tropical 35
Grevillea 110, **220**
   robusta 110, 220, 245
Guzmania 103, **221**
   lingulata 221
     'Broadview' 221

     'Splendens' 221
   minor 221
   monostachya 221
   tricolor 221
   zahnii 221
Guzmania hybrid 'Intermedia' 221
   'Magnifica' 221
Guzmania hybrids 221
Gymnocalycium 103, **222**
   denudatum 222
   friedrichii 222
   gibbosum 222
   mihanovichii 222
Gynura 104, **222**
   scandens 222, 223

H
Habit 17, 18
Haemanthus 101, **223**
   albiflos 223, 224
   katharinae 223, 224
     'King Albert' 223, 224
Handsprays 58
Hanging plants 25
Hardening off 74
Hardness, degrees of 57
Haworthia 107, **224**
   cymbiformis 225
   fasciata 224
   margaritifera 225
   papillosa 225
   planifolia 225
   reinwardtii 225
   tesselata 225
Heath 202
Heath family 115
Heating flex 53, 71
Hebe 111, **225**
   andersonii hybrid 225
     'Evelyn' 225
     'La Seduisante' 225
     'Variegata' 225
Hedera 89, 102, **226**
   canariensis 226
   colchica 226
   helix 207, 226
     'Crispa' 226
     'Glacier' 226
     'Goldheart' 226
     'Luzii' 226
     'Maple Queen' 226
     'Minima' 226
     ssp. *canariensis* 226, 227
       'Gloire de Marengo' 226, 228
     hybrid 'Pittsburgh' 228
Hedge cactus 155
Heel 354
Hemigraphis 100, **229**
   colorata 229
   repanda 229

Hibiscus 107, **229**
  abelmoschus 229
  cannabinus 229
  esculentus 229
  rosa-sinensis 229, 230
    'California Gold' 230
    'Cooperi' 230, 231
    'Flamingo' 230
    'Flore Pleno' 230
    'Laterita Variegata' 230
    'Miss Betty' 230
  schizopetalus 230
  syriacus 229
High pressure mercury lamps 48
Hillebrandia 102
Himantophyllum 166
Hippeastrum 46, 101, **231**
  aulicum 231
  vittatum 231
Hippeastrum hybrid 233
  'Apple Blossom' 231, 232
  'Fire Dance' 231
  'Happy Memory' 231
  'Picotee' 232
Hippeastrum hybrids 231
Holly fern 186
Horn meal 61
Hot house plants 55
House lime 337
Howeia 29, 109, **234**
  belmoreana 234, 235
  forsteriana 234
Hoya 71, 74, 102, **236**
  australis 236
  bella 236
  carnosa 236, 237
Humidifier 59
Hyacinth 238
  Roman 239
Hyacinthus 107, **238**
  orientalis 238
    'Amsterdam' 239
    'Anna Marie' 239
    'Delft Blue' 238, 239
    'Jan Bos' 239
    'L'Innocence' 238
    'Ostara' 239
    'Pink Pearl' 239
    'Tubergen's Scarlet' 239
Hybrid 354
Hydrangea 111, **240**
  macrophylla 240, 241
Hydroponics 43ff, 91ff
Hydro-pot 91, 92
Hypocyrta 106, **242**
  glabra 242
  nummularia 242
  strigillosa 242
Hypoestes 100, **242**
  dristata 242
  sanguinolenta 242
  taeniata 242

I
Illumination 47, 48
  period of 48
Impatiens 102, **243**
  holstii 243
  sultani 243
  wallerana 243
    'Ammerland' 243
    'Beauty of Klettgau' 243
    'Blaze' 243
    'Liegnitzia' 243
    nana 'Firelight' 243
    F_1 hybrids 243
    hybrid 'Orange Baby' 243
    'Rosa Baby' 243
Incarvillea 103
Individuality of a plant 12
Indoor propagator 27, 28
Interaction 10
Interior photographs 29
Internode 354
Iresine 101, **244**
  herbstii 24, 101, 244
  lindenii 244
Iridaceae 106
Iris family 106
Ivy family 102
Ixora 111, 217, **244**
  coccinea 244, 245
    'Biers Glory' 244
    'Morsai' 244
    'Shawii' 244

J
Jacobinia 100, **246**, 279
  carnea 246
  pauciflora 246
Jacaranda 103, **245**
  mimosifolia 245
Jardinière on landing 9
Jasmine 247
Jasminum 108, **247**
  officinale 247
  polyanthum 247
  sambac 247
    'Plena' 247
Jonquils 266
Justicia 246

K
Kalanchoë 22, 105, **247**, 248, 249
  blossfeldiana 105, 247, 248
    'Compacta Liliput' 248
    'Goldrand' 248
    'Orange Triumph' 248
    'Tom Thumb' 248
  jongmansii 248
  miniata 248
  tomentosa 248

Kalanchoë (Bryophyllum) 105, **249**
  daigremontiana 249
  laxiflora 249
  pinnata 249
  tubiflora 249

L
Labiatae 106
Leadwort 305
Leaf cutting 71
Leaf gloss 79
Leafmould 61, 353
Leathery foliage 17
Leguminosae 106
Lemon 163
Libonia floribunda 246
Light 49, 50, 51
  direction of 29
  meter 51
  physiology 41
  requirements 355
Liliaceae 106
Lilium 107, **250**
  auratum hybrids 250
  convallium 174
  longiflorum hybrids 250
Lilium hybrid 'Destiny' 250
  'Enchantment' 250
  'Golden Chalice' 250
  'Harmony' 250
  'Midcentury' 250
  'Paprika' 250
  'Rainbow' 250
Lillipot 174
Lily 250
Lily family 106
Lily of the valley 174, 175
Lime haters 57
Lime tree family 112
Liriope 107, **251**
  muscari 251
    'Variegata' 251
Lithops 101, **252**
  fulleri 252
Lobivia 103, **252**, 253
  densispina 252, 253
    f. *leucomalla* 253
  haageana 253
  jajoiana 253
  pentlandii 253
  tiegeliana 253
Long day plants 54, 354
Lophophora 103, **253**
  williamsii 253
Lucky clover 278
Ludwigia 108
Luwasa system 95
Lycaste 109, **254**
  skinneri 254
  virginalis 254

M

Madagascar jasmine 339
Madder family 111
Maidenhair fern 116
Malacocarpus 272
Mallow family 107
Malus punicus 110
Malvaceae 107
Mammillaria 103, 176, **254**, 313
  bocasana 255
  geminispina 255
  gigantea 255
  hahniana 255
  humboldtii 255
  pennispinosa 255
  plumosa 255
  prolifera 255
  rhodantha 254, 255
  schiedeana 255
  wildii- 255
  zeilmanniana 255
Manettia 111, **256**
  bicolor 256
  inflata 256
Manure 61
Maranta 107, 181, **256**
  bicolor 257
  leuconeura 257
    'Fascinator' 257
    'Kerchoveana' 256, 257
    'Massangeana' 257
Marantaceae 107
Marchantia polymorpha 84
Meadow saffron 171
Mealy bug 98
Medinilla 107, **258**, 336
  magnifica 21, 258
Melastomataceae 107
Melocactus 103, **259**
  maxonii 259
  obtusipetalus 259
  peruvianus 259
  violaceus 259
Mescalin 253
Mesembryanthum family 101
Mesembryanthamaceae 101
Microcoelum 109, **260**
  weddellianum 260, 261
Micro-climate 78
Mildew 99
Milkweed family 102
Miltonia 109, **262**, 274
  candida 262
  clowesii 262
  roezlii 262
  vexillaria 262
    rubella 262
Mimicry 252
Mimosa 106, 115, **263**
  pudica 245, 263
Miniature bottle garden 81
Mint family 106

Mirabilis jalapa 108
Monstera 89, 101, **264**, 294, 315, 341
  acuminata 264
  deliciosa 13, 41, 264, 294
    'Borsigiana' 264
  obliqua 264
    'Leichtlinii' 264
  pertusa 264
Moraceae 107
Moss 76, 88
Mother-in-law's tongue 324
Mother of thousands 137, 325, 343
Mountain palm 40
Moving 51
Mulberry family 107
Multiflora hyacinths 239
Mutation 354
Myrsinaceae 108
Myrtaceae 108
Myrtle 40, 265
Myrtle family 108
Myrtus 108, 147, **265**
  communis 265

N

Narcissus 101, **266**
  tazeta 266
    'Paperwhite' 266
    'Totus albus' 266
Nephrolepis 108, **269**
  cordifolia 269
  exaltata 'Bornstedt' 269
    'Dwarf Boston' 269
    'Rooseveltii' 269
    'Whitmannii' 269
    hybrids 269
Neoporteria 103, **267**
  gerocephala 267
  napina 267
  nidus 267
  senilis 267
Neoregelia 103, **267**
  binotii 267
  carolinae 268
    'Marechalii' 268
    'Meyendorfii' 268
    'Tricolor' 268
  concentrica 268
  princeps 268
  tristis 268
Neoregelia hybrid 'Volckhartii' 267
Nerium 101, **270**
  oleander 270, 271
    'Variegatum' 270
Nertera 111, **272**
  granadensis 272
Nettle family 112
Nidularium **267**
  innocentii 268
    'Maureanum' 268
    'Nana' 268

    'Paxianum' 268
    'Purpureum' 268
  striatum 268
Night cactus 330, 331
Nightshade family 112
Node 354
Notocactus 103, **272**, 273
  apricus 273
  concinnus 273
  haselbergii 273
  leninghausii 273
  ottonis 272
  submammulosus 273
  ubelmannianus 273
Nutrient solution 93
Nyctaginaceae 108

O

Oak fern family 110
Odontoglossum 109, 262, **274**, 275
  bictoniense 274
  cervantesii 274
  citrosmum 274
  grande 274
  pulchellum 274
  rossii 274
  uro-skinneri 274
Oenothera 108
Office plants 36, 41
Old man cactus 154
Oleaceae 108
Oleander 56, 270, 271
Oleandra 108
Oleandraceae 108
Olive family 108
Onagraceae 109
Oncidium 109, 262, 274, **275**
  bicallosum 275
  forbesii 275
  kramerianum 275
  ornithorhynchum 275
  tigrinum 275
  varicosum 275
Ophiopogon 251
  muscari 251
Ophthalmophyllum 219
Oplismenus 106, **276**
  hirtellus 18, 276
Opuntia 103, **276**
  basilaris 277
  bergeriana 277
  microdasys 277
    'Minima' 277
  rufida 277
Orange tree 163
Orchidaceae 108
Orchids 57, 108
Organic fertiliser 62
Origin 18
Ornamental pineapple 123
Ornamental pots 77
Oroya 104, **277**

neoperuviana 277
peruviana 277
Osmunda 354
Oxalidaceae 109
Oxygen 50

P
Pachyphytum 105, **278**, 280
  amethystinum 220
  bracteosum 278
  hookeri 278
  oviferum 278
Pachypodium 101, **278**
  geayl 278
  lamerei 278
Pachystachys 100, **279**
  lutea 279
× *Pachyveria* 105, 278, **280**
  clavifolia 280
  pachyphytoides 280
  scheideckeri 280
  spathulata 280
Palmae 109
Palms 17, 109
Palm pot 63
Pandanaceae 109
Pandanus 17, 29, 109, **280**
  sanderi 281
  utilis 280, 281
  veitchii 281
Panicum 276
Paphiopedilum 109, **282**
  barbatum 282
  callosum 282
  fairieanum 282
  glaucophyllum 282
  hirsutissimum 282, 283
  insigne 282, 283
    'Aureum' 282
    'Chantinii' 282
    'Harefield Hall' 282, 283
    'Sanderae' 282
  niveum 282, 283
  spicerianum 282
  villosum 282
× *Paphiopedilum harrisianum* 282
  lathamianum 282
  leeanum 282
Parasites 98
Parodia 104, **284**
  chrysacanthion 284
  maxima 284
  microsperma 284
  nivosa 284
  sanguiniflora 284
  schwebsiana 284
Pathenocissus 112
Passion flower 17, 40, 284, 285
Passion flower family 109
Passiflora 12, 109, **284**
  caerulea 56, 284, 285
    'Constance Elliot' 284

    'Empress Eugenie' 284
  gracilis 284
  incarnata 284
  maculifolia 286
  quadrangularis 109, 121, 286
  racemosa 286
Passifloraceae 109
Pavonia 107, **287**
  intermedia 287
  multiflora 287
Pea family 106
Peat fibre 61, 354
Pedilanthus 105, **287**
  tithymaloides 287
    'Variegatus' 287
Pelargonium 106, **288**
  capitatum 290
  crispum 290
  fragrans 290
  grandiflorum hybrids 288
  graveolens 290
  odoratissimum 290
  peltatum hybrids 289
  radens 290
  zonale F_1 hybrids 290
    hybrids 289, 290
Pellaea 84, 111, **290**
  atropurpurea 290
  cordata 290
  falcata 290
  hastata 290
  rotundifolia 290, 291
  viridis 290
Pepper 301
Peperomia 22, 110, **291**
  angulata 292
  argyreia 292
  arifolia 292
    'Princess Astrid' 292
  caperata 137, 291
  clusiifolia 292
    'Variegata' 292
  fraseri 292
  glabella 24
  griseo-argentea 292
  hederaefolia 292
  obtusifolia 292
    'Greengold' 292
  prostrata 292
  pulchella 292
  resediflora 292
  rotundifolia 292
  scandens 292
  serpens 292
  verticillata 292
Pereskia 316
Perlite 62, 354
Pesticide 96, 98
pH values 57
Phalaenopsis **293**
  amabalis 293
  aphrodite 293

  cornucervi 293
  schilleriana 293
  stuartina 293
  violaceae 293
Philadelphus 111
Philodendron 17, 22, 51, 74, 75
101, 264, **294**, 315, 341
  andreanum 295
  bipinnatifidum 295
  cuspidatum 295
  elegans 295
  erubescens 294
    'Green Emerald' 294
    'Red Emerald' 294
  ilsemannii 294
  laciniatum 295
  martianum 295, 296
  melanochrysum 295, 297
  panduriforme 295
  pertusum 294
  scandens 295
  selloum 296
  squamiferum 295
  verrucosum 295
Philodendron, climbing or creeping
  with undivided foliage 294
  climbing, with divided foliage 295
  non-climbing, with and without
  trunk formation 295
Phlebodium 110, **297**
  aureum 297
    'Glaucum Crispum' 297
    'Mandaianum' 297
Phoenix 109, **298**
  canariensis 298
  dactylifera 298
  roebelinii 298
Photosynthesis 49, 354
Phototropism 51
Phyllitis 84, 102, **299**
  scolopendrium 84, 299
    'Capitatum' 299
    'Crispum' 299
    'Digitatum' 299
    'Undulata' 299
Phyllocactus 55, 201
Pilea 112, **299**
  cadierei 299, 300
    'Minima' 300
  involucrata 300
  microphylla 300
  muscosa 300
  pubescens 300
  spruceana 299, 300
    'Norfolk' 300
    'Silvertree' 300
Pineapple plant 73, 123
Piper 110, **301**
  nigrum 301
  ornatum 301
  porphyrophyllum 301
  sylvaticum 301

Piperaceae 110
Pisonia 108, **301**
   alba 301
Pittosporaceae 110
Pittosporum 110, **302**
   coriaceum 302
   crassifolium 302
   tenuifolium 302
   tobira 302
     'Variegatum' 302
Pittosporum family 110
Plant care 36, 76
Plant case 28
Plant containers 64, 65
   for Luwasa soilless cultivation 95
   communities 40
   families 100
   furniture 27
   growth in artificial light 54
   window 30, 31, 33, 35, 85, 86, 90
Plants as presents 38
   borderline cases, light requirements 356
   for light shade 355
     moderately good light 355
     the plant window 356
   liking a light and cool situation in winter 356
   requiring little light 355
     plenty of light 356
   suitable for a north-facing window 355
   which overwinter in the cellar 356
Plastic bag 74
Plastic plant container 64, 68
Platycerium 110, **303**
   alcicorne 303
   bifurcatum 303
     bifurcatum 303
       'Hillii' 303
       'Majus' 303
   grande 303, 304
   wilhelminae-reginae 303
   willinckii 303
Plectranthus 106, **304**
   fruticosus 304
   oertendahlii 304
Plumbaginaceae 110
Plumbago 110, **305**
   auriculata 305
   capensis 305
   europaea 305
Poet narcissi 266
Poinsettia 54, 204
Poinsettia pulcherrima 204
Polypodiaceae 102, 110
Polypodium 297, 354
Polyscias 102, **306**
   balfouriana 306
     'Peacockii' 306
   filicifolia 306

   guilfoylei 306
   paniculata 306
Polystichum 84, 102, **306**
   auriculatum 306
   setigerum 307
     'Plumosum Densum' 307
     'Proliferum' 307
     'Proliferum Wollastonii' 307
   tsus-simense 306, 307
Polystyrene granules 62, 354
Polytrichum 84
Pomegranate 312
Pomegranate family 110
Positioning of house plants 10
Pot, correct for plant 62
Potbound 66
Pot frame 76
Potsherds 62, 66, 67
Prayer plant 256, 257
Pricking out 74
Primrose family 110
Primula 110, **307**
   acaulis 308
   × *kewensis* 308
   malacoides 307
     'Wädenswyler Typ' 308
   obconica 308, 309
     'Multiflora' 309
   praenitens 308
   sinensis 308
   vulgaris 308
     F_1 hybrids 308
Primulaceae 110
Propagating soil 72
Propagation 70
Protea 110
Protea family 110
Proteaceae 110
Pruning 80
Pruning to shape 80
Pseuderanthemum 100, **310**
   atropurpureum 310
   reticulatum 310
   sinuatum 310
Pteridaceae 110
Pteris 75, 84, 110, **310**
   cretica 310
     'Albo-lineata' 84, 310
     'Major' 310
     'Rivertoniana' 310
     'Wimsettii' 310
   ensiformis 310
     'Evergemiensis' 311
     'Victoriae' 311
   multifida 311
     'Cristata' 311
   serrulata 311
   tremula 311
   umbrosa 311
Pumice 46
Punica 110, **312**
   granatum 312

   'Nana' 312
Punicaceae 110
Pyrethrum 104

R
Rainwater 59
Rat's-tail cactus 126
Rauwolfia 101
Rebutia 104, **312**
   chrysacantha 313
   marsoneri 313
   minuscula grandiflora 313
     f. *violaciflora* 313
   pygmeae 312, 313
   senilis 313
   xanthocarpa 313
Rechsteineria 106, **314**
   cardinalis 314
   leucotricha 314
Red pepper 150
Red spider mite 98
Repotting 65
Rhaphidophora 89, 101, **314**, 328
   aurea 314, 315
     'Marble Queen' 315
Rhipsalidopsis 104, **316**
   gaertneri 316
   × *graeseri* 316
   rosea 316
Rhizomes 63
Rhododendron 105, **317**
   obtusum 317
   simsii 317, 318
Rhoeo 104, **319**, 333
   spathacea 319
     'Vittata' 319
Rhoicissus 112, **321**
   capensis 161, 320, 321
   rhomboidea 161
Robinia 87
Rochea 105, **321**
   coccinea 321
     'Gräsers Rote' 321
   falcata 178, 321
   jasminea 321
   versicolor 321
Root cuttings 71
Root formation—excessive 66
Root lice 98
Roots 68
Rosa 111, **322**
   'Baby Masquerade' 322
   'Coralin' 322
   'Doc/Degenhardt' 322
   'Dwarf King' 322
   'Dwarf Queen' 322
   'Happy/Alberich' 322
   'Little Buckaroo' 322
   'Mother's Day' 322
   'Pour toi' 322
   'Sleepy/Balduin' 322
   chinensis 322

'Rosina' 322
minima 322
lawrenceana 322
× *noisettiana* 322
noisettiana hybrids 322
roulettii 322
Rosaceae 111
Rose 322
Rose family 111
Rubber plant 210
Rubia tinctorum 111
Rubiaceae 111
Rue family 111
Rust 99
Rutaceae 111

S
Saintpaulia 17, 106, **323**
 ionantha 323
Sand 355
Sansevieria 17, 40, 71, 100, **324**
 trifasciata 'Craigii' 324
  'Golden Hahnii' 324
  'Hahnii' 324
  'Laurentii' 324
Sauromatum 101, **325**
 venosum 171, 325
Saxifraga 111, **325**
 cotyledon 325
 stolonifera 56, 325
 'Tricolor' 326
Saxifragaceae 111
Saxifrage family 111
Scale insect 97
Schefflera 29, 46, 102, **326**
 actinophylla 42, 326, 327
Scindapsus 74, 101, 314, **328**, 341
 aureus 314
 pictus 328
 'Argyraeus' 328
Scirpus 105, **328**
 cernuus 328
Scorched foliage 96
Scrophulariaceae 111
Sea lavender family 110
Sea urchin 131
Sedge family 105
Sedum 105, 220, **329**
 acre 329
 bellum 329
 morganianum 70, 329
 pachyphyllum 329
 rubrotinctum 329
 sieboldii 329
  'Mediovariegatum' 329
Seed bowl 70
Selaginella 84, 111, **330**
 apoda 330
 martensii 330
  'Watsoniana' 330
Selaginellaceae 111
Selenicereus 104, **330**

grandiflorus 12, 330, 331
pteranthus 331
Sempervivum 329
Senecio 104, **331**
 citriformis 331
 cruentus 332
  hybrids 332
  'Scarlet' 332
 haworthii 331
 herreianus 331
Sensitive plant 263
Setcreasea 104, **333**
 purpurea 333
 striata 146, 333
Sexual propagation 70
Shade plants 17
Sharp sand 61, 354
Short-day plants 54, 354
Shrimp plant 138
Side light 52
Siderasis 104, **333**
 fuscata 333
Silk bark oak 220
Sinningia 106, **334**
Sinningia hybrids 334
Sinopteridaceae 111
Situation 49
 in relation to the sun 53
 out of doors in summer 54
Skimmia 111, **335**
 japonica 335
Slipper flower 146
Small-cupped narcissi 266
Smithiantha 106, **335**
Smithiantha hybrid 335
 'Deschene's Strain' 335
 'Pink Domino' 335
 'Orange King' 335
 'Swan Lake' 335
 'Vespers' 335
Softening water 58
Soil-ball 66, 68
Soil, care of 76
Soilless cultivation 43, 45, 91
 history 44
 pots for 91, 92
 suitable for 46
Soil reaction 57, 353
 temperature 60
Solanaceae 112
Solanum 112, **336**
 capsicastrum 336
  'Variegatum' 336
 pseudocapsicum 336
  'Goldball' 336
  'New Patterson' 336
Sonerila 107, **336**
 margaritacea 336
 'Argentea' 336
Sooty mould 99
Spanish pepper 150
Sparmannia 112, **337**

africana 337
 'Flore Pleno' 337
Spadix 354
Spathe 354
Spathiphyllum 101, **338**
 floribundum 338
Spear flower 128
Sphagnum 61, 354
Spider plant 159
Spiderwort family 104
Spindletree family 104
Spurge family 105
Staghorn fern 303, 304
Stapelia 102, 156, **338**
 gigantea 338
 grandiflora 338
 variegata 338
Starfish flower 338
Stephanotis 74, 102, **339**
 floribunda 339
Star cactus 131
Stonebreak family 111
Stonecrop family 105
Stopping (tip removal) 354
Streptocarpus 106, **340**
Streptocarpus hybrid 'Constant Nymph' 14, 340
 'Wiesmoor' 340
Streptocarpus hybrids 340
Strophanthus 101
Succulents 17, 30, 103, 105
Sun 49
Surface roots 63
Swiss cheese plant 264
Syngonium 101, 295, **341**
 auritum 42, 341
 podophyllum 341
 vellozianum 341

T
Taproots 63
Tea plant 148
Technical terms 353
Terrestrial orchids 108
Tetrastigma 112, **341**
 volnierianum 341
Theaceae 112
Thrips 98, 99
Thunbergia 100, **342**
 alata 342
  'Alba' 342
  'Aurantiaca' 342
  'Fryeri' 342
  'Lutea' 342
 erecta 342
 laurifolia 342
 natalensis 342
Tiarella 111
Tiliaceae 112
Tillandsia 130, **342**
 lindenii 343
 usneoides 343

367

Ti-plant 194
Tolmiea 111, **343**
  menziesii 343
Top light 52
Top shoot 71, 74
Tradescantia 22, 56, 70, 104, 146, 276, 319, 333, **344**, 352
  albiflora 'Albo-vittata' 344
  blossfeldiana 344
  crassula 344
  fluminensis 344
    'Variegata' 344
  navicularis 344
Tree bark 90
Tree orchids 108
Trichocereus 104, **345**
  candicans 345
  grandiflorus 345
  pachanoi 345
  purpureopilosus 345
  schickendantzii 345
  spachianus 345
Tripod stand 56
Trumpet narcissi 266
Tub plants 56
Tuberous roots 63
Tulipa 107, **346**
  gesneriana 346
  'Keizerskroon' 346
Tulip 346
Turf loam 354
Tying 77
Type 18

U
Umbrella plant 183, 184

Urtica dioica 112
Urticaceae 112

V
Vallota 101, **231**
  speciosa 231, 232
Vanda 109, **347**
  caerulea 347
    'Hennisiana' 347
Variegated foliage 18, 353
Vegetative propagation 70
Veltheimia 107, **348**
  capensis 348
  viridifolia 348
Venus'-slipper 282
Verbena family 112
Verbeneceae 112
Veronica 225
Vigour 11
Vinca minor 152
  rosea 152
Vine family 61, 112
Vitaceae 61, 112
Vitis heterophylla 'Variegata' 122
  vinifera 112
Vriesea 103, **348**
  gigantea 348
  hieroglyphica 349
  imperialis 348, 349
  psittacina 348
  splendens 348
  'Flaming Sword' 348, 349
  'Major' 348
tessellata 348
Vriesea hybrid 'Rex' 348, 349

W
Wardian case 25
Water 58
Watering 58
Watering cans 58
Watering installation 89
Wax flower 17, 40, 50, 52, **236**
Whitefly 98, 99
Width container 85
Willow herb family 108
Winter temperature 355
Wood sorrel family 109
Woody plant, pruning 80

X
Xerophytes 355

Y
Yellow blotching 96 99
Yucca 100, **350**
  aloifolia 350, 351
    'Quadricolor' 350
    'Tricolor' 350
  gloriosa 350

Z
Zantedeschia 101, **350**
  aethipica 350
  elliottiana 350
Zebrina 104, **352**
  pendula 352
    'Quadricolor' 352
Zygocactus 104, 201, 316, **352**
  truncatus 352